Community, Environment and Local Governance in Indonesia

This book explores the forces reconfiguring local resource governance in Indonesia since 1998 by drawing together original field research undertaken in a decade of dramatic political change. Case studies from across Indonesia's diverse cultural and ecological landscapes focus on the most significant resource sectors – agriculture, fisheries, forestry, mining and tourism – providing a rare, in-depth view of the dynamics shaping social and environmental outcomes in these varied contexts.

Debates surrounding the 'tragedy of the commons' and environmental governance have focused on institutional considerations of how to craft resource management arrangements in order to further the policy objectives of economic efficiency, social equity and environmental sustainability. The studies in this volume reveal the complexity of resource security issues affecting local communities and user groups in Indonesia as they engage with wider institutional frameworks in a context driven simultaneously by decentralizing and globalizing forces. Through ground-up investigations of how local groups with different cultural backgrounds and resource bases are responding to the greater autonomy afforded by Indonesia's new political constellation, the authors appraise the prospects for rearticulating governance regimes toward a more equitable and sustainable 'commonweal'.

This volume offers valuable insights into questions of import to scholars as well as policy-makers concerned with decentralized governance and sustainable resource management.

Carol Warren is Associate Professor in Asian Studies and Research Fellow at the Asia Research Centre, Murdoch University.

John F. McCarthy is Senior Lecturer at the Crawford School of Economics and Government, Australian National University.

Routledge Contemporary Southeast Asia Series

Community, Environment and Local Governance in Indonesia

Locating the commonweal

**Edited by Carol Warren
and John F. McCarthy**

Routledge
Taylor & Francis Group

LONDON AND NEW YORK

First published 2009
by Routledge
4 Park Square, Milton Park, Abingdon, Oxon OX14 4RN
605 Third Avenue, New York, NY 10017

Routledge is an imprint of the Taylor & Francis Group, an informa business

First issued in paperback 2012

Typeset in Times New Roman by Pindar NZ, Auckland, New Zealand

British Library Cataloguing in Publication Data
A catalogue record for this book is available from the British Library

Library of Congress Cataloging-in-Publication Data
Community, environment and local governance in Indonesia: locating the commonweal / edited by Carol Warren and John F. McCarthy.
    p. cm.—(R4outledge contemporary Southeast Asia series)
  1. Natural resources—Indonesia—Management.
  2. Local government—Indonesia. I. Warren, Carol.
II. McCarthy, John F.
HC447.5.C66 2008
333.70959—dc22   2008018396

ISBN13: 978-0-415-43610-6 (hbk)
ISBN13: 978-0-415-54109-1 (pbk)
ISBN13: 978-0-203-88899-5 (ebk)

For the next generation –
Rusma, Niah, Paridah
and Aedan

Contents

Illustrations

Maps

Figures

Tables

Contributors

Greg Acciaioli currently lectures in Anthropology and Sociology at the University of Western Australia. He has held research fellowships at the Southeast Asia Research Centre (City University of Hong Kong), Asia Research Centre (Murdoch University), and Asia Research Institute (National University of Singapore). He is currently researching the interface of the indigenous people's movement with resource contestations in Indonesia.

Hidayat Alhamid conducted the research presented here as part of his doctoral studies at the School of Resources Environment & Society, The Australian National University. He is a forester by training, who continues to work towards sustainable and equitable management of natural resources in West Papua.

Laurens Bakker teaches anthropology and adat law at the Radboud University in Nijmegen, the Netherlands. His interests include customary normative systems and regional politics.

Chris Ballard is a Fellow in Pacific History at The Australian National University. He has conducted long-term research as an archaeologist, historian and anthropologist in Papua New Guinea, Indonesian Papua and Vanuatu. His current research interests include land reform in Vanuatu, the history of racial 'science' in Oceania, and indigenous cultural heritage in the Pacific.

Peter Kanowski is Professor of Forestry, and Deputy Director of the Fenner School of Environment and Society at The Australian National University. His principal interests are in forest policy and governance. He works extensively with colleagues in the Asia-Pacific region on a range of significant forest and society issues.

Anton Lucas is Associate Professor in the School of Political and International Studies at Flinders University in South Australia, where he teaches Asian Studies and Indonesian language. His research interests include decentralization, the politics of natural resource conflicts and agrarian change in post-Suharto Indonesia.

John F. McCarthy lectures at the Crawford School of Economics and Government at the Australian National University. He has worked with a number of non-government organizations, AusAID and the Centre for International Forestry Research (CIFOR). His research interests include environmental governance, political ecology, forest management, resource conflict and agrarian change.

Achmad Uzair Fauzan graduated from Gadjah Mada University, Yogyakarta in 2002, and has since worked for several organizations on issues concerning religion, politics and the urban poor. In 2007, he was involved in founding Lafadl Initiatives, which promotes alternative development ideas among youth. He is now pursuing postgraduate studies at the Institute of Social Studies, The Hague.

Jim Schiller lectures on Indonesian society and politics in the School of Political and International Studies at Flinders University. His research interests include the politics of elections, local state–civil society interaction and issues of accountability, transparency and participation in local governance.

Carol Warren is Associate Professor in the Asia Research Centre at Murdoch University. She works with a number of non-government organizations and has published widely on customary law, social change and environmental politics in Indonesia and Malaysia. Her current research interests include sustainable community development, indigenous rights, land tenure and agrarian policy.

Acknowledgements

This research project emerged out of a concern to explore the relevance of debates surrounding the 'tragedy of the commons' to questions of social equity, resource access and conservation affecting local communities in the period of governance reform in Indonesia since 1998. Recognizing the importance of a deeper understanding of socially embedded interrelationships between symbolic identities and practical interests for pursuit of the 'common good', we have adopted an expanded constructivist concept of the 'commonweal' as the framework for this fieldwork-focused comparative study of Indonesia's resource governance issues.

This research project was supported by an Australian Research Council grant (DP0211816) entitled 'Locating the Commonweal' and was hosted by Murdoch University's Asia Research Centre, where both editors were based. Long-term involvement with a number of other research groups sharing interests in these issues has contributed to the shaping of our research. Contributors have enjoyed the stimulation of participating in conferences and workshops sponsored by the Van Vollenhoven Institute of Law and Administration, Leiden University, Netherlands; the Centre for Maritime Research, Netherlands; the Max Planck Institute for Social Anthropology, Germany; the *Antropologi Indonesia* group at the University of Indonesia; the Asia Research Institute, National University of Singapore; the International Association for the Study of the Commons, U.S., and the Centre for International Forestry Research, Bogor, Indonesia. We are especially grateful to the Indonesian Institute of Sciences (LIPI), the regional university sponsors of our research, as well as the local collaborating researchers and NGO staff acknowledged separately in each of the case study chapters in this volume.

Among many others who have directly or indirectly contributed to our interpretation of the Indonesian cases presented here, we wish to mention: Dianto Bachriadi, Adriaan Bedner, Franz von Benda Beckmann, Keebet von Benda Beckmann, Noer Fauzi, James J. Fox, David Henley, Heru Komarudin, Tania Li, Zulkifli Lubis, Sandra Moniaga, Moira Moeliono, Jan-Michiel Otto, Nancy Peluso, Gerard Persoon, Lesley Potter, Kathy Robinson, Arianto Sangali, Leontine Visser, Yunita Winarto, and Zahari Zen. Thanks also to Bec Donaldson and Sherry Entus for editorial assistance.

Glossary of terms and abbreviations

Adat Customary law or practice

ADB Asian Development Bank

AMAN Aliansi Masyarakat Adat Nusantara (Alliance of Archipelagic Indigenous Peoples)

BAL Basic Agrarian Law (UUPA 5/1960)

BPD Badan Perwakilan Desa (Village Representative Body, a village-level institution established by regional autonomy reforms (UU 22/1999)

BTNLL Balai Taman Nasional Lore Lindu (Management Office of the Lore Lindu National Park)

Bupati Head of district level government (*kabupaten*)

CBNRM Community Based Natural Resource Management

Co-FISH Coastal Community Development and Fisheries Resource Management Project – (ADB funded)

CPR Common Pool Resource

CSIADCP Central Sulawesi Integrated Area Development and Conservation Project

Depsos Departmen Sosial (Department of Social Affairs, which was in charge of resettlement programs)

Desa Village level government under official (*dinas*) government structure

Dinas Official, state-based governance structures

DPR Dewan Perwakilan Rakyat (People's Representative Assembly – national parliament)

DPRD Dewan Perwakilan Rakyat Daerah (Regional People's Representative Assembly)

FWP Forum Wilayah Penyangga (Buffer Zone Forum)

HNSI Himpunan Nelayan Seluruh Indonesia (All Indonesia Fishers' Association)

HTI Hutan Tanaman Industri (timber plantations)

IUCN International Union for the Conservation of Nature and Natural Resources (now known as the World Conservation Union)

JapHama Jaringan Pembelaan Hak-Hak Masyarakat Adat (Network for the Defence of the Rights of Indigenous Peoples)

JED Jaringan Ekowisata Desa (Village Ecotourism Network)

Kabupaten   District level of government, below Provinces; headed by elected *bupati*

Kecamatan   Subdistrict level of government, headed by appointed *camat*

KMAN   Kongres Masyarakat Adat Nusantara (Congress of Archipelagic Indigenous Peoples)

KomNasHAM   National Human Rights Commission

Kopermas   Koperasi Masyarakat (community cooperatives)

KUD   Koperasi Unit Desa (Village Cooperative Unit)

LBH   Lembaga Bantuan Hukum Indonesia (Indonesian Legal Aid Institute)

LIPI   Lembaga Ilmu Pengetahuan Indonesia (Indonesian Institute of Sciences)

LKD   Lembaga Konservasi Desa (Village Conservation Organization)

LMDH   Lembaga Masyarakat Hutan Desa (Forest Village Community Council)

MPA   Marine Protected Area

NGO/LSM   Non-Government Organizations/Lembaga Sosial Masyarakat

Otda   Autonomi daerah (regional autonomy)

Otsus   Autonomi khusus (special autonomy)

PAD   Pendapatan Asli Daerah (Self-Generated Regional Income)

Pemda   Pemerintah daerah (regional government)

Perda   Peraturan daerah (regional regulation)

Perhutani   The State Forest Enterprise that manages commercial use of 2.5 million hectares of natural forest and forest plantations

PLN   Perusahan Listrik Negara (State Electricity Company)

PLTA   Proyek Listrik Tenaga Air (Hydro-electric Project)

Polhut   Polisi Hutan (Forest Police)

PPA   Perlindungan and Pengawetan Alam (Directorate of Nature Conservation)

Propinsi   (also provinsi) Province; the upper tier of regional government; headed by a governor

REPELITA   Rencana Pembangunan Lima Tahun (Five-Year Plan)

SFC   State Forestry Company (here refers to subunit of publicly owned forest enterprise – Perum Perhutani)

TNC   The Nature Conservancy

TNLL   Taman Nasional Lore Lindu (Lore Lindu National Park)

TPI   Tempat Pelelangan Ikan (fish auction centres)

UU   Undang Undang (national law)

UUPA   Undang Undang Pokok Agraria (Basic Agrarian Law)

WALHI   Wahana Lingkungan Hidup Indonesia (Indonesian Forum for the Environment)

YBHK   Yayasan Bantuan Hukum Rakyat (Foundation for Legal Assistance to the People)

YTM   Yayasan Tanah Merdeka (Foundation for a Free Land)

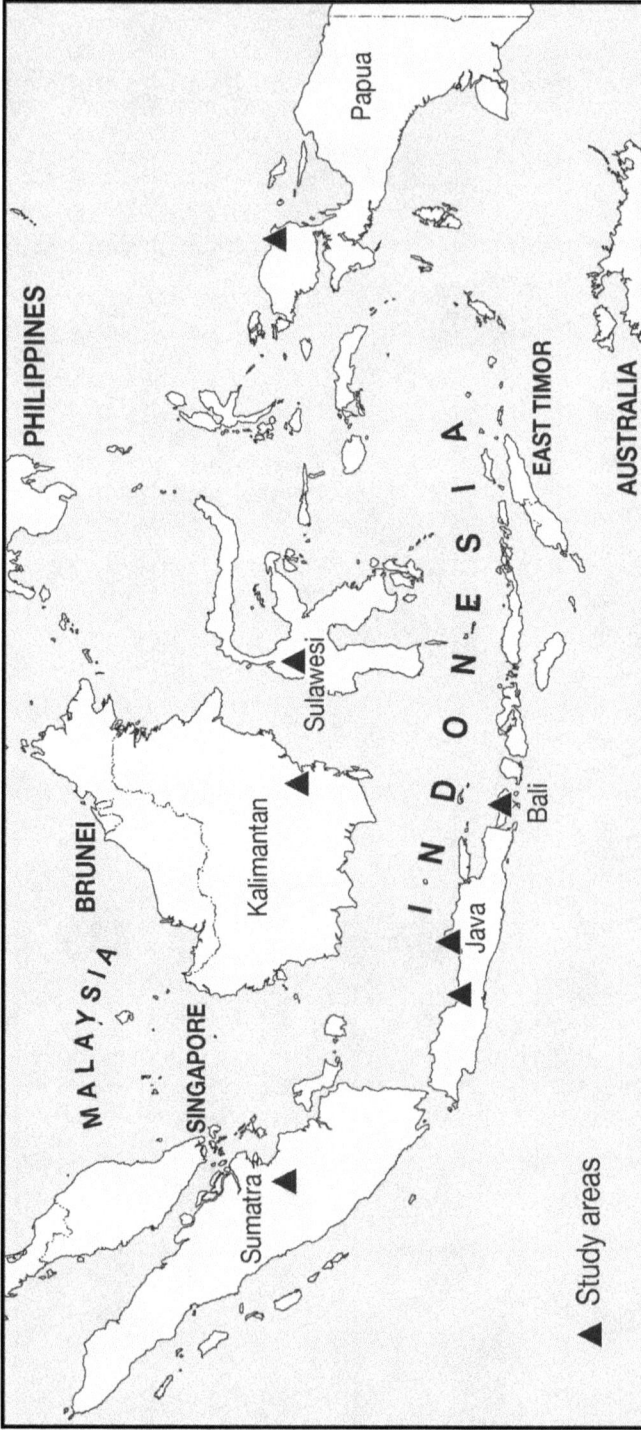

Map 1 Indonesia – case study locations.

1 Communities, environments and local governance in Reform Era Indonesia

John F. McCarthy and Carol Warren

Indonesia: crisis and reform

In the first decade of the new millennium Indonesia confronts daunting challenges posed by simultaneous economic, political and environmental crises. To a large degree, these crises and the reform movement (*reformasi*) they catalysed were precipitated by the excesses of President Suharto's 'New Order' regime (1966–1998), in which centralized authoritarian rule and political patronage facilitated unfettered exploitation of the nation's natural wealth and large population for capital-intensive development. In 1997–1998, Indonesia faced devastating forest fires,[1] which coincided with the unravelling of the country's economy, and finally the collapse of the Suharto regime after three decades in power. In the process, a sea change took place in the legitimacy accorded to the state's claims to control local resources and institutions in the name of 'the people' and the 'national interest'. In the post-Suharto 'Reform Era', the political space available to local actors and non-government organizations expanded dramatically. New political configurations and alliances emerged and alternative conceptions of common interest and identity were articulated, as actors engaged in an on-going struggle to build constituencies and reconstruct some form of 'commonweal'.

Exploring local contestations over resources across Indonesia's social and environmental landscape in the Reform Era, the case studies in this book foreground the concept of the 'commonweal' in an effort to broaden the treatment of resource security issues. Inspired by debates over the 'tragedy of the commons',[2] our use of the 'commonweal' concept encompasses the many facets of common interest and collective goods questions that have become apparent as analysts responded to Hardin's (1968, 1998) allegory and the subsequent common property and collective action literature. We use the term 'commonweal' to refer to the general welfare of the public, as well as the institutional, political, cultural and material domains through which that common welfare is pursued.[3] With this inclusive concept, we recognize that, if societies are to address social equity and human welfare needs within the limits set by the natural environment, they require new understandings of shared welfare and common interest, and new institutional arrangements. But we recognize that these efforts must increasingly address common welfare issues in a complex cross-scale governance context.

This conceptual framework gives due attention to social, political and symbolic 'capital', alongside their economic and natural counterparts, in analysing the sustainability and social security implications of decentralized governance and changing legal frameworks in contemporary Indonesia. At the same time, the concept of the 'commonweal' encourages us to resist reduction of meaningful to instrumental value in examining the struggles over Indonesia's natural, institutional and cultural resources. Working from a historically contextualized perspective regarding local ecologies, socio-economic and property relations (cf. Johnson 2004; Mosse 1997), this approach recognizes that communities of identity and interest may be constituted through processes at once informal and institutional, symbolic and material. It also recognizes that while 'identities' and 'interests' are mutually implicating, neither is reducible to or simply determined by the other. Governance regimes are ultimately expressions of values as well as interests and require moral legitimacy founded on conceptions of what is appropriate or good for the wider society (cf. Bellah *et al.* 1992; Rawls 1999; Arce 2003). Whether conceived on a local, regional, national or even global scale, a 'community' represents a potential locus for the pursuit of shared welfare and common interest – the 'commonweal' – in which the inter-connections between values, identities, institutional arrangements, power relations and resource rights must be explored.

The collective 'goods', which are organizationally produced and discursively conjured through such 'communities' are social, institutional and symbolic as well as material. In analysing the effects of decentralized governance in the new political constellations of Indonesia's Reform Era, we explore how the interplay between structures, identities and entitlement claims articulated across various local, regional and national scales and levels affects equity and sustainability outcomes. Entitlements are grounded in concepts of community and notions of collective good and right which are neither neutral nor fixed.[4] These are framed by group and individual identity, social position and interest, as well as the dynamics of alterity, and are affected by changing legal frameworks and shifting discursive grounds of legitimation. The rhetorical dimensions of governance (claims to represent 'the people' and the 'common interest') and the discursive struggles that situate collective identities are important dimensions of ongoing contests over the material resources that sustain communities, the socio-legal mechanisms that govern them, and the leaderships that claim to represent them.

Furthermore, locating the 'commonweal' necessarily engages the problem of articulating multiple forms of 'commonality' across scales. In the early nationalist period following the struggle for independence, a metaphoric and rhetorical slippage in the collective identities and claims surrounding citizenship underpinned the construction of the Indonesian nation-state as an 'imagined community' (Anderson 1991). From the outset, the conflation of local and national identities and associated resource regimes precipitated fierce contests of power and allegiance between the nation-state and its local constituents.

Following the collapse of the post-revolutionary experiment with democratic government and the removal of the populist Sukarno regime in 1966, the 'national interest' became the rationalizing principle for far more than a nation-building enterprise. Under Suharto's New Order regime, the project of building national out of local constructions of the commonweal could not be separated from the parallel political and legal divesting of local authority and resources through state policy. New Order legislation – in particular the Forestry Law (UU 11/1967), the Basic Mining Law (UU 5/1967), and the Village Government Law (UU 5/1979) – became instruments for the transfer of natural and institutional resources to serve the vested political and economic interests of Suharto's developmentalist regime. Even the explicitly socialist[5] Basic Agrarian Law (UUPA 5/1960), promulgated in the last years of the Sukarno presidency, was circumscribed and reinterpreted to deliver land and resources to private 'developers'. Through a perverse rhetorical slippage, conflating the nation (*kebangsaan*), the people (*rakyat*), and the common interest (*kepentingan umum*), the Indonesian state claimed unilateral authority over the nation's resources and presided over the transfer of public goods to private interests.

By the last decade of the New Order this once powerful populist-nationalist construction of a common Indonesian identity and interest faced increasing regional and local contestation. With the formal demise of the New Order regime in May 1998, the political constellation altered dramatically. These events had been preceded by the proliferation and growing strength of non-government organizations and a resurgence of customary *adat* groups in the most severely marginalized 'outer island' regions of Indonesia which had come to regard themselves as dominated by Javanese culture and Jakarta-based political and economic elites. Backed by activist students and given visibility by a revitalized and critical media, non-government organizations and *ad hoc* community groups across Indonesia began to take on the mantle of long repressed civil society.

In the decade since this crisis point, the Indonesian nation embarked upon wide-ranging reforms that aimed to democratize and improve governance systems, to address the grievances of regional societies marginalized by the previous regime and to attend to critical environmental issues. Following the sudden resignation of Suharto, reformers were eager to make the most of the fragile political base of the transitional Habibie government, demanding radical legal changes. They called for new laws that would guarantee democratic process, media freedom and recognition for the land and resource rights of ordinary Indonesians.[6] Among the most far-reaching responses to these political pressures, was the decentralization of significant aspects of Indonesian government.

Regional autonomy legislation and decentralized governance

A core problem facing reformers after 1998 was how locally embedded conceptions of commonweal – with their associated understandings of identity and entitlement – could be located within institutional arrangements being reconstructed at the national level. The political rationalities that now catalyse post-New Order Indonesia[7] are being forged from diffuse and rapidly realigning configurations of local, national and global interests and imaginings. In responding to this challenge, the state needed to (re-)create a political language and an institutionalized set of processes that could once again underpin the nation conceived of as a 'community' of interest and shared welfare. This coincided with a shift in global perspectives regarding state–society relationships: in policy narratives the state was no longer necessarily seen as the primary locus of political power and authority in the statist tradition of command and control. Rather, within the new governance paradigm, an 'enabling state' was now meant to work in concert with other actors to deal with collective action problems (Pierre and Guy 2000; Meynen and Doornbos 2004).

Under the good governance agenda, decentralization has been promoted as the technology of government available for achieving greater local participation in decision-making (Ribot 2001). In accord with this shift, the Indonesian state embarked on a process of reinventing institutional structures at the district (*kabupaten*) level and fashioning these domains as loci of accountability and representation through which a national commonweal could reinscribe itself in diverse localities. Revamped district institutions were expected to develop the capacity to 'bridge' state and civil society (Schönwälder 1997). In this way the reforms set out to rebuild governance systems, make Indonesia again attractive to investors and quell internal disaffection, especially among regional societies.

The outer island provinces that had been severely marginalized during the New Order period were now stridently demanding greater autonomy from the centre. Amid fears of national disintegration, legislators rushed through decentralization legislation. The Regional Government and Revenue Allocation Laws of 1999,[8] among the most sweeping of the legal changes introduced in the early Reform Era, gave greater powers to regional government in an attempt to assuage some of the resentment which has made local autonomy one of the driving issues of the political reform movement (*reformasi*). The legislation signalled major changes in the relative authority vested in representative bodies *vis-à-vis* the executive at every level, and a less hierarchical relationship among central, provincial, district, and village levels of administration.

In accord with the wider concepts and practices of 'good governance' circulating internationally, decentralization had long been advocated by the World Bank and other donors on the assumptions: that local governments are more accessible and therefore decision-making more transparent and accountable than is possible under strong centralized governments; that democratization will foster more direct community participation in development planning and resource management; and that tying decision-making processes more closely to the costs and benefits of resource allocation at the local level would produce better long-term outcomes for communities and the environment (MacAndrews 1986; World Bank 1994; Devas 1998; Manor 1999). Questions were immediately raised, however, regarding the extent of devolution of power and finance that would actually be implemented, and the practical potential of the new laws for achieving participation, equity or sustainable development. Given the political context in which the legislation was introduced, regional autonomy had to be understood primarily as a policy instrument directed towards national preservation, with questionable commitment from Indonesia's national elite (Aspinall and Fealy 2003; Erb *et al.* 2003; Schulte Nordholt and van Klinken 2007).[9]

Among contentious provisions of the new legislation was the vesting of autonomous authority at district/regency (*kabupaten*) level at the expense of the higher tier provincial (*provinsi*) government. The provinces had relatively greater authority under the previous structure, and theoretically offered a more qualified human resource base, broader policy perspectives, and greater capacity to redistribute public goods. Critics pointed to evidence of district policies favouring short-term, unsustainable resource extraction, and of local elites co-opting district government to private ends. But independent power in the hands of provincial governments had always been regarded with suspicion. Decentralization of authority to provincial level was perceived to threaten the integrity of the Indonesian state (Haris 1999; Juliantara 2000; Resosudarmo 2005; Schulte Nordholt and van Klinken 2007). Notwithstanding the transgressions of the Suharto era, and the ambivalent attraction of both national and local identities for most Indonesians, the unified nation-state held too powerful a place in popular-revolutionary concepts of an Indonesian 'commonweal' to be summarily abandoned.

Beyond reform of political structures, natural resource management was the next most urgent issue facing Indonesia's legislators. In 1999, shortly after approving the key decentralization legislation, parliament also passed a new framework Law on Forests (UU 41/1999). Although the framework regional autonomy law (UU 22/1999) had set out the fields of government where authority over natural resources was to be decentralized in rather an ambiguous way, it did emphasize the role of the districts. In contrast, the new forestry law allowed the Ministry of Forestry in Jakarta to remain the effective landlord of over 60 per cent of the nation's land area that was mapped as national 'forest estate' (*kawasan hutan*).[10] In contrast to the regional autonomy law, the forestry law assigned only limited authority to regional governments for issuing small-scale timber exploitation permits, retaining central ministerial authority over forest mapping and planning, the allocation of forest land uses, and licensing of large-scale timber concessions and industrial forest plantations.[11]

Legislative efforts to reconcile the conflicting claims of central and regional governments, and negotiate between popular demands and those of economic elites have been halting and inadequate. Acknowledging that the overlapping and contradictory state of existing natural resource and agrarian laws had led to conflict, exacerbated poverty and degraded the environment, the People's Consultative Assembly (MPR), the highest legislative body in the land, mandated the review and reorganization of all laws concerned with natural resources and land rights in its 2001 policy directive (TAP MPR IX/2001) (Lucas and Warren 2003; Contreras-Hermosilla and Fay 2005). In effect, legislators were pressed to recognize that a coherent policy response to Indonesia's many environmental crises and resource conflicts required reworking the nation's legal and institutional arrangements. This would necessarily entail reconsidering the ways in which the nation-state might serve as a locus for reconstructing the commonweal by working through the problems of nesting the local within the national resource management regime.

In subsequent discussions, reform of the critically important 1960 Basic Agrarian Law remained particularly contentious. Heavily influenced by the populist and socialist orientation of Indonesia's first President, Sukarno's, early post-revolutionary nationalist regime, the law restricted foreign ownership and paved the way for land redistribution and reform. At the same time, this fundamental law failed to grant rights in communal lands (*hak ulayat*) based on customary (*adat*) law a formal status equivalent to private (*hak milik*) title. Along with the Indonesian constitution, its privileging of 'national interest' in the disposition of land and resources had allowed the state to override the rights of regional minority cultures.

While the national land agency prepared a series of draft revisions to the 1960 Agrarian Law, other departments proceeded with draft frameworks for natural resources and spatial planning legislation. Competing departmental interests advanced their preferred concepts in the absence of an acceptable resolution of the conflict between investors, the state and the public, whose wellbeing these laws were supposed to protect. Conflicting interests among 'the people' themselves also proved difficult to broker.[12] In particular, the principle of prior collective rights claimed by minority indigenous cultures was pitted against the redistributive principle that had underpinned land reform under Sukarno and large-scale transmigration policies under Suharto.

With vying political interests represented in successive multi-party coalitions, and needing to shore up majority support in parliament, successive presidents have had limited capacity to coordinate policy in the Reform Era. While the notion of 'national interest' remains potent in political discourse and in legal formulations, the question of how to craft a coherent national policy framework for dealing with environmental and agrarian issues remains unresolved. In the meantime, the number of presidential decrees, government regulations, and ministerial decisions affecting land tenure and natural resources has continued to proliferate, in many cases with overlapping and contradictory provisions.[13]

Clearly, formal legal changes are only part of the story of Indonesia's changing governance scenario. To date legislative revisions have neither kept pace with popular demands for resource security, nor ensured protection of the nation's environmental 'commons'. Nevertheless, as the case studies in this book attest, the political dynamics surrounding regional autonomy have consolidated a dramatic transformation in the relationship between the regions and the centre, and with this, theoretically at least, the potential for better outcomes to emerge through local governmental processes. These detailed studies demonstrate the significance of local agency in reframing the practical operations governing communities and environments in the post-Suharto period. Local interests, both popular and elite, were not standing by waiting to see how far official implementation of regional autonomy policy would actually be taken by the state. On the ground, direct people's actions and the emergence or repositioning of local elites have generated changes in the distribution of resources, in many cases irrespective of legal niceties, and often exacerbating negative environmental impacts (Warren and McCarthy 2002; Lucas and Warren 2000; McCarthy 2004; Resosudarmo 2005).

Disillusioned with the corruption of state institutions – parliaments, courts, the bureaucracy – and their failure to represent the public interest or to fairly regulate the distribution of collective goods, local groups took matters into their own hands, staking their claims to a share in the nation's wealth. Land occupations, 'wild' logging and mining, people's 'justice', largely outside formal legal frameworks, became the primary avenue of public 'participation' in many parts of post-New Order Indonesia. This was especially apparent in regions where development of lucrative forest, mining and plantation sectors had dispossessed local communities in the long and vexed history of state and private 'enclosure' of traditional commons. In other cases, ordinary people had been forced to bear externalities in terms of environmental damage and infrastructure costs (especially in tourism, dam construction, mining and manufacturing), while reaping little benefit from the large-scale developments pursued by political cronies given privileged access to the nation's resources.

But even under the authoritarian New Order regime, settlers, investors, and agents of the state had sometimes found it necessary to pragmatically accommodate the *de facto* power of local actors and the underlying claims of local property systems. Informal alliances between military, regional official and local interests functioned in complex relationships to *adat* as well as official governance arrangements. All too often *ad hoc* local accommodations between parallel property systems and institutional arrangements came to serve the resource exploitation interests of colluding parties rather than sustainable management (Resosudarmo 2005; McCarthy 2006). In many respects the post-New Order state of affairs exacerbated these tendencies, creating a particular version of the 'tragedy of the commons' scenario.

Compounding the earlier disempowerments of local institutions, the loss of faith in the national promise of an Indonesian commonweal left a vacuum, with no legitimate institutional authority able to regulate resources in many parts of the archipelago. During the crisis that emerged after 1998, economic insecurity combined with the ambiguity in institutional arrangements and the failure of law enforcement to create a high degree of uncertainty for all resource users. The situation strongly favoured the rapid liquidation of resources, with actors working to take as much as they could as fast as they could, with little regard for future options (McCarthy 2004). Due to the erosion of its legitimacy, central government in Indonesia found it difficult to prevent the assertion of local interests well before the 1999 decentralization laws were implemented in 2001.

After 1998 district governments took the initiative to create new regulations on land and natural resource matters. In the midst of uncertainty, the districts created legal regimes that served their agendas by cherry picking from the complex matrix of national legislation.[14] In the first years districts enthusiastically extended their authority, granting small-scale exploitation permits to local clients, and contributing to the explosion in local (often extra-legal) logging operations occurring at this time (Barr *et al.* 2006). Districts and provinces also made use of their discretionary authority over spatial planning and new powers over land matters to grant plantation licences over large areas.[15] Although new provisions in agrarian, forestry and plantation laws and regulations allowed district and provincial government agencies some discretion to recognize *adat* rights,[16] few of these have been formally implemented. Because potent interests and incentives pushed in the other direction, the reforms failed to improve villagers' legal position significantly with respect to customary rights over resources (Colchester *et al.* 2006).

The need to iron out some of the discrepancies and put a break on the 'euphoria' of district opportunism provided the Megawati administration, supported by central government bureaucrats and legislators resistant to decentralization, the rationale to push back regional autonomy reforms. In late 2004 the national legislature replaced the earlier regional autonomy law by enacting Law No. 32/2004 on regional government. In clarifying the roles and authority of different levels of regional government, the new law emphasized the supervisory functions of provincial government.[17] Taken together with the other legal changes, this revised decentralization law muted the extent of authority and autonomy that districts had assumed with respect to natural resources under the 1999 law.[18] However, the state would not readily put newly formed local resource arrangements back into the 'national interest' bottle.

Critical to understanding the extent and implications of popular participation in decentralized institutions is an approach which links analysis of official government structures with studies of the informal governance practices that so often determine outcomes. As a preface to the case studies of complex resource contestation across the Indonesian archipelago in the Reform Era, the following section takes up theoretical treatments of the problems involved in efforts to construct and articulate sites of common interest – the commonweal – with a particular focus on the local domain. This discussion draws on the expanding literature on the 'commons' and collective action in dealing with resource management and entitlement (Young 2000, 2002; Berkes *et al.* 2003; Dolsak and Ostrom 2003; Ostrom and Nagendra 2006). Informed also by critiques from social and political ecology, we adopt an interdisciplinary and historically contextualized approach to analysing how struggles over natural resources are moulding social and environmental outcomes in today's Indonesia (cf. Peet and Watts 1996; Bryant and Bailey 1997; Mosse 1997; Ribot 1998; Johnson 2004; Robbins 2004).

Theoretical considerations

Despite the acute nature of Indonesia's problems, the issues the country currently faces are familiar ones. All societies confront a more or less analogous set of issues in constructing legitimate forms of governance for promoting the common interest and general welfare. The problem of building effective institutional arrangements for more sustainable use and fairer distribution of increasingly scarce resources has emerged as the ultimate challenge for governance from local to global scale in the new millennium. Over recent years a number of scholars have explored these 'commons' questions across a wide range of situations and contexts. In the process, the concept of the commons has expanded to incorporate public interest and collective action issues revolving around common ownership, community management, public space, and the accumulation of social capital. Here we focus on the dimensions of these debates explicitly related to the inter-linkages between questions of environmental sustainability and social equity.

The Commons

Hardin's (1968) classic exploration of the 'tragedy of the commons' explained eco-logical decline in terms of how the short-term economic incentives driving rational individual resource use work toward depletion and against the longer-term welfare and security of the larger social group. Hardin's account located the problem in the failure of the 'commons' as a common property resource system to contain the instrumental rationality driving unregulated individual behaviour. Subsequently the 'commons' has emerged as a central metaphor mobilized in discussions of environmental politics and efforts to define the conditions of effective resource and property regimes (Goldman 1998). Hardin's argument was taken by policy makers to support the neoliberal preference for privatization, since his conflation of 'the commons' with open access regimes had suggested that state or private manage-ment regimes were ultimately necessary to resolve the 'tragedy'. The assumed inefficiencies of state management left the private sphere the preferred model to produce more efficient, but debatably more sustainable or equitable outcomes.

Revisionist researchers responded to Hardin by examining how groups of users in different parts of the world have set up property regimes and institutional arrangements for resource management (McCay and Acheson 1987; Berkes 1989; Bromley 1989; Feeny *et al.* 1990). These discussions revealed that in many cases local institutions had indeed failed to ensure sustainable use, and, as described by Hardin, this has led to resource depletion. However, these authors cogently argued that this did not occur for the reason Hardin suggested. Rather, Hardin had mistaken an 'open access' situation, where there was an absence of any effective regulatory regime for the 'commons', where formal and informal systems evolved to control access and use. Contra Hardin, they showed how successful local level resource management regimes have worked effectively to exclude those without recognized rights or permissions, and to regulate use of resources by those with collectively recognized entitlements. Accordingly, rather than pointing to an intrinsic pathology of commons type management, Hardin's 'tragedy of the commons' was transformed into an allegory of resource depletion due to institutional failure. This failure was often ascribed to overlapping or ambiguous property systems, typically arising from state intervention and/or market penetration of common property regimes.

This discussion was taken further with the application of so-called 'collective action' approaches influenced by the New Institutionalist Economics with its focus on the relation between individual choice and social structure. Among others, Ostrom (1990) developed a general model of self-governing institutions, seeking to identify the conditions under which groups of people cooperate through institu-tions to successfully manage resources on the local level.[19] Ostrom particularly concentrated on how long-standing commons institutions functioned to overcome the 'tragedy of the commons' without the intervention of the state. The focal point of discussion was 'common pool resources' (CPRs) – resource systems from which it is difficult to control access and which can under certain conditions (population increase, commercial demand) be depleted. From this point a growing body of literature addressed environmental problems from an institutionalist perspective, discussing the 'design principles' that favour positive environmental outcomes.

These design principles encompass 'clearly demarcating boundaries, devising equitable rules for sharing benefits and costs, establishing effective monitoring arrangements for imposing graduated sanctions, and creating integrated systems by nesting smaller units within larger organizations' (Gibson *et al.* 2000: 228).

Ostrom (1990, 1992) persuasively contrasted the benefits of indigenous institutions evolved to suit local situations with the massive institutional failures that occur in highly centralized irrigation systems. This work demonstrated how large systems were unable to accommodate the interests of diverse groups and the variety of ecological and social variables involved. As Ostrom and Nagendra (2006: 1) have written, 'where users are genuinely engaged in decisions' regarding rules affecting use, there is a higher likelihood of users 'following the rules and monitoring others' than 'when an authority simply imposes rules'. This suggested that the solution might lie in the devolution of some responsibility to local user groups. Rather than a sovereign power (the state) making rules by fiat, Ostrom's model called for the development of multiple regimes where each organizational level would be relatively autonomous and could 'craft' rules to suit its own scale of management. In other words, each level should be 'nested' within a wider institutional framework responsible for a broad set of common interest concerns.

Ostrom's conclusions also resonated with current discussions of 'social capital' (Ostrom 2000; Putnam 2002; Dolsak and Ostrom 2003). For Ostrom, successful institutions (in this context defined as 'rules-in-use') heavily depended on shared values and norms. For example, she argued that local cultural traditions involve shared understandings of fairness, leadership responsibilities, members' rights and duties, and appropriate ways of doing things. Successful cooperation is connected to reciprocity[20] and the implicit expectation of benefits to be derived from future cooperation. Prior cooperation, institutional history, and past decision-making experience represent a resource (social capital) that can be used to forge future agreements. This leads Ostrom to argue that groups that do not develop these shared values may well find it difficult to craft institutions to manage CPRs effectively.

These revisionist analyses had a significant impact. Most notably, they reversed the polarity between public and private, state and local in assessing capacity to manage resources efficiently. Whereas the concept of the 'commons' had acquired a somewhat pejorative reputation after Hardin, now there was awareness that in a range of circumstances local institutions have sustainably managed CPRs. Consequently, scholars and policy makers became less enthusiastic about prescribing the transfer of property and resource management rights to central state control or to market mechanisms through privatization as solutions to commons problems.

Yet, a number of critical observations have emerged regarding the limitations of this revisionist 'commons' literature. Young (2002: 7) argues that in its focus on how to avoid the depletion and degradation predicted by Hardin's tragedy, this new model has been overstretched, with 'a range of situations that differ from one another in important ways' being forced into a single 'commons' conceptual box. In many situations where natural resources are shared, control is exercised by a range of different actors with diverse effects. In these situations analysis can typically identify 'a multiplicity of both discrete and overlapping property relations and a multiplicity of actors engaged in struggles over property rights' (Bruce *et al.* 1993: 628). Such situations tend to be too complex for the narrow institutional approach used in much of the commons literature. For instance, in real world situations, it is rarely possible to definitively distinguish public, private and common property, with complex bundles of property rights incorporating aspects of all three (Young 2002; Verdery and Humphrey 2004; Benda Beckmann *et al.* 2006).

The commons literature has tended to look at single CPR cases working on small spatial scales. However, most situations are more complex, with local institutions subject to various external pressures: management typically involves many heterogeneous actors, institutions and overarching levels of government. In addition, community management systems are affected by the state, the market and by other global forces. For these reasons communities are in many cases unable to effectively control the local environmental 'commons' on their own. This pointed to the need for multi-level and cross-scale analysis and management approaches (Berkes *et al.* 2003). In response, the institutionalist approach has adjusted, allowing that there is no single governance blueprint – in terms of institutional or tenurial arrangements – that would ensure sustainable management across different settings (Dolsak and Ostrom 2003; Ostrom and Nagendra 2006).

In explaining environmental outcomes in terms of individual actors' calculations regarding the costs and benefits of rule-governed behaviour, the collective action approaches ultimately derive from a disciplinary model of individual decision-making and rational choice. The central idea in collective action approaches is that institutions governing CPRs successfully provide structural incentives for individuals who, by behaving in accordance with them, reap benefits from restraint.

Many criticisms of this approach emanate from sociologists and anthropologists whose disciplines ultimately hold quite different epistemological assumptions. Rather than seeking to link individual psychological dispositions to structural incentives in explaining outcomes, the sociological tradition broadly assumes that social institutions channel individual action. In other words, holding that the normative obligations of social membership define the logic of individual action, sociological accounts focus more clearly on how the shaping of the normative world and the working of power and authority set the appropriate goals of individual action and specify the accepted means of pursuing those goals (Hechter 1981).

Seen from this perspective, rational choice analyses tend to rely on an ahistorical concept of self, abstracted from social and cultural context, without accounting for how subjectivities are constituted through specific social, cultural and discursive practices (Agrawal 1999). These theories also tend to depend upon an instrumental and historically de-contextualized understanding of common property (Mosse 1997). Further, as institutional approaches tend to be concerned with issues of efficiency and environmental conservation, they can insufficiently focus on distributive justice and entitlement issues (Johnson 2004). In other words, their concern is environmental governance rather than environmental justice (Zerner 2000).

Collective action approaches are also held to lack a mature approach to the issue of power. As a corrective Mosse (1997) has argued that analysis needs to more carefully attend to the 'historically-specific structures of power' that underlie the norms and conventions of collective resource use and the persistence of institutional arrangements. This analysis suggests the need to develop a more historically, culturally and politically grounded understanding of resources, rights and entitlements.

Furthermore, arguing for a broader conception of common property that recognizes symbolic as well as material dimensions, Mosse observes that common property resources are also repositories of symbolic resources. Following Bourdieu, he argues that conceptions of common property are expressive of social relations, status and prestige. As the maintenance of common property resources generates symbolic capital in the form of honour, authority and domains of influence, resource systems create economic capital, which is reinvested in symbolic capital via public institutions. This in turn can be deployed to effect social, economic and environmental outcomes. These synergies suggest that we need to look at uses of resources within a wider set of social and political relations, considering the intimate connection between institutional forms, cultural meanings, the articulation of interest groups, and the facts of ecology (Mosse 1997: 473).

One problem that the social and symbolic capital approaches tend to share with the commons literature in its many guises is an underlying tendency to instrumental reductionism. Even the expansions of the commons concept to include questions of symbolic value and identity politics, despite Bourdieu's warnings against economism, privilege instrumental readings of cultural meaning and social relations that leave little space for inherent or embedded value in nature, social or cultural life. Emotional attachment, intellectual commitment and other non-instrumental dimensions (however equivocal) of meaning and value are part of a 'moral economy' that infuses notions of commonweal with assumptions regarding justice, fairness and reciprocity.

Environmental governance: community, participation, decentralization

The emergence of revisionist commons and collective action approaches provided a shift in analytical focus toward practices of community management and stimulated an interest in local sites of intervention (Mosse 1997: 471). That is, the collective action approach came to form part of a discourse of governmentality (Agrawal 2005) that fits well with the dominant emphasis underpinning decentralization policy. Indeed, by valorizing the virtues of management by local user groups, this discussion also helped inspire the shift towards devolved governance of natural resources and new policy experiments with community-based resource management (Agrawal 2001; Persoon *et al.* 2003; Brosius *et al.* 2005; Dove *et al.* 2005).

However, the shift towards devolved management has generated its own set of issues. First, the community based natural resource management (CBNRM) approach tends to assume that the characteristics of 'community' promote desirable collective decisions.[21] For instance, there are assumptions in the CBNRM literature that because of local community dependence upon resources, they are more interested in sustainable use compared with distant state or corporate interests. In other words CBNRM posits a direct association between individual and collective, long- and short-term interests, and between community resource control and environmental protection. This assumption leads CBNRM frameworks to advocate returning rights of control to communities and privileging community management systems. But communities have diverse features, and in many cases local actors seek to secure their future through resource use and market involvement in ways that contradict the goals and assumptions of community-based conservation discourse (Li 2002; Persoon *et al.* 2003; Brosius *et al.* 2005).

This is connected with a second set of problems that derive from the way communities themselves are conceived. The CBNRM literature tends to assume that 'communities' are relatively homogeneous entities in which an uncontested 'public interest' can be readily discerned, collectively agreed and acted upon. Here the failure of some of the neo-institutional literature to recognize the import of structural considerations of power inequalities is a critical problem. To remedy such simplifications, analysis needs to bear in mind that unambiguously shared understandings stemming from face-to-face interaction within communities cannot be assumed. Community management does not guarantee the elimination of hierarchy, domination and marginalization that in varying degrees characterize power relations at all levels of social structure. As Agrawal (1999: 104) warns, 'one must turn to community as a form of social organization in which the concrete existence of difference, hierarchy, and conflict must be painfully and tediously negotiated if the political goals of development, conservation and democratic consolidation are to be meaningful'. Community-based management would necessarily need to engage with the troubling aspects of community complexities in negotiating 'the institutional arrangements that communities must cultivate to limit resource exploitation' (Agrawal 1999: 104). The emphasis on community reflects and reinforces an earlier move towards participation in government and development planning that began in the 1980s (Chambers 1983). Here participation was seen as both an instrument (it could facilitate the achievement of improved social and economic development objectives) and a goal (it might allow communities to have greater control over their lives and resources) (Little 1994). Broadly based public participation subsequently became a core principle in natural resource management, potentially contributing to economic efficiency, equity and development. Participation, it was argued, would make plans more relevant, give people more self-esteem, and help legitimize the decision-making process and the role of the state (Ribot 2001). Accordingly, 'participation' entered the vocabulary of public policy reform as theorists advocated the merits of decentralizing authority and resources within the state to lower levels of government. The argument was that decentralization could make the state more responsive to regional and local needs than when administrative powers were concentrated higher up within the bureaucracy (Webster 1992). Ribot (2001:13) argues that 'achieving many of the equity, efficiency, environmental and development benefits of participation is predicated on devolving decision-making powers and responsibilities' to local communities. 'This requires representative and accountable authorities or groups to whom powers can be devolved, or the need to create such authorities' if effective institutional forms do not already exist. In his view, 'locally accountable representation is a means for integrating across and mediating among' the differences found in communities.

The decentralized governance narrative conceived that a local government apparatus with devolved powers, by virtue of its position close to local populations, can perform a 'bridging function' between state and civil society (Schönwälder 1997). In other words, by linking the aspirations of local people with the political system, reformed local institutions would satisfy demands for more equitable and sustainable outcomes, and solve the legitimacy problem that faced the Indonesian state in 1997–1998. But effective reform requires a particular model of civil society that is capable of holding the state accountable and participating in decision-making. Whether the forms of participation and the type of civil society that the good governance concept requires are emerging throughout Indonesia remains an open question. In the wake of regional autonomy and democratic governance reforms since 1997, the studies in this volume consider how far transparency, accountability, and public participation in decision-making are contributing to a governance regime that would be able to address Indonesia's serious social and environmental problems.

Environmental justice, environmental governance and post-institutionalist approaches

We return now to the distinction noted earlier between environmental justice and environmental governance agendas (cf. Zerner 2000). On the one hand, the environmental justice agenda focuses on democratization, equitable resource access and sustainable development. Viewed from the perspective of a villager, perhaps, environmental justice might be understood in terms of moral economy: how should the governance of environmental resources be brought into line with the prevailing collective perception of fairness in the local and wider economic spheres? This raises culturally moulded questions of entitlement as well as distributive justice. On the other hand, the environmental governance agenda focuses on the effective management of environments, or more specifically on the institutional, political and economic means and processes through which environmental protection and sustainable use outcomes could be achieved, with little or no concern for distributive effects.

The collective action literature we discussed earlier is concerned primarily with the environmental governance agenda. A different literature analyses environmental problems in terms of questions of power, entitlement and 'moral economy' (Munro 1998; Agrawal 1999; Johnson 2004). This political and social ecology literature recognizes that individual and group rights are located within a community and/or a state regulated bundle of rights, the concrete exercise of which depends upon the power of actors to set, modify and enforce their claims within that domain. Furthermore, the ability of actors to benefit from their entitlements is also shaped by political and social conditionalities – including their capacity to organize to access and manage resources necessary for production, as well as their position within the circuits through which commodities circulate, whether reciprocal exchange arrangements or market mechanisms (cf. Robbins 2004; Bryant and Bailey1997; Mosse 1997; Ribot 1998; Ribot and Peluso 2002).

The fact that, as Johnson (2004) notes, these literatures tend to speak past each other, reflects the reality that the two agendas exist in tension: effective environmental governance may involve a trade-off with equity issues at some scale (Zerner 2000). This was implied by the tragedy of the commons parable. For instance, locking up a forest or a marine area for biodiversity conservation is likely to work against the immediate needs and wishes of local resource users. On the other hand, the long-term environmental services such protection provides may enhance future livelihood prospects and benefit a much wider group. Distributive justice and entitlement claims are not necessarily correspondent either. Citizens may assume an entitlement to benefit from all the nation's resources, while members of particular ethnic groups living in their traditional territories claim local resources by virtue of their indigenous status and customary rights. Further, these concepts are accommodated by potentially competing grounds of legitimacy in different discursive contexts. For example, Indonesia's indigenous minorities and poor transmigrants would appeal to different originary and redistributive principles in asserting their claims to the same land and resources.

While much of the environmental governance literature tends to overlook these perhaps irreconcilable distinctions, there is at least a tacit link between environmental justice and environmental governance agendas. The so-called 'four pillars of good governance' – accountability, transparency, legal predictability and widespread participation – entail democratic, rights-based decision-making. If these principles were materialized in resource governance processes, some of the worst excesses that have plagued decision-making dominated by vested interests in Indonesia could be constrained. Many environmental problems in Indonesia emerge from the transgression of locally embedded notions of justice and entitlement, where popular resentment helps fuel conflict and rapid liquidation of resources. On the other hand, these four pillars do not address sustainability, unless 'accountability' is expanded beyond its normal sense to internalize environmental costs and include ethical considerations of intergenerational equity and rights of 'nature'.

We argue that, despite the epistemological and methodological differences that draw these literatures apart, this bifurcation of environmental governance and environmental justice concerns is hardly beneficial. Bringing the problems addressed by the two literatures together helps to expose the nature of the practical trade-offs that otherwise remain implicit in policy discussions.

While institutionalist accounts do recognize the importance of social and informal processes, they tend to focus on 'crafting' institutions for better outcomes. Such accounts have been criticized for failing to fully consider the complexity and fluidity of the practical social arrangements that shape outcomes, often in ways that vary substantially from more public forms of decision-making and negotiation. Indeed, the *de facto* arrangements determining resource outcomes tend to be deeply embedded in social relations where there is often a great deal of ambiguity regarding rights of access and compliance with rules. Here outcomes occur through continuous processes of dispute, negotiation and 'the practical adaptation of customs, norms and the stimulus of everyday interactions' in decision-making (Cleaver 2001: 42). In other words, practical arrangements can also work as more informal 'patterned ways of living together', sets of mutual expectations that minimise anxiety and conflicts and provide tacit rather than explicitly formulated rules for organized action (cf. Bellah *et al.* 1992).

This 'post-institutionalist' approach questions the instrumental emphasis on 'crafting institutional arrangements' or 'getting the institutions right', instead suggesting the need to accept the dynamic nature of institutional practices, and accommodate 'a variety of partial and contingent solutions' (Cleaver and Franks 2003: 2).

If social and cultural embeddedness adds another dimension of intricacy, the multi-scaled character of identity formation and of governance in an era of intensifying global integration and interdependence complicates this further. Outcomes at the local level emerge within and as a result of socially embedded dynamics, at least as much as they are affected by the officially constituted institutional regimes associated with the state and with globalizing processes.[22] Socially embedded processes include local customary *adat* systems and other normative arrangements that order local property relations and are used to settle disputes. They also encompass social fields in which relations based on reciprocity and obligation including community, kinship and patron–client arrangements regulate social action. As the case studies in this book demonstrate, in many instances these processes are more important than state recognized structures and procedures set out in 'official' regional autonomy policies.

Contests over the powerfully linked resource, identity and authority domains represented in local, regional and national regimes are a prominent feature of the changing political constellation of post-New Order Indonesia. While the terms 'commons' and 'collective goods' have restricted reference to natural and institutional resources, the metaphoric glossing of 'commonweal' captures the wider range of senses in which access to resources is tied up with overlapping memberships and strategic positionings within and across these 'communities' of identity and interest. The studies in this book focus on the interdependencies between the natural, cultural and institutional resources that represent the most significant collective assets for local communities. They explore the articulations of practical interests and symbolic identities in the emerging local resource governance regimes of contemporary Indonesia.

Comparative case study themes

The case studies presented in this book investigate the concrete implications of Indonesia's changing governance regimes in the quest for democratic reform and for more equitable and sustainable resource use. They explore efforts to 'locate the commonweal' through studies carried out in a cross-section of local contexts. Cases have been selected to cover the spectrum of significant resource sectors – agriculture, fisheries, forestry, mining and tourism – and to represent the marked regional differences that characterize Indonesia's cultural-ecological landscape. The studies attempt to encompass the breadth and complexity of these issues by focusing on the relationship between local governance, environmental management and livelihood security as these intersect at particular regional sites. They investigate the complex relationships of contestation and accommodation within local communities and the increasingly fluid nature of the realignments taking place across scales and levels as regional autonomy policies and political democratization alter the power relationships that determine resource allocation.

The first case study, by Jim Schiller and Achmad Uzair Fauzan (Chapter 2), concerns forest, fishing and mineral resources in the Jepara region on the north coast of Java. In microcosm this account demonstrates the complexity of economic, social and ecological issues affecting local resource management in today's Indonesia. The Jepara case illustrates the vicious cycle of over-exploitation fuelling further resource decline. The study does show that increasing assertiveness of local groups over resources has in many but not all instances been able to force a more inclusive approach to decision-making on resource issues in the Reform Era. At the same time, Schiller questions the extent to which policy shifts remain captive to elite interests and sacrifice the long-term livelihood security of the wider population for immediate needs or gain.

Anton Lucas (Chapter 3) studies livelihood contestations over fisheries in Tegal district, on the north coast of Central Java. Here resource decline also fuels sea tenure claims and inter-village conflict over access to fishing grounds, as fishers turn to more intensive and damaging fishing technologies. Increasing pressures on already marginal livelihoods among artisanal fishers and hired workers in the fishing industry are a product of declining catches, rising costs, more intensive technologies and the profit margins of creditors and market intermediaries. Lucas examines some examples of conservation and development project interventions, which offer some potential for balancing livelihood and conservation values.

Greg Acciaioli (Chapter 4) analyses customary (*adat*) custodianship, resource entitlement, and multi-ethnic contestations in the management of Lore Lindu National Park, Sulawesi. Here an international environmental organization, The Nature Conservancy, has had long-standing involvement in conservation policy through its partnership with state agencies in the creation and monitoring of protected areas. This case explores the negotiations involving TNC and Lindu people toward accommodating *adat* claims in return for environmental commitments. It also explores the implications of indigenous co-management policies for different ethnic groups seeking to control land and resources in the area, and their deployment of developmentalist, environmentalist and indigenous (*adat*) rights discourses to these ends.

Laurens Bakker (Chapter 5), in contrast to some of the other case studies presented in the book, describes inclusive and foresighted community governance in two of the hinterland villages he studied in East Kalimantan. A strong sense of identity and resilient customary (*adat*) institutions continue to play an important role in decision-making and inter-community negotiations on forest protection and use in his case studies. While isolation and the high degree of integrity of local institutions has provided conditions conducive to good commons management in the two communities to date, different internal dynamics and relationships with outside groups and interests produce considerable diversity in their respective approaches to commons challenges.

Hidayat Alhamid, Chris Ballard and Peter Kanowski (Chapter 6) consider the history of disenfranchisement and the future prospects for indigenous forest management in Papua. Like minority groups across the country, the Rendani people had no say in the imposition of protected status on their forests. Now, with overlapping state and traditional management systems, effective regulation is lacking. This case study investigates the shortcomings of state responses to local demands for greater benefits, focusing in particular on the social and environmental implications of 'Special Autonomy' in Papua and the grafting of cooperatives onto 'reinvented' *adat* institutions.

John F. McCarthy (Chapter 7) describes the twin processes of deforestation and conversion to mono-crop plantation agriculture in Jambi, Sumatra. Focusing on the social and environmental impacts of the rapidly expanding oil palm plantation industry, this case study considers the extent to which reforms have increased popular participation in governance, and asks how decentralization and the revival of cooperatives have affected resource access and distribution, as well as patterns of conflict over agrarian resources.

Carol Warren (Chapter 8) traces the efforts of remote communities in Bali to reassert customary (*adat*) community control over land and resources through a community mapping programme sponsored by non-government organizations. Their attempts to constrain the alienation of land and the commodification of culture and environment associated with the mass tourism industry have met with some success. But, in the face of countervailing market forces, villagers and their NGO intermediaries have faced difficulties in realizing their ambitions to reduce dependence upon primary production through downstream processing and through small-scale community-based tourism.

In the final chapter, Warren and McCarthy return to the tragedy of the commons allegory and the debates it has generated on the role of communities, states and markets in confronting issues of equity and sustainability in contemporary Indonesia. They analyse the complexities revealed in these case studies of local responses to resource and conservation issues in the Reform Era, and consider the problems and possibilities of articulating common interests across local, national and global scales and levels.

These extended case studies explore how communities with different cultural backgrounds and resource bases are responding to greater autonomy, and to reformist pressures for popular participation in governance, more equitable resource access and distribution, and more sustainable management practices. The authors investigate the complex local dynamics within and among communities and local user groups as these engage with wider regional, national and international institutions and actors.

While the formal relationship between central and regional governments in Indonesia through decentralization reforms has attracted the most attention among political commentators (Erb *et al.* 2005; Schulte Nordholdt and van Klinken 2007), the effectiveness of decentralization and regional autonomy policies has as much to do with how these changes articulate to emerging formal and informal governance practices in villages and urban wards across the country. It is here that the real effects of reform must ultimately be measured. The contribution of the diverse case studies in this book lies in their detailed exploration of a range of important questions posed by the Indonesian case: Under what conditions have governance reforms led to greater public participation in decision-making, fairer resource distribution and better environmental management on the ground? To what extent do global rhetorics and processes of communication and economic integration influence the national/local political dynamic as these in turn frame public welfare and environmental sustainability debates? And how might a 'commonweal' articulate local communities within wider domains of authority and resources in building long-term common interests and identities? The studies that follow offer important insights into these urgent questions, of import beyond the Indonesian case.

Notes

1 The fires that ravaged Sumatra and Kalimantan, destroying up to 10 million hectares of forest, polluting the atmosphere and causing health problems at home and in neighbouring Singapore and Malaysia, were directly linked to government policies that had granted large timber concessions and plantation leases to national politico-business interests. While government officials and the timber industry blamed the exceptionally dry El Niño year and shifting cultivators' farming practices, evidence from satellite photographs showed that most of the fires originated in logging concessions and areas set aside for timber and oil palm plantations, with many of these licences held by Suharto cronies. With fire the cheapest means of clearing remnant vegetation for conversion to oil palm plantations, it was alleged that plantation owners had deliberately set many fires. Despite claims that those responsible for the fires would be fined heavily and have their permits cancelled, none of the major conglomerates with timber or oil palm interests were prosecuted (*Bisnis Indonesia*, 18 September 1997; *Kompas*, 22 March 1998; Reuters 15 April 1998).

2 See Hardin 1968, 1998; Feeny *et al.* 1990; Mosse 1997; Uphoff and Langholz 1998; Gibson *et al.* 2000; Dolsak and Ostrom 2003; Johnson 2004.

3 Derived from the Old English referring to the common welfare and common wealth, the term 'commonweal' is more widely used today in American English. See *New Oxford American Dictionary* and *Oxford Dictionary of English Etymology*.

4 Sen (1981) developed the concept of entitlement for the analysis of famine, but it has also been applied to wider patterns of natural resource use and property rights. Leach *et al.* (1999: 233) have defined entitlements as the 'legitimate effective command over alternative resource bundles'. Entitlements emerge from the processes where actors negotiate access to and use of resources, such as land and labour. Such negotiations necessarily involve relationships of power and identity, entail debates over meaning, and are mediated by prevailing institutional orders (cf. Leach *et al.*, 1999: 235–6).

5 Formulated in the years following the Indonesian Revolution under the country's first President, Sukarno, the Basic Agrarian Law declares the state to be 'based on Indonesian Socialism', and asserts that all rights in land have a social function

(UUPA 5/1960: §5,§6). Both the Basic Agrarian Law and the Indonesian Constitution declare all natural resources to be ultimately under the control of the state 'in order to provide for the greatest wellbeing of the people' (UU Dasar 1945 §33; UUPA 5/60, §2.1).

6 See statements from FKKM (1998); KPA Munas (1998); Latin (1998); AMAN (1999); INFID (1999).

7 See Warren and McCarthy (2002) for an in-depth discussion of the colonial and post-colonial resource and governance policies that had precipitated these challenges to the centralized Indonesian nationalist model.

8 *Undang Undang No. 22/1999 tentang Pemerintahan Daerah* and *Undang Undang No. 25/1999 tentang Perimbangan Keuangan antara Pemerintah Pusat dan Daerah.*

9 Indeed, as the crisis situation stabilized over the following decade, many of the early policy directions toward decentralized governance began to be restricted or reversed (see below).

10 Significantly, the new forestry law allowed for *hutan hak* (forest rights) areas within the forestry estate where private rights are recognized. While the law (UU 41/1999, §5) also allowed for *hutan adat* (customary forests), areas of the national forestry estate which the state would recognize and surrender to *adat* community management, up until the time of writing the implementing regulations required have yet to be passed.

11 The Ministry of Forestry's reluctance to support decentralization with respect to Forestry Law was reinforced by PP 34/2002. This regulation limited the power of districts to issue permits and levy taxes, and significantly departed from the spirit of UU 22/1999 (see Barr *et al.* 2006).

12 See Lucas and Warren (2003) for a detailed account of differing perspectives within government and academic circles, as well as among farmers' groups and NGOs, on revision of the Basic Agrarian Law.

13 Since the end of the Suharto period, key agencies such as the Forestry Ministry, Environment Ministry (KLH), the National Land Agency (BPN) and National Planning Agency (Bappenas) have championed particular legal conceptualizations that served their respective bureaucratic interests. So Indonesia's policy framework for natural resources remains contradictory. For instance, while the Forest Law (UU Kehutanan 41/1999) banned open-cut mining in protected areas (§38), mining companies are able to obtain licences from the Ministry of Energy and Mining. The latter ministry has been able to obtain exemptions under two executive regulations (Perpu No 1/2004 and KepPres 41/2004) which amend the 1999 Forestry Law, permitting mining in protected areas under certain conditions. As the former minister of the Environment, Emil Salim has noted when discussing the lack of policy coherence and an overall legal framework, 'now every ministry has their own framework law. It all becomes chaotic (*semrawut*), just like a forest of laws' ('RUU Tata Ruang punya lelemahan besar yang membahayakan', www.republika.co.id, *Republika*, 22 March 2006). For further discussion, see Contreras-Hermosilla and Fay 2005. See also McCarthy's chapter on Jambi in this volume for a discussion of the conflicts emerging from lack of clarity regarding spatial planning, plantation and forest concession laws.

14 For overviews see Contreras-Hermosilla and Fay (2005), and Thorburn (2004).

15 *Kompas*, 22 February 2005; *Suara Pembaruan*, 23 March 2007.

16 For example, Kepmen Agararia No 5/1999; Forestry Law 41/1999.

17 For relevant discussions see Patlis 2005; Barr *et al.* 2006.

18 Significantly, the new regional government law (34/2004) also emasculated the village assemblies (*Badan Perwakilan Desa* [*BPD*]), empowered under the earlier law.

19 For reviews of this literature, see Agrawal (2001), Johnson (2004).

20 See Rawls (1999: 190–224) for an extended discussion of the principles of reciprocity underpinning conceptions of justice and fairness. Rawls attempts to chart a position that draws together utilitarian, rational actor approaches with a relational understanding of how the reciprocal contributes to the formation of the 'common good' (1999: 193).

21 Dove *et al.* (2005) on the other hand point to what they call the reverse 'community bias' in the persisting tendency of conservation projects to treat local communities as 'the problem'. Against this their research project sought to 'identify both the ways in which local communities support conservation and the ways in which supra-community forces undermine it' (2005: xvi).

22 As Cleaver (2001) notes, the formal/informal distinction is misleading given that traditional social organizations often have formal aspects while formal, modern organizations are also affected by informal arrangements. Following her suggestion, we substitute 'organizational' and 'socially embedded' for 'formal' and 'informal' where appropriate.

Bibliography

Agrawal, A. (1999) 'Community-in-conservation: tracing the outlines of an enchanting concept', in R. Jeffery and N. Sundar (eds), *A New Moral Economy for India's Forests? Discourses of Community and Participation*. New Delhi: Sage Publications.

Agrawal, A. (2001) 'Common property institutions and sustainable governance of resources.' *World Development* 29(10): 1649–72.

Agrawal, A. (2005) *Environmentality: Technologies of Government and the Making of Subjects*. Durham, NC: Duke University Press.

AMAN (1999) Statement of the Congress of the Indigenous Peoples of the Archipelago, 14–22 March, unpublished ms, Jakarta: Aliansi Masyarakat Adat Nusantara.

Anderson, B. (1991) *Imagined Communities: Reflections on the Origin and Spread of Nationalism*. London: Verso.

Arce, A. (2003) 'Value contestations in development interventions: community development and sustainable livelihoods approaches', *Community Development Journal* 38 (3): 199–212.

Aspinall, E. and Fealy, G. (eds) (2003) *Local Power and Politics in Indonesia: Decentralisation and Democratisation*. Singapore: Institute of Southeast Asian Studies.

Barr, C., Resosudarmo, I., Dermawan, A. and McCarthy, J. (eds) (2006) *Decentralization of Forest Administration in Indonesia*, Bogor: CIFOR.

Bellah, R. N., Madsen, R., Sullivan, W. M., Swidler, A. and Tipton, S. M. (1992) *The Good Society*. New York: Alfred A. Knopf.

Benda Beckmann, F., Benda Beckmann, K. and Wiber, M. (2006) *Changing Properties of Property*. New York: Berghahn.

Berkes, F. (ed.) (1989) *Common Property Resources: Ecology and Community-Based Sustainable Development*, London: Belhaven Press.

Berkes, F., Colding J. and Folke, C. (eds) (2003) *Navigating Social-Economic Systems: Building Resilience for Complexity and Change*. New York: Cambridge.

Berkes, F. and Farvar, M. T. (1989) 'Introduction and Overview', in F. Berkes (ed.), *Common Property Resources: Ecology and Community-Based Sustainable Development*. London: Belhaven Press, pp.1–17.

Berry, S. (1989) 'Social Institutions and Access to Resources.' *Africa* 59 (1): 41–55.

Bourdieu, P. (1977) *Outline of a Theory of Practice*. Cambridge: Cambridge University Press.

Bourdieu, P. and Wacquant, L. (1992) *An Invitation to Reflexive Sociology*. Chicago: University of Chicago Press.

Bromley, D. (1989) 'Property relations and economic development: The other land reform.' *World Development* 17 (6): 867–77.

Brosius, J. P. Tsing, A. L. and Zerner, C. (2005) *Communities and Conservation: Histories and Politics of Community-Based Natural Resource Management*. Walnut Creek, CA: Altamira Press.

Bruce, J., Fortmann, L., and Nhira, C. (1993) 'Tenure in transition, tenures in conflict: examples from the Zimbabwe social forest', *Rural Sociology* 58 (4): 626–42.

Bryant, R. L. and Bailey, S. (1997) *Third World Political Ecology*. London: Routledge.

Butcher, J. (2004) *The Closing of the Frontier: A History of the Marine Fisheries of Southeast Asia c. 1850–2000.* Singapore: Institute of Southeast Asian Studies.

Chambers, R. (1983) *Rural Development: Putting the Last First.* Essex: Longman.

Cleaver, F. (2001) 'Institutions, agency and the limitations of participatory approaches to development', in B. Cooke and U. Kothari, *Participation: The New Tyranny.* London: Zed Books, pp. 36–55.

Cleaver, F. and Franks, T. (2003) How institutions elude design: river bank management and sustainable livelihoods. *BCID Research Paper No 12.* Alternative Water Forum, Bradford Centre for International Development.

Colchester, M., Jiwan, N., Andiko, Sirait, M., Firdaus, A.Y., Surambo, A. and Pane, H. (2006) *Promised Land: Palm Oil and Land Acquisition in Indonesia: Implications for Local Communities and Indigenous Peoples.* Bogor: FPP, Sawit Watch, ICRAF, HuMa.

Contreras-Hermosilla, A and Fay, C. (2005) *Strengthening Forest Management in Indonesia Through Land Tenure Reform: Issues and Framework for Action,* Forest Trends, http://www.forest-trends.org.

Davidson, J. S. and Henley, D. (2007) *The Revival of Tradition in Indonesian Politics: The Deployment of Adat from Colonialism to Indigenism.* London: Routledge.

Devas, N., (ed.) (1989) *Financing Local Government in Indonesia.* Athens, OH: Ohio University Center for International Studies.

Devas, N. (1998) 'Indonesia: what do we mean by decentralization?' *Public Administration and Development,* 17 (3): 351–67.

Dolsak, N. and Ostrom, E. (eds) (2003) *The Commons in the New Millennium: Challenges and Adaptation.* Cambridge, MA: MIT Press.

Dove, M., Sajise, P. and Doolittle, A. (eds) (2005) *Conserving Nature in Culture: Case Studies from Southeast Asia.* New Haven, CT: Yale University, Southeast Asia Studies.

Erb, M., Sulistiyanto, P., and Faucher, C. (eds) (2005) *Regionalism in Post-Suharto Indonesia.* London: Routledge Curzon.

FKKM (1998) 'New era for Indonesian forestry, forest resource management reformation', http://forests.org/gopher/indonesia/newindo2.txt (22 September 1998).

Feeny, D., Berkes, F., McCay, B. and Acheson, J. M. (1990) 'The tragedy of the commons twenty two years later', *Human Ecology* 18 (1): 1–19.

Gibson, C. C., McKean, M. A., and Ostrom, E. (2000) 'Forests, people, and governance: some initial theoretical lessons', in *People and Forests: Communities, Institutions, and Governance,* C. C. Gibson, M. A. McKean and E. Ostrom (eds). Cambridge, MA: MIT.

Goldman, M. (1998) 'Inventing the commons: theories and practices of the commons' professional', in *Privatizing Nature: Political Struggles for the Global Commons.* New Jersey: Rutgers University Press, pp. 20–53.

Greenough, P. and Tsing, A. L. (eds) (2003) *Nature in the Global South: Environmental projects in South and Southeast Asia.* Durham, NC: Duke University Press.

Hardin, G. (1968) 'The tragedy of the commons', *Science* 162: 1243–8.

Hardin, G. (1998) 'Extensions of the tragedy of the commons', *Science* 280, http://www.sciencemag.org.

Haris, S. (1999) *Reformasi Setengah Hati.* Jakarta: Penerbit Erlangga.

Hechter, Michael (1981) 'Karl Polanyi's social theory: a critique', *Politics and Society* 10 (4): 399–429.

INFID (1999) Statement of the National Congress of the NGO Forum on National Development INFID, July 1999, unpublished ms, Bali: Indonesian NGO Forum on National Development.

Juliantara, D. (ed.) (2000) *Arus Bawah Demokrasi: Otonomi dan Pemberdayaan Desa.* Yogyakarta: LPU.

Johnson, C. (2004) 'Uncommon ground: the 'poverty of history' in common property discourse', *Development and Change* 35 (3): 407–34.

KPA Munas [Musyawarah Nasional] (1998) *Deklarasi Musyawarah Nasional Konsorsium*

Pembaruan Agraria, Yogyakarta, December, unpublished ms. Bandung: Konsortium Pembaruan Agraria.

Latin (1998) 'Reorientasi Sektor Kehutanan Untuk Menduking Pemberdayaan Ekonomi Rakyat', http://www.latin.or.id/berita_biotrop.htm.

Leach, M., Mearns, R., Scoones, I. (1999) 'Environmental entitlements: dynamics and institutions in community-based natural resource management', *World Development* 27 (2): 225–47.

Li, T. M. (2000) 'Articulating indigenous identity in Indonesia: resource politics and the tribal slot', *Comparative Studies in Society and History* 42 (1): 149–79.

Li, T.M. (2002) 'Local histories, global markets: cocoa and class in upland Sulawesi', *Development and Change* 33 (3):415–37.

Little, P. D. (1994) 'The link between local participation and improved conservation: a review of issues and experiences', in D. Western, R. M. Wright and S. C. Strum (eds), *Natural Connections: Perspectives in Community-based Conservation*, Washington, DC: Island Press, pp. 347–72.

Lucas, A. and Warren, C. (2000) 'Agrarian reform in the *era reformasi*', in C. Manning and P. van Diermen (eds) *Social Dimensions of Reformasi and Crisis*. London, Zed Books, pp. 220–38.

Lucas, A. and Warren, C. (2003) 'The state, the people, and their mediators: the struggle over agrarian law reform in post-New Order Indonesia', *Indonesia* 76: 87–126.

Manor, J. (1999) *The Political Economy of Democratic Decentralization*, Washington, DC: World Bank.

MacAndrews, C. (ed.) (1986) *Central Government and Local Development in Indonesia.* Singapore: Oxford University Press.

McCarthy, J. F. (2004) 'Changing to gray: decentralization and the emergence of volatile socio-legal configurations in central Kalimantan, Indonesia', *World Development* 32 (7): 1199–223.

McCarthy, J. F. (2006) *The Fourth Circle: A Political Ecology of Sumatra's Rainforest Frontier.* Stanford, CA: Stanford University Press.

McCay, B. and Acheson, J. (eds) (1989) *The Question of the Commons: The Culture, and Ecology of Communal Resources.* Tucson: University of Arizona Press.

Meynen, W. and Doornbos, M. (2004) 'Decentralising natural resource management: a recipe for sustainability and equity?', *European Journal of Development Research* 16 (1): 235–54.

Mosse, D. (1997) 'The symbolic making of a common property resource: history, ecology and locality in a tank-irrigated landscape in South India', *Development and Change* 28 (3): 467–504.

Mosse, D. (2001) '"People's knowledge", participation and patronage: operations and representations in rural development', in B. Cooke and U. Kothari (eds), *Participation: The New Tyranny?* London: Zed Books, pp. 16–35.

Munro, W. A. (1998) *The Moral Economy of the State: Conservation, Community Development, and State Making in Zimbabwe.* Athens, OH: Ohio University Center for International Studies Publications.

Ostrom, E. (1990) *Governing the Commons: The Evolution of Institutions for Collective Action.* Cambridge: Cambridge University Press.

Ostrom, E. (1992) *Crafting Institutions for Self-Governing Irrigation Systems.* San Francisco: ICS Press, Institute for Contermporary Studies.

Ostrom (2000) 'Private and common property rights.' in B. Bouckaert and G. D. Geest (eds.) *Encyclopedia of Law and Economics.* Cheltenham, UK: Edward Elgar, University of Ghent.

Ostrom, E. and Nagendra, H. (2006) 'Insights on linking forests, trees, and people from the air, on the ground, and in the laboratory'. *Proceedings of the National Academy of Sciences of the United States of America* (6 November 2006): 1–8.

Patlis, J. M. (2005) 'The role of law and legal institutions in determining the sustainability of integrated coastal management projects in Indonesia', *Ocean & Coastal Management* 48: 450–67.

Peet, R. and Watts, M. (1996) 'Liberation ecology: development, sustainability, and environment in an age of market triumphalism', in R. Peet and M. Watts (eds), *Liberation Ecologies: Environment, Development, Social Movements*. London: Routledge, pp. 3–37.

Persoon, G., van Est, D. and Sajise, P. (2003) *Co-Management of Natural Resources in Asia: A Comparative Perspective*. Copenhagen: NIAS Press.

Pierre, J. P. and B. Guy (2000) *Governance, Politics and the State*. New York: St. Martins Press.

Putnam, R. (ed.) (2002) *Democracies in Flux: The Evolution of Social Capital in Contemporary Society*. New York: Oxford University Press.

Rawls, John (1999) *John Rawls: Collected Papers*. Cambridge, MA: Harvard University Press.

Resosudarmo, B. (ed.) (2005) *The Politics and Economics of Indonesia's Natural Resources*. Singapore: ISEAS.

Ribot, J. C. (1998) 'Theorizing access: forest profits along Senegal's charcoal commodity chain', *Development and Change* 29(2): 307–41.

Ribot, J. C. (2001) Local Actors, Powers and Accountability in African Decentralizations: A Review of Issues, Paper prepared for International Development Research Centre of Canada Assessment of Social Policy Reforms Initiative, to be published by United Nations Research Institute for Social Development (UNRISD).

Ribot, J. C. (2002) African Decentralization: Local Actors, Powers and Accountability. UNRISD Programme on Democracy, Governance and Human Rights Paper, No. 8, December.

Ribot, J. C. and Peluso, N. L. (2002) 'A theory of access', *Rural Sociology* 62 (2): 153–81.

Robbins, P. (2004) *Political Ecology: A Critical Introduction*. Malden, MA: Blackwell.

Schulte Nordholt, H. and van Klinken, G. (2007) *Renegotiating Boundaries: Agency, Access, and Identity in Post-Suharto Indonesia*. Leiden: KITLV Press.

Schönwälder, G. (1997) 'New democratic spaces at the grassroots? Popular participation in Latin American local governments', *Development and Change* 28 (4): 753–70.

Scott, J. (1976) *The Moral Economy of the Peasant*. New Haven, CT: Yale University Press.

Sen, A. K. (1981) *Poverty and Famines: An Essay on Entitlement and Deprivation*. Oxford: Oxford University Press.

Thorburn, C. C. (2002) 'The Plot Thickens: Decentralisation and Land Administration in Indonesia.' http://dlc.dlib.indiana.edu/archive/00001161/00/Thorburn.pdf.

Uphoff, N., and Langholz, J. (1998) 'Lessons for avoiding the tragedy of the commons', *Environmental Conservation* 25 (3): 251–61.

Verdery, K., and Humphrey, C. (eds) (2004) *Property in Question: Value Transformation in the Global Economy*. Oxford: Berg.

Warren, C. and McCarthy, J. (2002) 'Adat regimes and collective goods in the changing political constellation of Indonesia', in S. Sargeson (ed.) *Shaping Common Futures: Case Studies of Collective Goods, Collective Actions in East and Southeast Asia*. London, Routledge, pp. 75–102.

Webster, N. (1992) 'Panchayati Raj in West Bengal: popular participation for the people or the party?', *Development and Change* 23 (4): 129–63

World Bank (1994) *Governance: The World Bank's Experience*, Washington, DC: World Bank.

Young, O. R. (2000) 'Institutional Interplay: The Environmental Consequences of Cross-Scale Interactions. Constituting the Commons: Crafting Sustainable Commons in the New Millenium', the Eighth Conference of the International Association for the Study of Common Property, Bloomington, IN.

Young, O. R. (2002) 'Why is There No Unified Theory of Environmental Governance?', presented at 'The Commons in an Age of Globalisation', the Ninth Conference of the

International Association for the Study of Common Property, Victoria Falls, Zimbabwe, 17–21 June.

Zerner, C. (2000) 'Towards a broader vision of justice and nature conservation', in C. Zerner (ed.), *People, Plants, & Justice*. New York: Columbia University Press.

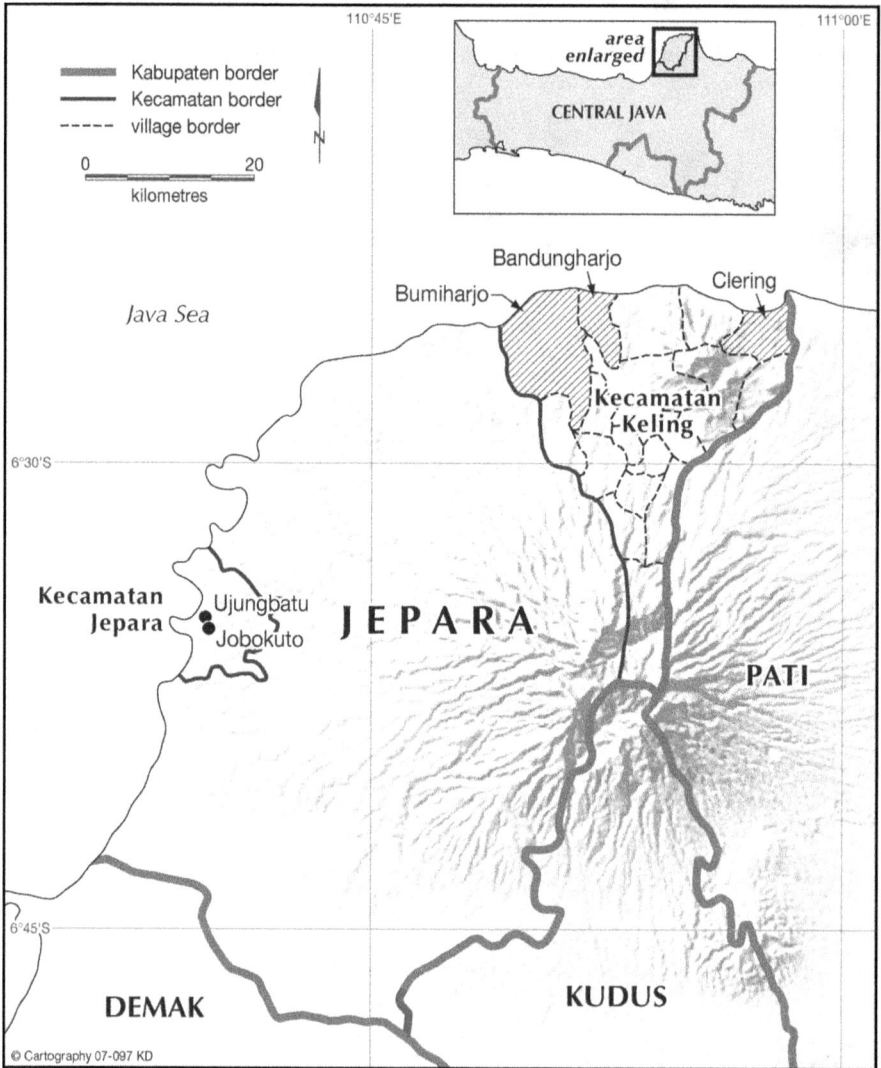

Map 2 Jepara District (kabupaten), Central Java Province.

2 Local resource politics in Reform Era Indonesia

Three village studies from Jepara

Jim Schiller and Achmad Uzair Fauzan

This study explores natural resource politics and state–society relations in Jepara, Central Java. It examines villagers' efforts to influence the way resources are used or distributed. It asks how villagers perceive and pursue common interests in cases of resource conflict and how the local state (Pemda) considers and responds to community conflict over resource use. The chapter will pay particular attention to the impact of changes in the distribution of authority between levels of government on the ability of citizens to express grievances and pursue aspirations, as well as the capacity of local government and civil society to resolve resource conflicts. The study traces natural resource conflicts in three sectors of the Jepara economy, forestry (and the related furniture industry), fisheries and mining. These cases illustrate changing and uncertain distributions of power between levels of government that affect resource contestation.

Jepara and its ecosystem

> The term amphibian is rather fitting for this district [Jepara] which is on the Muria peninsula. Citizens can earn a living from the sea, the lowlands or the mountain.[1]

Jepara is a district (*kabupaten*) on the north coast of Central Java. Its administrative centre, the town of Jepara, is located 70 kilometres from the provincial capital, Semarang. Most of Jepara's million people live on 1,000 square kilometres of land between Mount Muria and the Java Sea.

The marine, forest and mountain environments have long oriented Jepara's economic activity. In its 'golden age', 300 to 400 years ago, location and its sheltered harbour made Jepara a successful trading port. Reduced forest cover and the resulting siltation gradually pushed the coastline seaward and made Jepara's port too shallow for large ships. The Dutch colonial forces that razed Jepara (in 1701) in reprisal for trading with other Europeans, concentrated their settlement in the more accessible town of Semarang. This combination of political and ecological factors left Jepara a backwater until the 1980s when first Jepara's domestic and then its export furniture industry boomed.

Mount Muria protects Jepara's Java Sea fish nurseries from damaging easterly winds and waves, and its watershed forests slow runoff and thus provide year-round water for farming and other human needs while protecting against flooding. They also have provided much of the quality timber which Jeparans used to gain their national and international reputation as woodcarvers, carpenters, and furniture makers.

Both the forest and marine environments have come under increasing pressure since the 1980s. Approximately 24 per cent of Jepara is classified as forested: some 17 per cent is or was mainly teak forest controlled by the State Forestry Company (SFC); 4 per cent is planted in rubber and coconut trees and controlled by a state plantation company, PTPN Nusantara IX; 3 per cent is protected forest (*hutan lindung*) or privately owned or managed 'people's forest' (*hutan rakyat*)[2] and planted in a variety of trees including teak, mahogany and acacia. Following the 1998 collapse of the Suharto government the entire teak forest as well as some of the estate crops and nature reserves were cut down illegally. Since 1998 most teak forests have been replanted and looted again. By 2003 more than 27,000 hectares of land were declared degraded (*lahan kritis*). The decline in forest cover has been followed by increasingly frequent and severe flooding. Looting of teak forests in Jepara and across Java has also led to shortages of quality teak. This has weakened the furniture industry's international reputation. Poorer quality and increasingly expensive teak have led to job losses in an industry which directly employs 35 per cent of the local workforce.[3]

Data on the marine environment in the Java Sea near Jepara is scarce. What we do know suggests intensifying environmental stress from over-fishing and coral destruction, as well as declining biodiversity and fish stocks. Squires *et al.* (2003: 109) write that total fishing effort 'has most likely exceeded sustainable yield since at least 1985'. Jepara's growing population and industry have contributed to increased pollutants in the Java Sea. Extensive mangroves provided ideal conditions for fish nurseries and protected the coast from strong waves. Senior fisheries officials estimate that since independence more than 90 per cent of mangrove forests have been cleared for settlement, lucrative brackish water fish ponds (*tambak*), and firewood. This has led to coastal abrasion and reduced fish nurseries. Largely uncontrolled mining of coastal sands, riverbeds and mountain areas has also contributed to environmental degradation and flooding.

These environmental stresses interact with serious impacts on the economy of Jepara. The disappearance of Jepara's (and Java's) teak forest reduces the capacity of the furniture industry to offer alternative or supplementary employment for fishers. The over-fishing of the Java Sea reduces yields and encourages fishers to adopt more intensive fishing methods, which deplete fish stocks still further, eventually forcing fishers to search for other employment. Illegal and environmentally damaging mining is driven by declining opportunities in the marine, forestry and furniture sectors. Deforestation and mining-related erosion have led to flooding and coastal abrasion, which in turn put pressure on livelihoods in agriculture, fishing and aquaculture.

Jepara: state and civil society

The Jepara district government (Pemda) in the 1980s was a 'powerhouse' (Schiller and Martin-Schiller 1997), confident of the increasing scope and weight of its authority over its citizens. It was also a relatively responsive, 'developmentalist' regional authority whose senior officials sometimes found it easier to cooperate with local leaders than to enthusiastically pursue central government demands.

This responsiveness of regional government was largely due to the prominence of Nahdatul Ulama (NU, the Awakening of the Religious Scholars) as the paramount social organisation[4] in the area, as well as the growth of a largely indigenous Indonesian (*pribumi*) and locally owned teak furniture industry. NU provided a hegemonic social and cultural organisation whose leaders could deny the state legitimacy or cooperate to make local authorities successful. The growth of a confident indigenous business class, and of an industry that employed 85,000 workers in more than 2,000 small businesses, providing more than 28 per cent of the Gross Regional Domestic Product (GRDP) and 35 per cent of employment, encouraged community leaders to be more demanding and state officials to be more responsive, open, and cautious in dealing with Jepara's local elite. It also provided openings for 'ordinary' Jepara citizens to be more openly critical and to expect more from local government.

The fall of Suharto, combined with the promise of a new '*Era Reformasi*' – free from corruption, collusion and nepotism (KKN), with a reduced role for the military, and an unnerved government trying to prove it was more accountable – had a dramatic effect on the already vexatious political, economic and social climate in Jepara. In the 'euphoria' that followed Suharto's fall,[5] people felt completely free to vent their grievances against the state. At its peak in 1999, there were 1,080 cases of public protest in the villages of Jepara.[6] More than 100 village officials were forced to resign. The district government library was burnt down and other government buildings were the target of demonstrations and riots.

The regional government's response to the general loss of prestige of state authority was to try to appear yet more responsive and participatory. In mid-1998 the district head (*bupati*) issued a Plan for Implementing Reformasi in Every Field as proof of his pro-reform credentials. The plan specified the procedures, costs, times and offices involved in the delivery of common public services. It threatened to dismiss any village official or civil servant who did not follow procedures and urged citizens to report complaints.[7]

There have been several other participatory innovations. One of these involves travel by senior officials each Friday to attend mosque and engage in dialogue with villagers. Another is a regular talk-back radio show in which the district head and senior officials answer public questions or criticisms about government services. Still another is the evaluation of public services by NGOs, which includes a quarterly dialogue between local officials and NGOs at the Inter-actor Forum (*Forum Lintas Pelaku*) established by Pemda. Since the 1999 election the DPRD has been divided, with no political party dominating. The DPRD is at the same time an arena where local people can make demands and attempt to have their grievances redressed – through appeals to representatives, commissions or political parties – and a part of local government that is a target for critical oversight by NGOs.[8]

Civil society's response to the changed basis of state authority is visible in the mushrooming of advocacy NGOs and *ad hoc* citizens' organisations. More than 60 NGOs are registered, but the number of unregistered NGOs, movements, foundations and forums is much greater. Most of these organisations are local, but some outside NGOs operate in Jepara and many local NGOs work in conjunction with national and international agencies and donors. Some of these organisations emerge to uncover and combat injustices or poor governance, while others cooperate closely with Pemda to monitor or deliver its services.

Jepara's dense network of NGOs has been recognised by the district head Hendro Martojo as a useful source of feedback on government performance.[9] Of course, the government also tries to co-opt community leaders and to influence how organisations and the public see its actions.[10] However, in post-Suharto Jepara there is no question that Pemda faces more public scrutiny and pays more attention to public demands than ever before.

All three of the resource conflicts – regarding teak forests, fishing nets and feldspar mining – involved villagers less than 20 kilometres apart, in the Keling subdistrict in the north of Jepara (see map 2). Keling is the closest thing mainland Jepara has to a frontier, being as far from the district capital as you can travel by road or track. Some villages are barely 100 years old and population density is relatively low.[11] Keling contains most of Jepara's forests and mountainous areas. It has the highest percentage of non-Muslims, 9.5 per cent (Biro Pusat Statistik Jepara 2002: 184),[12] and is widely acknowledged to contain the district's highest percentage of nominal Muslims. It also has something of a reputation as a lawless area.

The three resource disputes discussed here involve villagers, the local state and private or state-owned companies. In each case we will outline the grievances, aims, actions, organisation, skills and social standing of the villagers trying to influence resource politics. We then look at the other contestants involved and discuss how and why Pemda (District Government) has responded to those disputes, and how its legal authority is understood. Finally, we analyse resource conflict outcomes and consider the current state of local resource politics.

Forests and furniture

... furniture is almost like the breath of life for Jepara people.[13]

Nationally and globally Jepara is best known for its furniture. Legends date carving skills back to the sixteenth century. Exports of classical Jepara teak furniture date back at least to the beginning of the twentieth century, but the antique reproduction and garden furniture exports for which Jepara is famous have expanded rapidly since the 1980s.[14] By 1999 exports and domestic sales were worth more than US$300 million annually.[15] Furniture and allied industries (sawmilling, logging, upholstery, packaging, transport, and accommodation) supplied more than 100,000 jobs and were easily the largest contributors to GRDP.

Illegal logging of state forests broke out over wide areas of Indonesia in 1998 and continued on a massive scale until 2001. Looting was especially severe in Jepara.[16] The high price of teak – on the market in 2001 a 3 m log 15–25 cm in diameter was worth Rp 750,000 or more than three months' wages for an agricultural worker – made it an alluring target.[17]

How could the State Forestry Company (Perhutani), Jepara's local government and owners and workers in its valuable furniture industry ignore the devastation of the forest upon which they depended? How did they view and respond to the forest resource crisis? Before we can answer these questions we need to understand the villagers' views of the resource use dispute.

Bumiharjo village is surrounded by the Indonesian State Forestry Company (SFC), Perhutani's Forest Sub-Unit, and by the PTPN IX Nusantara rubber and coconut plantation. The village's 7,500 people occupy 2,858 hectares. Some 75 per cent of the village land is owned by the two state forest companies, and only 25 per cent by villagers, 39 per cent of whom are landless agricultural labourers.[18] This shortage of land and of jobs increases Bumiharjo's dependency on the SFC. Peluso (1992) vividly describes forest villagers' resistance to efforts by the SFC and its employees to enforce compliance with company cultivation practices, rules and interests. Her study of peasant exploitation and resistance in the state teak forests of Java fits well with the Bumiharjo villagers' view of their relations with SFC.

One way that Bumiharjo villagers tried to overcome the shortage of land was to grow rice along the river bank (*lambiran*) that borders the plantation.[19] More than 400 villagers work the approximately 100 hectares of *lambiran*. SFC officials disputed the villagers' right to use this land, and claimed that villagers widened the river banks and encroached on teak forest land. Villagers accused the SFC officials of maliciously felling trees on to their rice crops just before the rice was to be harvested. They also claimed that they had to leave 5 per cent of any harvest in bribes for SFC employees.

Villagers also objected to the unequal allocation of land below the replanted teak trees to grow crops, a practice called *tumpang sari*. Only 380 quarter-hectare plots were set aside for *tumpang sari* and less than half of these were distributed to Bumiharjo villagers. The village had 842 landless agricultural labourers while less than 190 villagers had been allocated *tumpang sari* rights. Only 15 per cent of villagers had been given 2 hectares or more. Villagers complained that SFC employees unfairly benefited from their control over access to land, trees and labour. There were other grievances, but the central problems were that teak was a very valuable timber, that little of that wealth was flowing to those who lived near the forest, and that villagers believed that they were exploited by SFC and its employees.

All these local grievances might have been enough to motivate the 'wild logging' (*penebangan liar*) that occurred. However, notions of moral economy also came into play. It was widely believed that President Suharto, his family and cronies had accumulated great wealth, and much of it from Indonesia's forests.

Many saw the breakdown of the security apparatus as an opportunity to do what their leaders had done and have their turn at amassing wealth.[20] No wonder villagers talked of the looting as 'timber demonstrations' (*demo kayu*), not *penjarahan hutan* (forest looting), and joked about the forest as *hutan negoro* (literally, state forest, but in Javanese this can also mean 'forest to be cut down').[21]

It is clear that villagers had strong motives for demonstrating their unhappiness with Perhutani and with their impoverished circumstances. A representative in Jepara's district assembly saw the forestry villagers' relations with Perhutani as 'a time bomb waiting to explode'.[22] All that was needed was opportunity, and that was provided by the paralysis of the security forces that followed Suharto's fall. It is interesting to note, however, that 'wild logging' in the Bumiharjo forest did not occur until January 1999, after months of forest looting elsewhere in Java. Some of the villagers claimed that they joined in the looting only after they found that they could not stop outsiders from looting their forests. Reportedly, when Bumiharjo villagers blocked the road into the forest, outsiders blockaded the only other road to the village. Allegedly afraid of being isolated from the outside world and of being attacked, the locals joined the looting.

Whatever we make of the initial reluctance to participate in the looting, it did not last. One village official estimated that 75 per cent of the villagers were detained for forest looting, although most were released without being tried. Looting on a massive scale – involving hundreds, sometimes thousands, of villagers – continued for about three months,[23] and on a lesser scale until May 2000, when the entire forest was logged. Some replanting of teak continued but many young teak trees were stolen for firewood or pulled up to allow planting of peanuts, cassava and other crops. One stand of teak, plot 151, which was replanted in 2000, has been protected and six years later has 15 cm trunks and is the pride of the forest (see below). In 2005 a nearby 61-hectare nature reserve (*cagar alam*) which had been spared from earlier logging – reportedly because the rangers had good relations with the villagers – was entirely felled and the land planted in rice. Also in 2005 some of the low-value rubber trees from the Nusantara plantation were stolen to sell for firewood.

So what has been the outcome of the 'wild logging'? The short-term gain for the village was substantial.[24] The number of four-wheel vehicles owned by locals rose from 23 in 1997 to 79 in 2000. The number of motorbikes increased from 216 in 1997 to 683 in that year. The number of Type A (permanent, walled and tile roofed) houses rose from 32 to 342 over the three-year period. Village per capita income increased five-fold in current rupiah between 1997 and 2000.[25]

By 2003, however, the boom was over. Respondents reported that many villagers were unemployed and in financial difficulty. Because the forest was gone, villagers had access to more agricultural land. However, they discovered that in the absence of teak trees water was in short supply. The forest's root system had helped to retain water near the surface, and the result of wholesale felling was a substantial decline in crop yields. Cassava production yielded only 20 tons per year, worth Rp 4 million. It cost more than Rp 2 million to clear the teak roots and pay for seeds, fertilizer and labour expenses. One measure of economic decline was the sale of *nemer* (cultivation rights during the teak replanting process). Another was the village's own contribution to its development funds (*dana swadaya*), which fell from Rp 170 million in 2000 to 39 million in 2003.[26]

How did the SFC, the Jepara furniture industry and regional government respond? Nationally, SFC went through a change of status from autonomous state company (PTPN) to full public company status (*Perum*). It was also forced to deliver embarrassing annual announcements and forecasts of reductions in its timber production and profit. Having few trees left to log, SFC emphasised its role as a protector of the forest and environment rather than its role as a profit-making producer of timber. Much more could be written about the national SFC; however, this chapter will focus on SFC's local response in Bumiharjo village and Jepara district.

We need to consider both Perhutani's immediate action to try to stop or limit the forest looting and its longer term 'fix' for the security of the teak forests. The use of Perhutani's forest police, other local employees, and the state police to frighten away would-be illegal loggers was unsuccessful. They were too few, too fearful for their own and their families' lives – and, it is widely believed, too willing to be paid off – to catch many illegal loggers. Both villagers and Jepara furniture makers said that some SFC employees were themselves involved in arranging the transport and purchase of stolen timber. It was reported also that some were paid to certify illegally logged teak as legal and available for sale.[27]

Village informants said that the forest police and officials made patrols, but that the looters simply waited until they went home. Illegal loggers paid villagers to monitor and report the movements of forest police and security officials. Some local forestry officials tried (unsuccessfully) to curb looting by telling villagers they could fell a few trees for themselves and for their relatives. Some also wrote the names of those who logged trees on the remaining roots. One informant said that this delayed illegal logging for a few days while villagers waited to see if the loggers were detained. Police set up as many as nine roadblocks on the road to Jepara. However, these check points did not prevent the passage of illegal timber. Instead, police collected tolls of Rp 100,000 or more for each vehicle. On forest roads, villagers unable to log or carry timber charged fees to allow trucks or utility vehicles to pass.[28]

Perhutani's long-term effort to restore and safeguard the forest through community involvement – the Joint Community Forest Management scheme (*Pengelolaan Hutan Bersama Masyarakat* [PHBM]) – has not been well received in Bumiharjo. In 2002 a Forest Village Community Council (*Lembaga Masyarakat Desa Hutan* [LMDH]) was established to involve villagers in forest co-management. An informant on the executive board said it was 'theoretically formed for [village] prosperity, but in practice only so that the people do not steal the forest'.[29] The LMDH is supposed to have four sections, but only the security section is active.[30]

The SFC has reclaimed very little of the deforested land now being used for agriculture. The LMDH was told that when replanting occurs, the space between trees would be doubled allowing villagers to grow more crops for longer before the shade of the teak trees block growth. SFC officials had said villagers would receive 100 per cent of the proceeds of the first thinning of trees, which normally occurs about five years after planting. Plot 151 was replanted in 2000 and should have been thinned by 2005. This was not done, allegedly because SFC feared that the entire plot would be cut down. Now SFC is saying that the first thinning of Plot 151 will count as a repeat thinning so the LMDH will receive only 25 per cent of the sale price. Estimates are that it will be worth at least Rp 10 million to the LMDH. The LMDH is supposed to receive 25 per cent of the proceeds of each additional thinning and of the final sale of the teak trees.

There are several problems with Perhutani's solution. One is that very few in the village trust SFC to keep its promises. The change in the rules for Plot 151 reinforces the mistrust. A second is that there is lack of clarity and lack of agreement about who will receive the money. Some villagers believe that only the *nemer* holders who cultivate crops under the trees – in exchange for tending and protecting them – are entitled to a share.

Whether the money goes to the *nemer* owners or to the LMDH, distribution is still a problem. People who occupied the forest land immediately after the looting have received additional *nemer* from Perhutani, while those not involved in the looting or occupation of the forest did not. As the forest has disappeared the water and nutrients available for other crops has declined. Similarly, as incomes have fallen most *nemer* rights have been traded to more wealthy Bumiharjo residents and to outsiders. Distributing the profits to *nemer* holders would only increase inequality in Bumiharjo and alienate the landless agricultural labourers who see the proceeds of the forest going to Bumiharjo elites and outsiders.

If the money is handed over to the LMDH, the chairman plans to divide the receipts of the thinning: 35 per cent to the LMDH executive; 35 per cent to social groups (e.g. prayer groups, mosques and youth groups); and 30 per cent to *nemer* holders. This allocation is also problematic. People who were not given *nemer* have not joined the LMDH and may receive nothing. There is also certain to be conflict within the various groups and between them over the distribution of a large amount of money in a poor community.

A final problem is that while 25 per cent of a teak tree will be worth at least several thousand dollars in 60–80 years, as one villager put it, 'How can we eat until then?'[31] Teak may be an attractive long-term investment (in 2005 a first class 80 cm diameter log was worth $830 per cubic metre),[32] but a tree planted in 2005 might produce first-class logs in 2065 – a long wait for a poor farmer. It remains an unanswered question whether it is possible to convince poor people that it is better to look after an asset (teak) that will probably have a very good return in 60–80 years' time rather than using the land and labour to produce a small, but quick, cash or food crop return.

The furniture industry has certainly not assisted in resolving the resource use dispute. The industry, or to be more specific, the post-devaluation global demand for cheap Jepara furniture, helped to fuel the 'wild logging'. Most of the illegally logged trees from Bumiharjo, Jepara's 18 other forest villages and much of central and East Java became Jepara furniture. In this competitive industry illegal logs could be purchased for half to two-thirds of the official SFC auction price.

Furthermore, illegal logs could often be purchased on credit from sellers anxious to unload their cargo and return to the forest for more timber.

Furniture entrepreneurs wanted to maximise profits in the short term and the illegal logging made that possible.[33] The furniture makers and their association (Asmindo) assumed that other fast-growing timbers could replace teak forests (e.g. acacia), and that quality 'super-teak' could be made to grow quickly (15 years instead of 60–80).[34] They believed that they could substitute imported, rainforest mahogany for plantation teak, and that, if necessary, they could relocate their business or find another way to make money.

Pemda has long been supportive of development in the furniture industry.[35] If you talk to senior local officials it is obvious that they view the problem of forest management being able to provide a sustainable supply of quality timber for Jepara's industry as important, but well beyond its control. By 1997, Jepara's furniture industry was using about 600,000 square metres of teak per year. Two years later, in 1999, Jepara required in excess of 1 million square metres of teak annually. This was much more than the legal teak production for the entire island of Java, and clearly unsustainable. By 1997 Jepara's forests could provide only 1 per cent of its furniture industry's teak needs.

To maintain even Jepara's 1997 level of furniture exports would require the maintenance of environmentally sustainable and socially equitable conditions for teak production across Java. The chances of achieving this in the near future seem slim. Within Jepara the authority to manage the teak forests rests clearly with the SFC. As Zainuri, a representative in the Jepara District Assembly put it, 'the authority is with them (Jakarta). Our capability is the capability of the fax [to pass on messages]. Let them (SFC) manage, we should not intervene.'[36]

The district head, Hendro Martojo, has acknowledged that teak furniture production at 1999 levels is unsustainable. To increase the availability of timber for the furniture industry he proposed planting fast-growing teak, acacia and other fast-maturing trees, as well as importing other timbers. He has also promoted domestic tourism to help reduce the dependence on furniture.

Privately, many Pemda officials are disparaging of the way SFC manages the forests, deals with the forest villagers, and supplies timber to the furniture industry. They note that there was no looting of trees in the *hutan rakyat* (people's forest), privately owned land in which people grow teak and other trees, often provided by the District Estate Crop and Forestry Service. They argue that Pemda's own reforestation efforts are much more successful. Pemda has introduced a small-scale agricultural extension and community reafforestation programme providing teak seedlings and other tree crops for use by villagers on their own land. In the past this programme has given priority to villages with degraded land in river watershed areas. In 2006 the District Forest and Estate Crop Service began providing teak and other seeds to Bumiharjo.

The regional autonomy laws have done nothing to make it easier for Jepara's district government to resolve these forest resource disputes. SFC is at least promising to pay more attention to the welfare and rights of forestry villagers. The Jepara District Government is interested in the problem, but only has the power to advise the SFC and central government. Their message would be that extension work with forestry villagers on private land might increase production and be more sustainable.[37] But Pemda's small efforts at reafforestation to date apply only on private land outside the national forest estate.

The recent illegal logging of a nature reserve in Bumiharjo suggests that this more anarchic means of contesting resource use might be the most dangerous for the sustainability of the forest and the furniture industry. There was a degree of legitimacy about taking trees from the widely despised Perhutani, which was expressed in the idea of *demo kayu* (timber demonstrations). However, the act of looting one of Central Java's oldest nature reserves whose managers had maintained good relations with the villagers suggests that the villagers believed that the reserve would be logged by outsiders if they failed to log it first.

Jepara fisheries: *jaring cotok* conflict

> It is hard to be a fisher nowadays. The cost of going to sea is high but the value of the catch does not amount to much … [Costs of] almost all daily needs go up. But I still believe that I can live for as long as the sea is still there and boats can sail.[38]

Fishers have responded to declining yields and increasing costs in a number of ways.[39] One response is to move to other occupations to reduce dependence on fishing income. Until its recent decline, Jepara's furniture industry had provided more reliable and better paid employment alternatives for many. A second response is to invest more capital to acquire new technology, larger engines or more fuel, and to spend extra time at sea. All of these methods have been employed in Jepara, including the use of sonar, powerful lights and divers to find, attract and herd fish into nets. One technique adopted widely in the 1990s was to modify traditional *lampara* or *cantrang* nets, making the holes smaller and adding a weight to pull the net to the bottom. This net, called a *jaring cotok*, is trawled. Because the net is fine mesh it collects everything in its path, large fish and small, fish eggs, prawns and shellfish. And because it has a heavy plank or weight that is dragged across the bottom, it destroys coral. The weight on the *cotok* also frequently cuts stationary nets as it passes over them.[40]

As in the Tegal case (see Chapter 3), a more conservative and potentially more sustainable path to dealing with declining yields is to restrict access to fishing grounds and to prohibit the use of damaging new technology in designated zones. The aim is to restore fish stocks and achieve sustainable fish yields – in short, to optimise future income at the expense of current production. It is this effort by the fishers of north Jepara to keep these *cotok* mini-trawlers out of their traditional fishing zone that is the focus of this second case study.

The *cotok* is an innovation by fishers and *juragan* (fishing boat owners) in response to falling incomes, rising fuel and other costs, and to government policy. In the 1970s the government promoted the use of trawlers (called *pukat harimau*), which began to increase rapidly in Indonesian waters. In 1978, however, the central government banned trawlers from traditional fishing zones.[41] In the late 1980s and 1990s, in response to the ban and falling yields, some boat-owners began to modify their traditional nets.

The adoption of the *cotok* occurred mostly in central and south Jepara, where investors in fishing had to compete against the booming furniture industry for capital and labour. Those supplying capital and fishers in central and south Jepara had more exposure to consumer demands and more options for investing their capital and labour than those in the less commercial north.[42] In north Jepara there tends to be a larger percentage of small boats crewed by the owner and one crew member, while the central and south Jeparan boats tend to be larger, with multiple crew members. To put that another way, north Jeparans are artisanal fishers and they are in conflict with a capital intensive commercial industry in central and south Jepara.

The result of widespread adoption of the *cotok* was increased fish catches for the first few years, followed by falling yields and incomes. In response to a decline in income, *cotok* fishers, mainly from central and south Jepara, began to spend longer at sea and to range further outside their usual fishing grounds. This brought them into areas that the coastal fishers of north Jepara considered to be their traditional waters, which reportedly had not experienced as much coral damage or over-fishing as other Jepara coastal waters.

In Jepara, fishers make up only a small fraction of the population of most 'fishing' villages. Previous studies by Mubyarto *et al.* (1984) note that Jepara fishing communities tend to be poor, have lower levels of education and more health problems than those of other villages in Jepara. Fishing incomes fluctuate greatly. The west monsoon (September to March) brings rain and storms, which frequently prevent fishing. It is estimated that only 20 days a month are available for fishing. The full moon also reduces catches. Some fish populations are seasonal. Regardless of catches or prices, which rise and fall dramatically, fishers have to eat as well as buy and maintain boats, nets, engines and other equipment. They also have to purchase fuel for their boats.

These uncertainties have led to complex patron–client relations to share risks and the fish catch. Boat owners called *juragan* take a varying share of the catch, after subtracting running costs, and providing a smaller share to crew members (called *pendega*). *Pendega* rely on boat-owners, and both rely on *bakul* (fish-buyers) to provide the credit that will feed their families during times of hunger (*musim paceklik*) and to provide the money for fuel, repairs and the nets and other equipment needed for fishing.[43]

Jepara has about 3,500 fishing boats, 3,500 *juragan* and 8,950 *pendega*. The Jepara Marine and Fisheries Service claims that approximately 420 of those boats are equipped with *cotok*. North Jepara has only an estimated eight[44] *cotok* out of nearly 800 fishing boats. A larger percentage of fishers own their own boats than in the rest of Jepara. In Bandungharjo – the north Jepara fishing community at the heart of the effort to ban *cotok* – there were 112 fishing boats usually crewed by two, and only about 20 per cent of village households were regularly engaged in fishing. Their boats had small outboard engines and rarely ranged outside the four nautical mile Jepara district zone.

From the early 1990s the fishers of Bandunghardjo saw more and more *cotok* mini-trawls in their local waters. By 1996, after growing despair that the national ban on trawlers would not be applied to the mini-trawlers, and after reports of declining catches, some of the local fishers decided to take their own action. They seized and destroyed three mini-trawl boats from a much more populous village in central Jepara. Their action occurred just before Indonesia's national economic crisis. With the breakdown of authority that accompanied the collapse of the Suharto regime, and because Bandungharjo had a small number of fishers, village leaders (none of whom were fishers) feared retribution either at sea by the more numerous central Jepara fishers with faster and larger boats, or on land by hired gangs who might raid the village or block its only road.

It was in this state of anxiety, made worse by a violent conflict at sea in August 2000, that Toifuri, a relative newcomer, but locally influential fisher, attended an informal discussion of local social problems organised by several NGOs from neighbouring districts. Toifuri mentioned the fishing conflict, as did some leaders from other villages. Toifuri said that he thought his village would be safer and stronger if other fishers' groups from northern Jepara were enlisted into a larger organisation. With substantial assistance from the Kudus-based NGO, PPHM,[45] the Forum Nelayan Jepara Utara (North Jepara Fishers' Forum [NJFF]) was created. It brought together fishers' cooperatives and organisations from four subdistricts (*kecamatan*). Its first goal was to negotiate an end to the conflict with central and southern Jepara. Its second aim was to lobby for banning the *cotok* and for enforcement of that ban. Its third goal – at least from Toifuri's perspective – was to make the conflict less personal and less focused on hostilities between his isolated and outnumbered village and the more populous and better-connected villages of Ujungbatu and Jobokuto in central Jepara.[46]

The first goal was formally achieved after eight months and several meetings between north Jepara fishers and village heads and their counterparts from central Jepara. The first meetings were called by the Jepara police commander, but the police and fishers soon announced that the matter was being returned to the fishers for settlement. PPHM brokered the meetings. They had earlier gained the confidence of central Jepara fishers by mediating an internal financial dispute within a fishers' cooperative. The PPHM leader, Lala, played a major role in gaining trust between the two groups of fishers. Leaders on both sides of the conflict were impressed by her efforts. One crucial difference from government-organised meetings was that they involved real fishing leaders, not just the village officials.

On 18 April 2001 an agreement ending the conflict between north and central Jepara fishers was signed by chairmen of five fishers' groups from north Jepara and six fishers' groups from central Jepara. North Jepara agreed to return the six boats it confiscated and central Jepara its three. Costs of any repairs were to be the responsibility of the owners. The agreement went on to state that the two parties 'agree in principle to commit themselves not to use mini-trawlers ... which have been banned by the government'. The central Jepara fishers agreed to remove the mini-trawlers from Jepara waters in stages over six months. The fishers agreed to tolerate all fishers in Jepara waters as long as their nets 'did not violate laws and were environmentally friendly'. During the six-month withdrawal period, the central Jepara fishers agreed that *cotok* boats would 'only be tolerated ... as far as the north Jepara sub-district boundary'.

Perhaps the most contentious part of the agreement was in a section labelled 'Regarding Risks and Sanctions'. It notes that the north Jepara fishers will conduct 'operations' against fishers who use *cotok*. It states that *cotok* users do so at their own risk and that any boats seized in 'operations' will become the property of the north Jepara fishers. Finally, both sides agreed to surrender anyone violating this agreement to law enforcement authorities. The agreement closes with a promise to consult and to refer conflicts to the courts in Jepara if necessary. The agreement was signed and sealed by the 11 fishing group leaders and witnessed by the Jepara police commander, the Jepara Marine and Fisheries Service, and the district head. Attached to the agreement was the 1999 Jepara district head decree banning *cotok* nets.

It would seem that in the eight months since the clash the forum had achieved most of its aims. However, negotiating an agreement between fishers and gaining the endorsement of a ban from the Jepara district government was not the end of the matter. There was doubt about whether the local government had the will or resources to enforce the agreement. Past government efforts to enforce the ban had led to demonstrations and threats of violence by *cotok* fishers. Jepara's Polisi Airud (Water and Air Police) had only one boat to patrol 72 kilometres of shoreline and more than 1,500 square kilometres of Jepara waters.

The forum decided that they would have to enforce the ban themselves. To do that they purchased a Rp 35 million boat with a large inboard motor. To pay for the boat they collected contributions from each of the constituent fishing groups. In the Bandungharjo group dues were Rp 5,000 per boat per month. Toifuri says that the boat frightened the mini-trawlers away. Unfortunately, the running costs were high, the boat needed expensive repairs, and it proved difficult to sustain the collection of dues, so the boat was kept in service for less than a year.

The other aim of the forum was to strengthen the commitment of district government to a ban on mini-trawls. Again the PPHM played a role as community advisor and lobbyist for the forum.[47] Over the space of two years, four forum meetings were held to plan a draft Regional Regulation (Perda) banning mini-trawls, and to suggest revisions to a draft prepared by the Jepara Marine and Fisheries Service (Dinas Kelautan dan Perikanan). Forum meetings were lively, with differences of opinion about which types of nets should be banned. The main division was between those who wanted the widest possible range of nets banned and those who argued that they should not try to 'win alone' (*menang sendiri*) because that would create more opposition to the passage of the legislation. The May 2003 meeting ended with the approval of a draft regulation and agreement to let the forum executive decide which nets to suggest for banning.

In July 2003, after nearly three years without an incident in north Jepara, a mini-trawler from central Jepara (Jobokuto) was seized and its crew injured. In the same month the fisheries service submitted a draft proposal for a ban on mini-trawls to the regional assembly for consideration. Although there had been long delays, the informal leader of the North Jepara Fishers' Forum remained optimistic that the mini-trawlers would be banned, their fishing grounds protected, and that his village would not be singled out for retribution because north Jepara would be seen as united.

That optimism was not entirely justified. Of the three goals only one has been fully achieved. The conflict was brought under control, perhaps because of the trust established between the fishing group leaders or perhaps because the creation of a North Jepara Fishers Forum deflected some attention away from the individual villages involved. Bandungharjo has not been attacked. However, on the main point, the NJFF appears to have lost. Pemda, in the end, did not introduce legislation or continue to support the *cotok* ban. Surprisingly, even without the local state endorsement of the ban the NJFF leaders report that there have been no further incidents of outsider or locally owned mini-trawlers fishing in their waters.

Why did Pemda fail to deliver the promised *cotok* ban? There was a mix of factors involved. Although the Java Sea and its fishers are an important part of local identity, ocean fisheries are not a leading sector in the economy. It is true that Pemda Jepara established its own Marine and Fisheries Service, but that was in response to the central government's creation of a Department of Marine and Fisheries, and its emphasis was on ocean resources and aquaculture, not fishing. It is also true that the district head's speech at the Marine and Fisheries Service noted Jepara's long relationship with the sea, its annual fishers' festival (*Iomban*), and the potential of its seas; but he was more enthusiastic about coastal development, ocean tourism and brackish water ponds than fishing. Fishing is not a major source of Jepara income or employment. The ocean fisheries sector's percentage of GRDP declined from 1.6 per cent in 1993 to 0.8 per cent in 1998.[48] In 2004 the industrial, agricultural, forestry and services sectors were all 20 to 30 times larger. The industrial sector alone employed eight to ten times as many workers as fishing.[49] It is likely, therefore, that banning mini-trawlers was perceived to be less of a priority than other problems.

Pemda's attention to the marine environment and the problems of fishers is influenced by a number of sources. Jepara is home to a campus of Diponegoro University's Oceanography Faculty and marine laboratories, whose staff and students have sometimes led campaigns to protect coral reefs and mangrove forests. Diponegoro University researchers were ambivalent about the ban. They argued that fishers needed the mini-trawler to increase yields, and that they would not do much damage to the environment if they were kept away from the coral reefs. The newspapers (mainly *Suara Merdeka* and *Kompas*) that Pemda officials read assiduously, presented a picture of declining incomes and angry fishers playing out a tragedy of the commons, taking violent action to protect their interests. The negative reporting of fishing conflicts and the government's inability to control them was an important motivator for Pemda's back-down. Senior Pemda officials did not want to enact a regulation that they did not have the will or resources to enforce. This would make them seem even more incompetent, while risking conflict with the province over local authority to regulate fishing practices.

By 2004, Pemda officials had had three years to gauge the effectiveness of efforts at controlling *cotok* and fishers' reactions to these regulations. They found the resistance to the ban strong, and government capacity to police the ban very limited. When the police confiscated *cotok* nets or detained fishers using *cotok*, the police and the Fisheries Service faced large groups of unruly demonstrators threatening to burn the police headquarters and fisheries office if the nets and fishers were not returned immediately. These pro-*cotok* fishers lived only a few kilometres from town and numbered about 9,000.[50] Large crowds could be mobilised in town at short notice, and with little chance to organise a defence.

Also in 2004 the central and southern Jepara fishers groups began to organise resistance to the ban. In April 2005, when the NJFF called a meeting with all Jepara's fishing groups and concerned state officials, only 11 people of the 30–40 expected turned up to discuss the final draft of the ban. Hired thugs (*preman*) reported outside the hall warned people that bad things would happen if the group approved the draft. In the end the meeting adjourned without voting on the ban.

The Fisheries Service recommended to the district head that the ban be withdrawn because it would involve fishers from other districts and was therefore a provincial matter. They also claimed rather lamely that the national law banned mini-trawlers anyway. They promised that fishers caught using the *cotok* net would be warned.

Feldspar mining dispute

Clering village is located 60 kilometres northeast of Jepara and 20 kilometres off the North Muria road, the 'back road' that links Jepara with Pati. It is about as far away from the administrative centre of Jepara as you can be on the mainland. This area has few connections to the bureaucratic and political elite in Pemda.

Clering has a population of about 5,500. Two-thirds of village land is state owned (*tanah negara*). Villagers scratch out a living on the remaining one-third. Brackish water fish ponds, rice fields, dry fields (*tegalan*) and residences occupy the remaining third. The fish ponds are rapidly being eaten up by coastal abrasion. Fishing catches are down, and agriculture is also said by locals to be in decline. Even the village's labour-intensive roof tile industry faces stiff competition from new mechanised production methods.

Clering's main visible source of wealth is in its mountains. Mount Ragas, a part of the Muria Range, has a large working deposit of feldspar. Some of the ore is also rich in kaolin, a valuable mineral used in ceramics, glass and glazing.[51] Other 'mountains', hills, and beaches in Clering and surrounding villages are also believed to have a high concentration of useful mineral sands.

The value of feldspar and kaolin seems to fluctuate and to depend on the point of purchase. The Semarang Mining Company (Semarang Mineral Pembangunan or SMP), reportedly pays Rp 96,000 (A$15) for a truckload (+ 4 tons) at the minefield, or Rp 220,000 (A$34) for a truckload at its Pati depot. It sells processed feldspar at Rp 2.4 million (A$350) a truckload.

When feldspar was discovered in the village in the early 1990s,[52] the discoverer set out to buy as much of the mountain as he could. At the time villagers regarded the land as useless (*tanah pemajekan*) and were happy to sell it cheaply. Later, of course, they felt swindled. By 1994 the pioneer of feldspar mining was pushed out and his lease was given to the company that later became SMP. The then governor of Central Java province, Soewardi, took a direct interest in the granting of mining rights. His daughter, Dona Sophianti, was reported to be a director of SMP.[53]

SMP is by far the largest miner on the mountain.[54] However, other leaseholders operate as well, and from the beginning there have been unauthorised mining operations by villagers and disputes over who has the right to mine. There have also been intermittent arrangements between SMP and the village in which the company paid some of the salaries of village officials and contributed to village income. So, villagers' distrust of outsiders, as well as their long-standing misgivings about how the authorities (the police and local government) and the company would deal with villagers' efforts to increase their income from feldspar all contribute to the current conflict.

Local people described their grievance with SMP in some detail in interviews, discussions and active protests. The company signed an agreement with the village on 21 January 2001 at the district government offices, granting the village compensation including unspecified mining rights.[55] Arrangements with the company for monthly payments to be made to village officials in lieu of salary were never fully met. Fixed payments, to the Village Development Fund of Rp 15,000 (A$2) for each truckload of feldspar ceased after 2004. And village requests that the road to the mine, damaged by heavy mining traffic of up to 100 dump trucks per day, be resurfaced were ignored.

Complaints about dust from open-cut mining resulting in respiratory problems for the students at a nearby *madrasah* (Muslim non-residential school) received no response. Villagers had hoped the company would pay to move the school. Because the feldspar is transported in open dump trucks, the dust has been a health problem affecting all areas from the site along the road to the depot in Pati. Another complaint was that the company hired more outsiders than villagers. It was also said that field directors who cultivated good relations with the village were dismissed without notice.

The year 2003 marked the beginning of the current crisis. In that year police arrested two villagers for stealing SMP feldspar. After the arrests, villagers were too afraid to mine either the village site or other deposits on the mountain. The company mine area was sometimes locked and patrolled by armed police. At least one village protest leader received late night phone calls warning him to burn his archives of village protest action.

The village elite seems to be firmly supportive of the action against the company.[56] Village leaders and local supporters see the need for unity in order to stand up to the company. They believe the police and the legal system are likely to serve the company, but see Pemda Jepara (Jepara district government) and the DPRD as potential allies. Certainly, the Jepara government and assembly have taken an interest in the case and have been critical of the company. The media is generally seen as friendly to village interests. For example, *Suara Merdeka*, Central Java's leading newspaper, has devoted 16 stories in 2005 to the mining dispute and almost all of these are supportive of the villagers or critical of the company.[57]

In the mining case, unlike the fishing dispute, NGOs played only a minor role in organising or acting for the village.[58] Instead, village organisations (government and non-government) were used or created for the campaign. The Village Council (*Badan Perwakilan Desa*) sponsored meetings with Jepara state officials. A Village Community Enterprise Board (*Badan Usaha Masyarakat Desa*) was established to overcome legal objections to village management of a mine. A Village Assets Concern Forum (*Forum Peduli Asset Desa*) and a 'Little People's Forum' (*Forum Wong Cilik*) were created to assist in carrying out the public campaign. These informal village organisations could organise demonstrations, send protest letters, issue pamphlets, convey an image of mass support in media coverage, provide a podium for village elites and give those elites a seat at negotiations with the company and with the state. They also allowed the village government to deny responsibility if any laws were broken during the campaign. Finally, the militia-like village Ansor youth group was given the task of providing security at meetings.[59]

The village strategy to convince the authorities of their righteousness and determination shows considerable sophistication and some understanding of contemporary Indonesian laws and local politics. This comes out clearly in our research notes from a meeting with Pemda Jepara officials organised by the village on 26 May 2005.

The meeting site was the village volleyball field. To gain entrance the guests from the provincial and district governments had to pass through a coffee stall (*warung*). In the stall and throughout the volleyball field, numerous posters and banners were hung stating the villagers' complaints about the mine and their demands. Around the perimeter more than 50 uniformed village security militia were posted, perhaps as a signal to officials that the village leadership was trying to maintain law and order.

The meeting place was crowded with more than 100 seated villagers. Three speakers were appointed to speak for the village. The Head of the Village Council (BPD) began the meeting by stating why the village was angry with PT SMP and announcing the village's three demands. These were: (1) Recognition that SMP's 10 hectare mining lease (SIPD) had expired on 17 May 2005; (2) The lease would be turned over to the village owned industry unit (BUMD); (3) SMP's other 2.7 hectare mining lease was to be continued on condition that all non-technical workers must come from Clering, and that the same fee negotiated in 2001 of Rp 15,000 (A$2) for each truckload of feldspar departing the mine site would be paid to the village.

The aim of the Pemda negotiators appeared to be to try to convince the villagers to soften their demands to increase the chance that the company and the governor would grant them some mining rights and other compensation. After the presentation of the grievance letter and the three demands by the Head of the Village Council, it was the turn of the Jepara and Central Java state officials to respond. District officials put considerable effort into trying to convince the villagers that

Pemda Jepara was on their side. The main speakers were both from Jepara: Hermin Priyanto, the Head of the Energy, Mining and Environment Service, and Mulyaji, the Head of the Government Division, which is in charge of subdistrict and village government affairs. Boediono, the regional manager of the provincial government's Energy and Mining Management Board (BPPE), spoke only briefly.

Hermin Priyanto began by saying that he had been delegated by the district head, Hendro Martojo, who was absent because he had to attend a public discussion about the upcoming district head election in Jepara. Hermin Priyanto explained that the district head was very supportive of the Clering people. He said the proof of this was that Hendro had recently met with the managers of PT SMP and 'challenged' (*ditating*) them on their commitment to the village. Furthermore, the district head told the directors of SMP that he would not sign the letter of recommendation required by the province before granting an extension to the SMP mining lease until SMP met two conditions: the dispute with the people of Clering had to be resolved; and their compensation claim had to be met. It was reported that Hendro had already refused a request from the provincial mining service for his signature approving a continuation of the mining licence. Hermin Priyanto claimed that at their meeting, the SMP directors had told the district head that they were willing to pay a fee for reforestation, to recruit Clering mine workers, and to respond to the request for mining rights if the village communicated their request to the governor. After suggesting a compromise, which will be discussed later, he ended his presentation by once again stressing the Jepara Pemda's support for Clering. He faced the audience and said, 'We have a moral debt to help you.'

The next Pemda speaker was the Head of the Government Section from the district government. Mulyaji traced the history of the conflict and noted the expiry dates of SMP's 10 and 2.7 hectare leases. He urged the Clering Village Head to send a letter to the governor as soon as the Jepara district regulation (Perda) on mining was enacted. The letter should set out the claim for payments and the request for a mining license for the 2.7 hectare site. The smaller, but kaolin rich and therefore more valuable per ton, 2.7 hectare site could still be negotiated, but not the 10 hectare one, he said.[60] He stressed that they should not put the district head in a corner (*menyudutkan*) or make it difficult for the district head to convince the governor – who still held the power – to meet the villagers' request. He noted that the district head was fully committed to supporting Clering, demonstrated by his immediate response to the BPN mapping of the village, which the villagers claimed had unfairly decreased the villagers' land ownership. He had ordered BPN to remap the village. If Hendro gave a recommendation for approval, the villagers' claim had a good chance of success. He pleaded with the citizens to feel pity for the district head who would have to face up (*berhadapan*) to the provincial government if the people continued to demand the 10 hectare site. At one point he called on Boediono, the representative of the provincial government: 'try to have the [mining] of PT SMP stopped'. This was greeted with clapping by the participants.

Mulyaji claimed that the district government held a trump card in the negotiations with the province because an extension of the mining permit required the recommendation of the district head. One of the appointed village speakers interjected: 'If the district government actually has a trump card why not use it to ask for the 10 hectares?' Mulyadi said that the company and the provincial government also have trumps. He seemed to be suggesting that the company's existing lease and the governor's ultimate decision-making authority on this matter put them in a strong position. He also mentioned that the Jepara District Government was still very dependent on finance from the provincial and central governments, so Pemda could not just do anything the villagers wanted. The villagers then noted that if they were given 5 hectares they could provide as much in mining fees per year (Rp 80,000,000 equals A$11,000) as the company provided for 10 hectares, effectively doubling the tax rate. The public meeting ended without agreeing on a compromise offer from the village.

Negotiations between the government officials and the village leaders[61] started in earnest after the meeting broke up. Mulyaji explained again that they had little chance of success with their demand for 10 hectares. However, if they asked for 2.7 hectares at any location they chose on the mountain, they could count on a recommendation letter with the support of the district head. The village leaders quickly seemed to agree to Mulyaji's request. The officials left after Mulyaji asked the village secretary to bring a letter and the village seal to his office the following day so that he could immediately draft the district head's recommendation letter.

After the officials departed, the village held another meeting which decided to hold out for the 10 hectare rather than accept the 2.7 hectare compromise offer. A senior government official interviewed on 20 June was not very happy. He commented:

> If that's the way it is, in the end we will return the problem to the citizens. Let them write their own letter. Those of us who know a little about the regulations wouldn't ask for so much. We [Pemda Jepara] know even a district head's District Regulation (Perda) can be annulled, let alone a Village Regulation or citizens' demands.

Still, he intended to invite the village head to his office and to try to negotiate an alternative set of demands from the people of Clering.

Almost a year later (April 2006) there has been no final decision on Clering's request for payments and mining rights. Jepara's officials are still hoping that the villagers will reduce their demands and that a deal with the governor and the company will prove possible. The district head has still not signed the recommendation letter required for the ten-year extension of the mining permit. The company and the village are still threatening each other and locking each other out of the mining area.

There may be reasonable cause for villagers' expectations of a substantial victory. They realise, as one village official said, that this is no longer 'the New Order period when district heads did not dare to stand up to the governor'.[62] Jepara Pemda's response to the villagers' claim is affected by a larger resource issue, the question of which level of regional government controls mining in Jepara. Jepara is facing a widespread outbreak of unlicensed and uncontrolled mining of beach sands, river sands, and other minerals, causing erosion and damage to fishponds and to river banks. Pemda officials see this mining as both an environmental problem and as a potential source of income. They have even introduced a new District Regulation controlling all category C mining in Jepara. But this district level regulation clashes with recent provincial mining regulation. A senior Pemda official told us that it will take several months for the District Regulation to be reviewed and approved by the provincial government and by the Ministry of Home Affairs in Jakarta. The new Regional Government Law 32/2004 would seem to make the district's chances of gaining that authority weaker. At any rate, the province controls other sources of district income, including large infrastructure projects like bridge-building, road construction, ferry transport and infrastructure for the Karimun Jawa national park, which might be at risk if the district offends the provincial governor

The district head has to weigh up the relative costs and benefits of opposing the governor. Certainly, he will be considering the consequences for the government budget and the Jepara economy. If he is as effective a politician as local people believe, he will also be weighing up the patronage powers that he would gain from greater control over mining rights, against the patronage opportunities he would lose from a reduction in provincial subsidies.

Another change from the New Order that the villagers understand is the sharing of the executive's power at the district and provincial levels with the DPRD at both levels. Clering villagers have travelled to Jepara and Semarang on several occasions to demonstrate outside the assemblies. Both the Jepara and Central Java legislatures have sent commissions (*komisi*) to visit Clering and have invited villagers to Jepara or Semarang to discuss their grievances. Jepara assembly members have visited other districts to compare how mining permits are monitored and controlled. District assembly members from Keling have advised the village leadership on how to approach the local government and how to bargain with the company. The draft mining regulation produced by the Energy, Mining and Natural Environment Service of the Jepara District was reportedly largely rewritten by its assembly. The changes were mainly to widen the scope of district control over mining permits (*Surat Ijin Pertambangan Daerah*).

Another consideration for the district head in his reaction to villagers' concerns was political support. The district head was running for re-election in November 2006 and would benefit from support in Jepara's less Islamic north. His local reputation as a fighter for local interests against outside companies and provincial bureaucracies could be enhanced if he successfully challenges the company and the governor.

Comparing cases – resource conflicts, contestants and outcomes

This chapter has argued that village citizens' participation in resource conflicts has recently become bolder, more sophisticated and more organisationally impressive. There is more space for collaboration in the representation of local interests and for the negotiation of resource conflicts. The local government has become more transparent, more open to participation, and more concerned with accountability. However, this has not necessarily resulted in better environmental governance, greater access to resources by those who challenge existing resource arrangements, or successful, workable compromises.

The three village-level conflicts (forestry, fishing and mining) all involve long standing issues that pre-date the fall of Suharto. Tables 2.1 and 2.2 compare some features of those conflicts.

The first dispute is over the management and distribution of wealth generated by the teak forests. The collapse of the state and security apparatus provided an opportunity for villagers to protest against their poverty and dependence amidst SFC's control of enormous teak forest wealth. They asserted their claims by logging out the teak forests.

Since 1999 the SFC has been trying to re-establish the forest as a sustainable resource for the furniture industry. They have tried to control the forest by promoting, and ostensibly pursuing, a more equitable social forestry programme that provides more incentives for villagers to replant and protect the forest. Seven years later there is little evidence of success at gaining villagers' trust or compliance, or at re-establishing the forest.

Table 2.1 Resource conflicts in three Jepara villages

Village:	Bumiharjo	Bandungharjo	Clering
Resource:	Teak forest	Fishing waters	Feldspar mine
Principal adversary	forest villagers	north Jepara fishers	Clering villagers
	Perhutani/State Forest Company	other fishers	SMP Semarang Mining Co.
Goal	greater access to forest and land	protected fishing resources	greater access to jobs and revenue
Moral claim	protest injustices	sustainable fishing	agreed compensation not delivered
Social standing/ leadership	low	low-medium	village wide support including village elite
Village solidarity	low	only 20% of villagers fish	high
Dispute outcomes	temporary wealth gain	ban efforts failed	still contested
	promised fairer share of forest wealth	some Pemda and NGO aid to protest villages	village elite and some others seem likely to win some benefits
Social capital outcomes	no benefits evident	political contacts and skills	political contacts and skills
		confidence in citizen rights	confidence in citizen rights
Environmental outcomes	loss of teak forest, erosion, flooding	deterioration of fishery	loss of tree cover and deterioration of air quality

Table 2.2 Pemda response to resource conflicts in three Jepara villages

Village:	Bumiharjo	Bandungharjo	Clering
Resource:	*Teak forest*	*Fishing waters* (north Jepara)	*Feldspar mine*
Principal contestants	villagers State Forest Company	north Jepara fishers central and south Jepara fishers and creditors	villagers SMP – Semarang Mining Co.
Actions	spontaneous forest looting	seized boats formed forum cooperating with NGO negotiated end to hostilities drafted mini-trawler ban lobbied Pemda and fishers	illegal mining blocked mine access lobbied Pemda and Province negotiated with SMP mine strong media campaign
District government (PEMDA) legal authority	no authority over *Perhutani (SFC)* state forest land	Pemda first claimed authority to ban nets then denied authority or wisdom of exercising such authority	Pemda is challenging provincial control
PEMDA response	none beyond forwarding village complaints	initially sympathetic to proposal for ban, then lost interest	sympathy for village claim refused to renew SMP mining permit
Outcomes	temporary wealth gain *Perhutani* promises fairer distribution intensive forest use unsustainable	ban efforts appear to have failed some state aid to protest villages fishing unsustainable	negotiations continue village likely to win some benefits mining unsustainable

The fishing dispute pits traditional coastal fishers from north Jepara, who are trying to protect their customary fishing grounds and small incomes, against fishers from central and southern Jepara, and beyond, trying to introduce new technology. The outside fishers have larger boats, larger engines, fish over wider areas, and use *cotok* nets to trawl for fish. The north Jepara fishers argue that the finely meshed *cotok* nets quickly reduce fish populations. Furthermore, the weight attached to the *cotok* damages coral reef fish nurseries and frequently cuts traditional fishers' stationary nets.

The north Jepara fishers have been unsuccessful in their efforts to obtain an official ban from the Jepara district government. However, they have succeeded in keeping their waters free of *cotok* nets for four years. They have also created a community action and lobby group, the North Jepara Fishers Forum, which spans more than a dozen villages in four subdistricts. It seems likely that the process of establishing and maintaining the NJFF, cooperating with the NGO community, gaining media coverage, lobbying the local government, and negotiating with leaders from *cotok* villages have all provided some social capital, in the form of skills, confidence, personal contacts and trust.

The negotiation of a fishers' agreement between villages seems to have lowered hostilities and developed a degree of trust between fishing community leaders. These face-to-face meetings between pro and anti-*cotok* leaders may help to explain why there has been no *cotok* fishing and no violent incidents in north Jepara waters since the initial conflict.[63] The NJFF leaders also gained contacts and earned goodwill with the Fisheries Service, the Pemda bureaucracy and the Jepara District Assembly. Their skills in dealing with government and good relations with Pemda and the NGO community have enabled NJFF villages to obtain small aid grants.[64] In 2005 they also influenced a Pemda decision to oppose an effort by the All Indonesian Fishers' Association (HNSI) to become the sole organisation for Jepara fishers.

The mining dispute is focused on efforts by a north Jepara village to gain compensation for environmental impacts (respiratory problems and damage to roads and bridges), and to obtain a 'fair' share of the income from the quarrying of feldspar and kaolin in a large open-cut mine managed by PT SMP. The villagers claim that the company agreed to village demands at a 2001 meeting arranged by Pemda Jepara. However, the company claims that their representatives at that meeting did not have authority to make a deal with the villagers.

Since 2003 the company has used police, the courts and, allegedly, gangs (*preman*), to try to maintain its control over the mine.[65] Villagers have engaged in illegal mining, blocked the only road to the mine, protested to the district assembly in Jepara and the provincial assembly in Semarang, negotiated with the company, and carefully staged meetings to lobby for the support of district and provincial officials. Almost the entire village elite have been involved in establishing and mobilising various associations, committees, forums, and movements.

There is still no agreement with the mining company, but the village elite remain confident that they are in a position to achieve a generous settlement from SMP, making operation of the mine difficult and expensive. They could take feldspar, block the entrance to the mine or sabotage the dump trucks that collect the ore. Certainly, their organisational skills and sustained effort have won some respect from district government (Pemda) officials. Pemda Jepara's desire to wrest control over mining from the provincial government, and the district head's dissatisfaction with the company, also reinforce that optimism.

Villagers, local government and resource governance

What differences were there in the nature of the collective action taken and the government responses in these three cases? And what were the implications for resource governance?

The teak conflict was the simplest and most straight forward of the resource conflicts. Bumiharjo villagers, like those from hundreds of other forest villages in Java, cut down nearly all of the contested teak forests. Since 1999 they have continued to cut down or pull out most of the replanted seedlings. The resource conflict with Perhutani remains at a standoff. Perhutani has established a reforestation and forest protection programme that may offer better terms for the villagers. However, villagers do not have faith in the SFC's good intentions, and their direct action response – looting – was largely anarchic and counter-productive. The looting seems to have left the village with depleted resources, more tensions, and little capacity to cooperate for common goals. Perhaps because of that tension and suspicion, NGOs have found it difficult to create links or establish programs in Bandungharjo.

The north Jepara fisheries case, on the other hand, is an example of local leadership and collective action in the longer term interest of local fishers and the resource upon which they depend. Toifuri, a fisher who migrated to Bandungharjo in the 1980s, has played a key role in organising the campaign to keep mini-trawl nets out of local waters. The campaign for a District Regulation enforcing a ban seems to have been unsuccessful. After initially siding with the fishers' group supporting the ban, Pemda decided that it was too difficult and too dangerous to enforce the law and that its effort to assume authority over fishing might cause conflict with other districts and with the provincial government.

Toifuri's alliance with the respected legal aid NGO field worker, Lala, made it possible to establish the North Jepara Fishers Forum, thus networking the six local fishers' groups of north Jepara. They succeeded in calming a violent conflict with a central Jepara fisher group. If it had been endorsed by legislation the agreement negotiated between fishers would have been a model for community initiated and negotiated resource regimes backed by the state. However, even without state power or legitimacy the ban has held for four years. Pemda's retreat from support for the trawler ban may have been rationalized on democratic grounds, since there are three times as many fishers in southern and central Jepara fishing villages as in the north. It is more likely though, that fear of violence by the *cotok* fishers living near the Jepara capital was the strongest motivator. Still, the local government developed a link to village civil society that encouraged villagers to make demands on the local state and encouraged some local government officials to see responding to public demands as part of their duties. That more organised and more activist village level society and its more responsive local state will be sorely tested by central government plans to build a nuclear reactor near the fishing villages of north Jepara.

The mining case is probably the most remarkable example of a campaign led entirely by villagers to gain resource rights. The organising, lobbying, demonstrating, media campaign and stagecraft brought to bear were impressive. The villagers seem to have strong support from the district head, the district assembly, and the most important regional newspaper, *Suara Merdeka*. Perhaps Pemda and the assembly are more supportive because the amount of money at stake is far greater than in the fishing conflict, at least over the short term. This means the potential for government revenues or 'rents' for local state and village officials is far greater. However, the district head's concerns about his coming re-election campaign, his anger at the SMP directors for reneging on an agreement his office brokered, and his desire to control environmental problems caused by mining in Jepara are also factors. Shared interests with regional authorities improved the chances of local (albeit mainly elite-led), forces gaining from resource struggles in this case.

So what can village protagonists, Jepara government officials and other onlookers learn from these cases of village engagements in resource politics? The first lesson is that connections and solidarity matter. The mining village had the full support of the village elite, while the fishers' forum made up only a small percentage of their village population and had no village officials in its leadership.[66] A second lesson might be that violence or threats of force are still an important part of the capacity to gain attention and influence policy. The Jepara forest villagers were just one of many mobs that struck Perhutani across Java/Indonesia. One result was a new, purportedly, more villager-favourable reforestation programme from Perhutani. The north Jepara fishers' violence at sea gained the attention of the district government, and the power of the more numerous, centrally located pro-*cotok* mobs helped spur a Pemda back-down on the mini-trawl ban. The mining villagers' capacity to block mining or steal feldspar, and their use of village security groups to combat the company's own thugs, forced the company and Pemda to pay attention. Power and connections with the local government, civil society organisations and the underworld of thugs are all important.

A third important lesson is that none of these outcomes looks promising for the environment. Jepara has to deal with a rapidly degrading natural environment. Only the north Jepara fishers who argued for sustainable fishing had a strong environmental position. The mining villagers complained about the respiratory problems caused by the mining, but campaigned for compensation, not an end to mining. The forest villagers have recently logged out the last nature reserve in north Jepara. Any programme that could lead to sustainable forestry would have to be sensitive to local conditions. Perhutani has so far proved unable to build those conditions, and the chances for increased local autonomy allowing Jepara and other districts to achieve that seem poor.

Finally, whether a more participatory local resource politics will in the long run be more capable of dealing with environmental and resource security issues depends on whether more villagers can see the benefits of collective public participation and whether more local and national government officials can see the benefits of improved governance. If there are no perceived benefits for villagers from efforts to redistribute resources or protect their environment, then the appeal of violent collective action may be stronger. As the teak looting indicates, this would not be good for the environment or ultimately for local livelihoods.[67] If both local civil society and local authorities can see the process of participatory resource governance as empowering and satisfying, then Jepara's complex, environmental challenges may yet prove manageable.[68]

Acknowledgements

The research for this study is based on periodic field research in Jepara from 2003 to 2006. The research was made possible by a grant from Flinders University and by funding from the Australia Research Council. We conducted more than 80 formal and informal interviews, discussions and focus group meetings.

Notes

1 Hendro Martojo, district head of Jepara, quoted on the 456th anniversary of the founding of Jepara. *Suara Merdeka*, 9 April 2005.
2 See Portal Dinas Kehutanan Propinsi Jawa Tengah for details on *hutan rakyat*, http://www. dinashut-jateng.go.id/mod.php?mod=userpage&page_id=28 (accessed 4 March 2006).
3 See Kantor Biro Pusat Statistik Kabupaten Jepara, *Jepara Dalam Angka 2006*.
4 Its vote in the free elections of 1955 was 55 per cent and despite repression and vote-buying by the state party, Golkar, NU's political party, PPP, obtained more than 40 per cent of the votes in each of the New Order's controlled elections. See Schiller (2004).
5 Jepara district government (hereafter referred to as Pemda): officials used the term to explain public enthusiasm for protest and street politics.
6 Pemerintah Daerah Dati II Jepara (2001).
7 The plan was not entirely rhetoric. Many village officials and some state employees were compulsorily retired.
8 A long campaign by local and national NGOs eventually led to the prosecution and conviction of the 1999–2004 DPRD Chairman and one Vice-Chairman and the return of millions of rupiah by assembly members.
9 Personal communication, Hetifah Sjaifudian, 20 February 2006.
10 A recent field trip to Kalimantan and Malaysia organised by the district assembly and some senior officials is a case in point. All the Jepara-based journalists and several of the NGO leaders were invited as well. Presumably, they could not criticise the event if they were part of it. In the end, some of the NGO leaders, the district head, his deputy and several legislators dropped out because of negative publicity. Those NGO activists who went along with it negotiated a district regulation (*Perda*) on transparency in government and public participation as the price of their cooperation.
11 Some 404 per square kilometre vs. approximately 1,000 for the rest of Jepara.
12 Jepara district is 97.2 per cent Muslim.
13 Interview with Indonesian Furniture Association official, 25 March 2003.
14 Schiller and Martin-Schiller (1997) attributed that growth to Jepara's reputation as a centre of fine carpentry, its industry-supportive local government, deregulation of the export process, the availability of teak and skilled, inexpensive labour.
15 This conservative estimate is based upon discussion with industry association and local government officials.
16 The Head of Planning at the Indonesian Ministry of Forestry claimed that the Pati forest unit, which includes Jepara, lost 30,000 m3 of teak annually between 1998 and 2001. By 2004 only 200 ha of productive teak forest remained out of 14,000 ha. (*Suara Merdeka*, 10 February 2004). A 2003 outline paper delivered by Jepara's district head to the Central Java Forum on Joint Management of Forest Resources with the Community claims that more than 300,000 Perhutani trees with a value of just over Rp 20 billion were illegally logged between 1998 and 2000.
17 Interviews with Bumiharjo villagers, 11 and 12 August 2003 and 30 May 2007.
18 Bumiharjo Village 2004, *Badan Pembangunan Desa, Daftar Terinci Pembangunan Desa* [Bumiharjo Village Development Board Detailed List of Village Development] unpublished report [posted on village hall information boards].
19 Any unreferenced information on Bumiharjo is from anonymous interviews and informal discussions conducted between 2003 and 2007.

20 This amoral taking advantage of a situation for personal gain is called *mumpung* (because the opportunity presents itself) in Javanese.

21 However, this is not to deny that they took the act of felling teak from the forest seriously. Some forest villagers ritually slaughtered water buffalo as offerings to the spirits that protect the forest before logging began.

22 Interview, 16 August 2003.

23 For at least those three months chainsaws could be heard in the forest 24 hours a day and sawmills were operating continuously. There was even a market where you could buy food, saw blades, and all the loggers' needs at two or three times the Jepara price.

24 Fauzan (2004: 215).

25 These results were calculated from 1997 and 2000 Bumiharjo village data recorded in the Bumiharjo village government office.

26 Bumiharjo Village 2004 *Daftar Terinci Pembangunan Desa* [Detailed List of Village Development], unpublished report, Bumiharjo: Badan Perwakilan Desa (BPD).

27 The furniture makers of Jepara frequently joke about '*kayu spanyol*' not 'Spanish timber' but half stolen timber (*kayu separuh nyolong*) and (*kayu Kartini*) named after the Jepara-born pioneer in women's education, whose famous writing (*Habis Gelap, Terbitlah Terang* [After the Darkness, Emerges the Light] is a euphemism for illegal logs certified as legal.

28 Money went to the individuals involved and sometimes to mosques, village road-building funds and the poor.

29 Interview, 21 January 2006.

30 Ibid.

31 Interview 12 August 2003.

32 Cifor (2005) 'Java teak: a livelihood hardwood', http://www.cifor.cgiar.org/docs/_pf/1_ref/publications/newsonline/33/hardword.htm (accessed 15 October 2005).

33 This short-term outlook is apparent in the 30 per cent rejection rate that Jepara furniture exporters had on their 1999–2001 sales: the exporters rushed production. Overseas buyers complained that teak furniture was not properly dried (and therefore subject to warping or cracking), poorly built or carved, hastily varnished or painted, and badly packed.

34 Monsanto Chemicals was marketing *jati emas* (gold teak), a genetically modified teak, which it claimed matured in 15 years. Source: interview with Perhutani official, Jepara, 31 March 2006.

35 See Schiller and Martin-Schiller (1997).

36 Interview, 16 August 2003.

37 Both local government officials and assembly representatives expressed this view.

38 Jepara fisher, quoted on 456th anniversary of the founding of Jepara (*Suara Merdeka*, 8 April 2005).

39 According to official figures, between 1995 and 2001 yields fell from 3.2 to 1.3 million kilograms while the value of that catch rose from Rp 2.2 to 5.3 billion, considerably less than the increase in the cost of living associated with the dramatic currency devaluations over that period. Data supplied by the Jepara Marine and Fisheries Service (Dinas Perikanan dan Kelautan Pemda Jepara).

40 Another damaging means of achieving short-term income gains at a high environmental cost involves illegal coral mining. Coral theft for construction – a boat-full yields Rp 150,000, equivalent to three to five days' income for a fisherman – and coral destruction by trawling have decimated the reefs. Estimates are that more than half the reefs in the Karimun Jawa marine national park are damaged and up to 90 per cent in the four nautical mile Jepara coastal zone. See Marine Diving Club Diponegoro University, '2002 Karimunjawa Reef Check Accountability Report' (unpublished). Still other fishers resort to bombing and poisoning fish to improve their catch.

41 See Simorangkir 2000: 111–19.

42 Why would an experienced fisher continue to fish when he could earn a more regular and

substantial income, even in an unskilled job, in the furniture industry? Furniture workers also face less danger and are less affected by rising fuel costs than fishers, although, as we have seen, declining timber as a result of over-exploitation is now affecting the availability of employment in this sector.

43 See the Lucas study in this volume for a detailed analysis of the impact of credit arrangements in the fishing industry on the north coast of Java.

44 Sub-district figures categorise *cotok* with some other nets.

45 Pusat Pelayanan Hukum dan Masyarakat [The Community and Legal Service Centre].

46 Some of the boats destroyed and crews attacked were from those villages.

47 PPHM met with fisheries department officials and drafted legislation after gaining approval from the forum.

48 Badan Perencanaan Pembangunan Daerah Kabupaten Jepara with the Faculty of Economics of the Diponegoro University, *Jepara Your Future Business Destination.* Semarang (2000: 8).

49 It would be unfair to say that state officials in the Marine and Fisheries Service have no interest in the welfare of the fishers. There seemed to be both a respect for the hard and dangerous work that fishers did and a fear that if they were aggrieved they could cause a great deal of difficulty for government officials.

50 About three times the number of north Jepara fishers.

51 Feldspar is a clay widely used in pottery production. Personal communication, Liz Morrell 17 March 2004.

52 A man from Pati collected a sample of the sand to be analysed by his son who was studying geology at Gadjah Mada University.

53 *Gatra*, 22 July 1995, 'Bisnis Anak Pejabat: Menang Melawan Pemda' (Officials' Children's Businesses: Winning against Regional Government), http://www.hamline. edu/apakabar/basisdata/1995/07/20/0005/html (accessed 9 March 2006).

54 It sued its field director for stealing from its mine site for three years. It claimed the loss was just short of Rp 13 billion (nearly A$ 2 million) Data was compiled from several interview sources.

55 'Kata Pak Suwito, Penambangan oleh Desa Legal' [According to Suwito the Mining by the Village Is Legal], *Suara Merdeka*, 21 June 2005. The arrangements were reported in detail in 'Mendulang Emas dari Gunung Ragas' [Spoon-Fed Gold from Mount Ragas] in a Jepara local government magazine, *Gelora Bumi Kartini* (April 2001): 28–9.

56 Perhaps the village leadership was motivated because the village has no land used to pay village officials (*tanah bengkok*).

57 This inclination was confirmed by interviews and informal discussions with journalists.

58 LPBHNU, the legal aid arm of Nahdatul Ulama, however, is providing legal assistance to Clering residents charged with stealing feldspar from SMP's mine site.

59 Ansor is the youth wing of NU, Jepara's hegemonic Muslim social organisation.

60 His argument about why the larger site was not negotiable touched on technicalities of mining leases, five-yearly reviews and expiry dates. It was not very convincing to the meeting or to the authors.

61 The *carik*, the BPD head, three of the village forum leaders, but not the village head.

62 Interview with Clering, BPD head.

63 However, another important factor is that rising fuel costs have reduced the distance that fishermen can afford to travel. Fish yields and prices have not kept up with costs.

64 For example materials and money to protect houses and boats from ocean waves.

65 However, the courts have not always sided with the company. On 26 September 2006 the Jepara court ordered compulsory mediation between the company and three villagers over disputed claims to 6 hectares of feldspar mine. 'Sengketa Gunung Ragas Dimediasi PN Jepara' [Mt. Ragas Dispute Mediated by Jepara State Court] *Suara Merdeka*, 28 September 2006.

66 Although the creation of a North Jepara Fishers Forum aggregated the voices of the

fisher minorities and encouraged Pemda and the NGOs to pay greater attention to its demands.
67 Indeed Jepara furniture exports have declined from US$ 141 million in 2004 to US$ 127 in 2005 and to US$ 119 million in 2006: *Suara Merdeka*, 4 July 2007, http://www.suaramerdeka.com/harian/0503/12/eko12.htm (accessed 4 July 2007).
68 The plan to begin construction of Indonesia's first nuclear power plant in north Jepara in 2008 will meet considerable local concern. However, it is even less likely to be a matter that local government or civil society will be able to influence.

Bibliography

Biro Pusat Statistik Jepara (2002–2006) *Jepara Dalam Angka (2002–2006)* Jepara Statistics Centre.
'Bisnis anak pejabat: Menang melawan Pemda' (Senior Officials' children's businesses: Winning against regional government), *Gatra*, 22 July 1995, http://www.hamline.edu/apakabar/basisdata/1995/07/20/0005/html (accessed 9 March 2006).
Butcher, J. (2004) *The Closing of the Frontier: A History of the Marine Fisheries of Southeast Asia c. 1850–2000*. Singapore: Institute of Southeast Asian Studies.
Development Planning Bureau with the Faculty of Economics of Diponegoro University (2000) *Jepara Your Future Business Destination*. Semarang.
Diponegoro University Marine Diving Club (2002) *Karimunjawa Reef Check Accountability Report*. Semarang: unpublished paper.
Fauzan, Achmad Uzair (2004) 'Poor Forest (Still the Same) Poor People: The Practice of New Forest Management in Jepara, Indonesia', unpublished manuscript.
Foley, M. and Edwards, B. (1996) 'The paradox of civil society', *Journal of Democracy* 7 (3): 38–52.
Mubyarto, Seotrisno, L. and Dove, M. (1984) *Nelayan dan Kemiskinan Studi Ekonomi Antropologi di Dua Desa Pantai* [Fishers and Poverty – An Economic Anthropological Study of Two Coastal Villages]. Jakarta: Rajawali.
Peluso, N. (1992) *Rich Forests: Poor People*. Berkeley, CA: University of California Press.
Pemerintah Daerah Dati II Jepara (1998) *Laporan Hasil Pelaksanaan Reformasi di Segala Bidang* [Jepara Second Level Regional Government Report on the Results of Implementing 'Reformasi' in Every Field]. Jepara.
Pemerintah Daerah Dati II Jepara (2002) *Laporan Pertanggungjawaban Akhir Masa Jabatan Bupati Jepara kepada Dewan Perwakilan Rakyat Daerah Kabupaten Jepara 1997–2002* [Jepara Second Level Regional Government End Term Accountability Report for the District Head of Jepara to the Jepara Regional Peoples Representative Assembly 1997–2002]. Jepara.
Pemerintah Daerah Dati II Jepara (2001) 'Mendulang Emas dari Gunung Ragas' [Spoon-Fed Gold from Mount Ragas] in *Gelora Bumi Kartini* (April 2001): 28–9.
Pitcher, T. and Hart, P. (1982) *Fisheries Ecology*. London: Chapman and Hall.
Putnam, R. (1993) *Making Democracy Work*. Princeton, NJ: Princeton University Press.
Pye, L. (1998) 'Civil Society, Social Capital and Civility', *Journal of Interdisciplinary History* 29 (4): 763–82.
Sahidin (2004) *Kala Demokrasi Melahirkan Anarki Potret Tragedi Politik di Dongos* [When Democracy Gives Birth to Anarchy: a Portrait of a Political Tragedy in Dongos]. Yogyakarta: Logung Pustaka.
Schiller, J. (2002) 'Looking For Civil Society: The 1999 Election in Jepara', *RENAI, Jurnal Politik Lokal & Sosial-Humaniora*, Year II, No. 1.
Schiller, J. (2004) 'What is an Election Supposed to Do? A local perspective on the 2004 Indonesian Elections', 19th Biennial Conference of the Asian Studies Association of Australia, Canberra 29 June–2 July.

Schiller, J. (1996) *Developing Jepara: State and Society in New Order Indonesia*. Melbourne: Monash Asia Institute.
Schiller, J. and Martin-Schiller, B. (1997) 'Market, culture, and state in the emergence of an Indonesian export furniture industry', *Journal of Asian Business* 13 (1): 1–24.
Simorangkir, S. (2000) *Perikanan Indonesia [Indonesia Fisheries]*. Denpasar: Bali Post.
Squires, D., Omar, I.H., Jeon, Y., Kirkley, J.,Kuperan, K. and Susilowati, I. (2003) 'Excess Capacity and Sustainable Development in Java Sea Fisheries', *Environment and Development Economics* 8: 105–127.
Suara Merdeka (1997–2007), http://www.suaramerdeka.com.

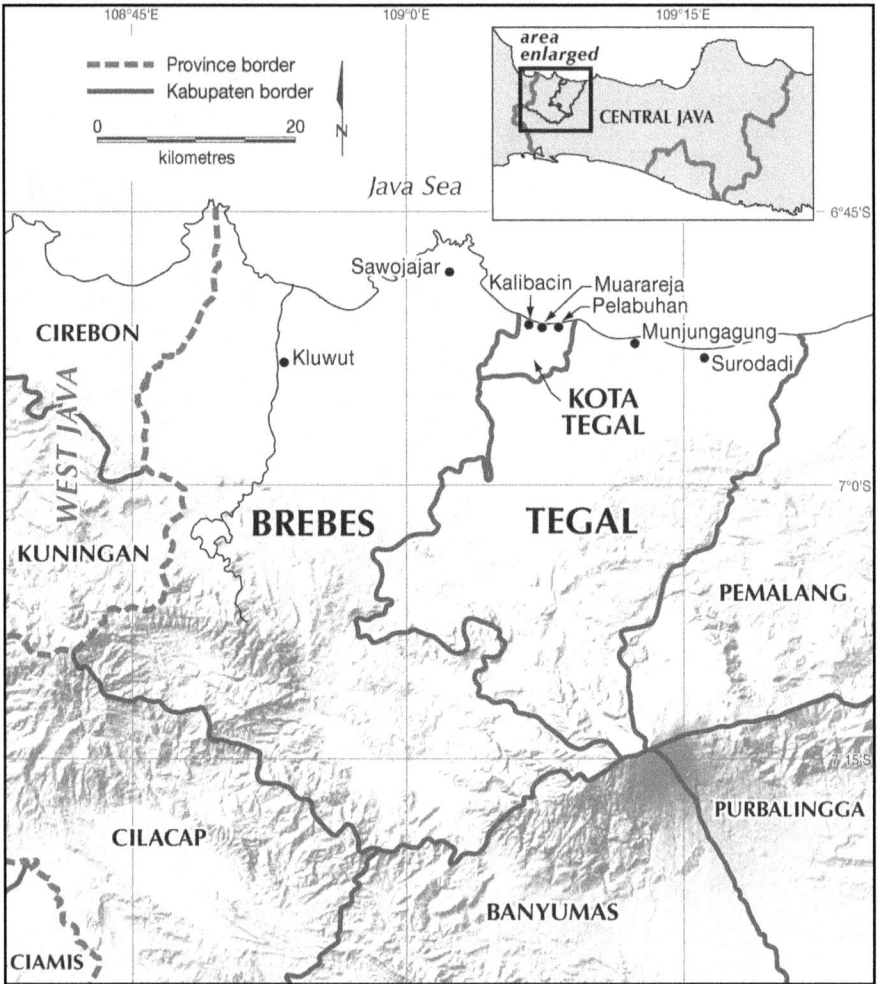

Map 3 Brebes and Tegal Districts (kabupaten), Central Java Province.

3 *Berjuang diatas perahu*[1]

Livelihood, contestation and declining marine resources on Java's north coast

Anton Lucas

[The meaning of] sustainable fisheries ... must be broadened [to be] *environmentally friendly, economically sound* and *socially just* [English original]. These three dimensions are the spirit of fisheries development planning in this era of regional autonomy.

(Minister for Fisheries and Marine Affairs, 2002)[2]

Introduction

The deteriorating situation of the fisheries in the Tegal region of Java's north coast is connected with a national and global intensification that is driving down catches in most parts of the world. The FAO estimates that 75 per cent of the world's marine fisheries are fully exploited, over exploited or already depleted (Mous and Pet *et al.* 2005: 260). Overfishing caused by a combination of factors – new technologies, too many boats and too many fishers – has led to a steady decline in the Java Sea fisheries since the mid-1980s, and to the current crisis in fishers' livelihoods in the Tegal region, with up to half the fleet unable to put to sea.[3] The reasons for this are complex and not solely due to the worldwide decline in fish stocks. The rising price of fuel has become prohibitive for many fishers, and indebtedness to trader moneylenders (*bakul-pengijon*) makes it difficult for fishers to move out of their industry even if there were alternative employment options. Increasing competition from both local and foreign-owned vessels is putting more pressure on larger boat operators, who in turn put pressure on small fishers. The challenge now is 'finding a way to exploit the seas ... that preserves habitats and species while providing the people of the region with an essential source of protein ...' (Butcher 2004: 291).

One of the concerns expressed in this chapter is the classic commons dilemma of how to create a fair and effective set of institutional arrangements to manage this common-pool resource (CPR) to avoid destruction of the Java Sea fishery. Both large and small fishers, faced with declining catches per unit of effort,[4] are now increasingly in conflict, challenging central and regional government authority in the process. Early in the Reform Era, several local fisher communities tried to limit access to 'their' section of the Java Sea fishery. Although there is no record of any territorial system of resource control in the Tegal area, amongst the communities involved in these conflicts the concept of 'open access' to a shrinking commons is now being challenged.

The conflicts over fishing rights between communities early in the Reform Era, indicates the plight of artisanal fishers trying to maintain a livelihood in the face of increasing competition from foreign as well as other more highly capitalized Indonesian fishers. In the context of decentralization and regional autonomy, this chapter explores these conflicts and the problem of maintaining livelihoods as overfishing and indebtedness drive ecological and economic decline.

The chapter also considers how various interest groups (government, NGOs and different sectors among local communities) in the Tegal region are responding to these challenges.

The political economy of fishing in the Tegal region needs to be sketched here. There is not one 'common' interest among fishers. Different competing actors with different economic means and hence access to more effective, and often more destructive, technology are in conflict since some actors are able to invest in the new technology (larger boats, mini-trawl nets, or *arad*), with adverse impacts on other users of the resource. The main users of these fisheries fall into two groups: at the bottom end are small fishers working on boats under 10 GT, employing one to five deckhands. Fishers (including owners, captains and deckhands) do not usually own or finance their own boats, equipment or supplies. They obtain credit from trader-moneylenders, to whom they must sell their catch through a relationship that ensures an indefinite state of indebtedness. Thus *bakul-pengijon* can insist that fishers use mini-trawl nets, which catch more fish but damage the environment (see Table 3.1 for impacts of different fishing technologies).

The second group, the medium (10–30 GT) and large (above 30 GT) purse seiners, are often financed or owned and operated by Chinese, some of whom are also *bakul-pengijon*. The bigger boats sell most of their catch to traders via the two Tegal municipal fish auction centres (TPI, *tempat pelelangan ikan*), which are bypassed when indebted small fishers sell their catch direct to *bakul-pengijon*. Central government agencies run the harbours and issue licences to boats above 30 GT. The provincial government is responsible for issuing licences to boats under 30 GT, running the (often defunct) TPI and the municipal Fisheries Department keeps limited statistics on boats, technologies and catches.

Other actors are NGOs working with fisher communities, and Co-FISH,[5] the ongoing Asian Development Bank project which has funded the construction of a new harbour and fish auction facilities in Tegal, and created the Karang Jeruk Fish Sanctuary as a marine protected area (MPA). Fisher communities have weak political power, as the municipal elected assembly, Dewan Perwakilan Rakyat Daerah (DPRD), considers the fishing industry, which contributes little to regional income, an unprofitable investment.[6]

Table 3.1 shows a selection of fishing technologies (boats and nets), type of fish caught, fishing zones that boats/nets are supposed to operate within, and environmental impacts of fishing technologies in the Tegal region during 2005.[7] The impact of fishing nets varies according to mesh size and whether the

Table 3.1 Selected fishing technologies and their impacts in the Tegal Region in 2005

Type of net S = set T = trawled	Type of boat, tonnage (GT), and horsepower of boats using this technology	Net mesh width (inches)[a]	Net specifications length (m) × depth (D) (m); demersal (D); pelagic (P)	Type of fish caught	Date introduced	Fishing Zone permitted to use this technology[b]	Social and environmental impacts
1 Fish lines (*pancing*) (S)	*Compreng* 0.5 GT, 10 hp	no mesh	300 × 12 (P)	*cracas, kuro, manyung*	1971	1a	Damages the Karang Jeruk reef
3 *Kejer* (S)	*Jukang* 4 GT 3–10 hp	3.5	800 × 1[c] (D)	prawns (*rajungan*)	1980	1a	Often left attached to a floating marker, can become tangled with other nets
2 Gill net *Gemplo* (S)[d]	*Sopek* 4 GT, 16–20 hp	0.75	150 × 20 (P)	*teri nasi, teri jawa*	1970–1994	1b	Operates on the edge of reefs, small mesh prevents undersized fish escaping
3 Millennium gill net (S) *Gilnet milenium*	As above	4	1000 × 12 (D)	*mayung, kakap, tongkol*	2002	1b	Wide meshed nets allows immature fish to escape
4 Mini trawl *Garuk* (T)	As above	1	27 × 2 (D)	*rajungan, ikan rucah*	2005	1b	Takes immature shrimp stocks
5 Fish traps *Badong* (S)[f]	*Sopek* 5 GT 16–30 hp	1.5	Cage nets 40 cm × 500cm attached to a line 20–30m deep (D)	*rajungan*	2003	1b	Are operated 24 hours a day, with long term impacts on stocks; compete for space with gill and *gemplo*

(*cont.*)

Type of net S = set T = trawled	Type of boat, tonnage (GT), and horsepower of boats using this technology	Net mesh width (inches)[a]	Net specifications length (m) × depth (m); demersal (D); pelagic (P)	Type of fish caught	Date introduced	Fishing Zone permitted to use this technology[b]	Social and environmental impacts
6 Mini-trawl *Arad* (T)	As above but also used with *dogol* boats	1	27 × 2 (D)	*udang, gembung, bambangan, cumi-cumi, ikan rucah*	1980	Ib	Trawls the sea bed, damages fish habitats and small mesh of net captures undersized/juvenile fish
7 Mini purse seine *Pukat cincin* (S)	As above	1	150 × 30 (D)	*tongkol, tengiri, bawal*	Early 1980s	Ib	Small mesh net takes young fish stock; Lamp fish lures to attract fish cause 'overfishing'
8 Modified trawl *Cantrang* (T)	*Dogol* 20 GT 100 hp	1	200 × 5 (D)	same + *kembung, kakap, pare*	1980	II	Uses mechanized sweep action; main cause of overfishing
9 Large purse seine (S) *Pursin besar*	Purse seine boats (KM – *kapal motor*) >28 GT, 100–200hp	1	600 × 70 (D)	*pare, tongkol, bawal, kakap, tengiri*		II	Causes overfishing; Use of ice instead of refrigerator ships results in loss of a substantial proportion of catch on 15 day trips

Source: Data from interviews (September 2004) and personal communications with Mardiyono (December 2006)

Notes:

a Fishers say net mesh of less than 2.5 inches destroys fisheries by taking too many undersized small species. The Indonesian Fisheries Department and local fishers use imperial standard for net mesh measurement.

b See n. 19 for the technology and boat size to be used in each zone.

c *Kejer* nets consist of 14 separate lines (*ting-ting*), each 60 metres long.

d *Gemplo* nets have a fine mesh (0.75 inches) called a *waring* which prevents the small *teri nasi* from escaping.

e A modified trawl net similar to *arad*; with rising fuel prices, *garuk* are popular because cheap to buy and operate, and take in larger catches.

f Muarareja *badong* fishers set up to 500 traps, seven metres apart and stretching 3,500 metres; they cook the catch on their boats to get a higher price (Rp 30,000 per kg, Rp 10,000 more than raw shrimp).

net is trawled (T) or set in a fixed location (S). The nets that create the biggest environmental problem are modified trawl nets – *arad* and its close relatives *garuk* and *cantrang*. These small mesh nets take everything from the sea floor, with serious ecological consequences. Because of declining catches and livelihoods, fishers continue using modified trawl nets, or create new nets, such as the *badong* (cage trap nets) to make ends meet.

Since 1999 the state has tried to resolve conflicts between net users by regulating different uses within declared fishing zones. The data in Table 3.1 show in which zones different fishing technologies and boat sizes are permitted to operate in the Tegal region. In the context of declining catches, this zoning system is proving difficult to maintain. For example, modified trawl nets (*cantrang*) used on large boats (*dogol* of 20 GT) are supposed to operate in zone II. But because of decreasing catches and increasing costs, they are going on shorter trips (one day instead of two weeks) and many boats are now operating 'illegally' in zone Ib (restricted to boats of less than five GT). Zoning has not prevented smaller boats using modified nets (longer nets with smaller mesh width) catching more undersized fish 'illegally' in zone Ia;[8] Neither has zoning prevented conflicts between different net users, because the zones are not adequately patrolled by the Fisheries Department. There is also conflict between those fishers using mini-trawl nets and others using stationary *kejer* nets or fish traps (*badong*).[9] Boats using modified trawl nets are not supposed to operate inside Zone Ia (*jalur* Ia) (the three nautical mile limit), the zone which is meant to be reserved for set nets, other nets which are not modified, and traditional boats (no more than 10 metres long) (Direktorat Jenderal Perikanan Tangkap and Co-FISH 2002: 7).

Declining small fisher livelihoods and overfishing in the Tegal region

Despite the general unreliability of fisheries statistics, it is possible to find indicators of overfishing by purse seine fishers in particular. Purse seines, which replaced the banned trawl boats, have been intensively studied since the late 1980s. Recent research has confirmed earlier data (McElroy 1991) about their contribution to declining Java Sea pelagic fisheries. Suherman and Nugroho (2004: 5–7), in an analysis of data from Pekalongan harbour from 1976 to 2001, show that the rapid growth in adoption of large purse seine nets in pelagic fisheries has been accompanied by use of longer nets, more powerful underwater lamp lures, larger engines, and a capacity to travel longer distances to fishing grounds in East Java waters and the Straits of Makassar. Despite this intensification, the total pelagic fishery yields and catch per unit of effort have declined since 1998 from a high point in 1992.[10]

Table 3.1 shows some of the fishing technologies used in Tegal over the last decade. The introduction of a new technology, namely fish traps (*badong*), has caused the most recent conflicts between fishers. *Badong* are cheap to buy and profitable to operate, but compete with other technologies operating within the three-mile zone. *Badong* are small mesh fish traps, which are not liked by gill net (*gemplo*) operators whose nets get tangled in the traps, while fishers using *kejer* nets find they get fewer prawns (*rajungan*) because of competition from fish traps.[11] *Badong* users are also in conflict with *garuk* mini-trawl operators, whose nets also get caught in the fish traps. They in turn complain that shrimp stocks are declining because of *garuk* operators.

Thus the fishing industry in Tegal is marked by increasing conflict between users of different technologies, namely trawl versus set nets, and between different types of set nets, which intensifies social tension as livelihoods from fishing come under increasing pressure. Furthermore, there is no longer any traditional fisheries resource management regime to mediate these conflicts in the Tegal region, although there apparently had been territorialized systems of limiting open access to marine resources in some East Java coastal communities in the past.[12] So far the state too has failed to keep the large purse-seiners from entering the six nautical mile fishing zones (Ia and Ib) that are supposed to be reserved for small fishers.

A study by the transnational Co-FISH project found that in Tegalsari fishing village (Tegal municipality) there has been a drastic decline in the fish catch between 1999 and 2003, leading to a decrease in the number of families whose livelihood was obtained from fishing, implying that fishers have been leaving the industry in large numbers. In the same period other occupations have also declined in that village, suggesting a general exodus to find work elsewhere, while in Muarareja the catch sold at auction declined, but numbers of fishers rose slightly,[13] suggesting more fishers were selling their catches through trader-moneylenders.

Biodiversity decline is being observed in the increasing prevalence of a particular species in the catch, and in the number of species that are now officially endangered. A Co-FISH survey found that between 2002 and 2003, at seven TPI in the Tegal-Brebes-Pemalang region, 14 species were no longer being caught, while a further 10 species were being caught in very small numbers.[14]

The impact of overfishing has translated into lower shares of the catch for boat crews, for both owners and deckhands of boats of five GT tons in particular. In four harbours in Tegal region (Kluwut Sawojajar, Larangan and TPI no. 1-Surodadi), 25–50 per cent of the local fishing fleets are lying idle. According to a small boat owner from Kluwut, 25 per cent of the fleet of 623 boats are not going to sea because of declining catches, a reduction in the size of fishing grounds,[15] stagnant prices, and increasing operating costs, particularly fuel, which has quadrupled in four years.[16] As a result of fuel price increases, many fishers now use fuel mixing (sump oil with diesel) to reduce costs. While this does not make a big impact on the size of crew shares, it reduces the life of boat engines by two to three years. These are indications of increasing economic hardship due to the open access nature of what should be a more strictly managed common-pool resource.

Regional autonomy and contestation over open access

Rokhmin Dahuri, Minister of Marine Affairs and Fisheries in Megawati Sukarnoputri's cabinet, wanted regional autonomy to change a centralist management system, where marine resources were controlled by large-scale fishers, to a system which would 'raise economic welfare and equalize disparities among regions'. He proposed to create a sustainable Indonesian fishery that would be 'environmentally friendly, economically sound and socially just' (Dahuri in Satria 2002: xv).

Articles 3 and 10 of the Regional Autonomy Law (UU22/1999) gave districts and municipalities 'authority' (*kewenangan*) to manage the seas to a four-mile limit, and provincial government management of the seas to a 12-mile limit. Some regional governments and some fisher communities interpreted 'authority' to mean 'sovereignty'. So attempts were made to territorialize fisheries, that is, to exclude 'outsiders' from fishing in waters over which fishers from particular districts and provinces now claimed exclusive rights (Satria *et al.* 2002: 1–2, 49). During 1998–2000 fishers tried to establish territorial rights over local fishing grounds, declaring areas offshore of their home harbours/districts/municipalities as single user areas.[17] This territorialization phenomenon, known as *pengkapling laut* (parcelization of the sea), led to violent conflicts between users of marine and coastal resources (see below).

Alongside the new four- and 12-mile management zones created by the decentralization law (previously the national government was responsible for management of all Indonesian waters), fishing zones (*jalur*) have been reactivated, adapted from the 1976 legislation (Minister of Agriculture Regulation no. 607/1976) to reduce conflict between trawlers and small fishers. As we have seen, under Minister of Agriculture Regulation No. 392/1999, Indonesia's fishing grounds were divided into three fishing zones with fishing gear, boat size and capacity specified for each zone. Boats not complying face heavy fines, having their licences revoked, or both, if they are caught.[18]

Neither regional autonomy maritime management zones, which aimed to give provincial and district governments more control over maritime resources, nor the resurrected fishing zones, which aimed to manage open access to preserve the common-pool resource, will preserve sustainability of fisheries in the Java Sea without effective enforcement.[19] Implementation is not facilitated by the fact that specified zones of regional authority do not correspond with fishing zones.

Despite provision to decentralize the issuing of fishing licences to districts, provincial authorities still issue licences for boats under 30 GT (and motors of less than 90 hp), and boat owners still go to the Directorate General of Marine Fisheries in Jakarta for boat licences over 30 GT.[20] Then they wait long periods before being issued with a Fishing Licence (SPI, *Surat Penangkapan Ikan*); the Fisheries Business Licence (IUP, *Izin Usaha Perikanan*) is issued by the provincial government (for boats under 30 GT, otherwise in Jakarta). On the SPI licence, a Tegal ex-*juragan* (owner who operated four purse seine boats before going broke) says:

Local businessmen (*pengusaha pribumi*) feel under great pressure arranging an SPI because there are many difficulties. You have to go to the office of the Director General of Fisheries in Jakarta to apply. It takes three to four months to get the license, and costs four to five million rupiah for 'administration'.

It's valid for one year, so you only have a license for the eight to nine months remaining. Fishers can't wait that long. They should issue temporary SPI. The system makes people crazy (*gebleg*).[21]

Many boats take a risk and go to sea without the large number of licences or certificates required. In mid 2003 three fishing boats – ironically owned by the Tegal municipality – were impounded and their captains given jail sentences for not having the necessary paperwork.[22] Neither the Tegal water and air police nor the municipal Fisheries Department have the patrol boats or security officers to patrol 'their' region of the Java Sea, given the hundreds of boats at sea at any one time. The Tegal harbour master has only eight marine inspectors for the 125 large purse seiners (over 30 GT) that unload their catch every day.[23] This situation demonstrates the lack of state capacity to create an effective resource management regime.[24]

Another local issue is which level of government runs the harbour. The national Department of Communications refuses to allow district level administrations to take over the control of local harbours.[25] Provincial level administrations still control the fish auction system and infrastructure. The Central Java provincial government collects a levy of 5 per cent via the TPI of which only 0.95 per cent goes to the municipal government.[26] The latter says that without funds it cannot improve the livelihood of fishing communities in Tegal municipality.[27] This suggests that despite decentralization, insufficient authority and resources have been devolved to local government for them to create an effective resource management regime.

Since *reformasi* began, contestation over access to the marine resource has been caused by increasing use of modified trawl nets, and by fishers excluding from their fishing grounds boats which they say bring no economic benefits and cause environmental damage to local communities. We will now look at one such conflict between two fisher communities within the Tegal region.

Conflicts over marine resources

Throughout 2002 Tegal fisher communities clashed along the north coast of Java and with other island communities in the Java Sea and in the Banka Belitung region of South Sumatra. A brief analysis of these conflicts demonstrates the complex environmental, political and administrative/legal issues behind these disputes.

The conflict between Muarareja and Surodadi (15 km to the east of Tegal municipality – see Map 3) shows the actions small fishers took early in the Reform Era to protect their livelihoods inside the three-mile fishing zone from over-exploitation by boats using banned trawl nets. On 7 March 2000 a *sopek* boat with three crew from Muarareja was fishing with a mini-trawl inside the three-mile zone near the coastal village of Surodadi.

A group of suspicious *kejer* net fishers from Surodadi was watching the Muarareja boat closely. Not long after this four Surodadi boats appeared and surrounded the Muarareja boat. They caught the crew using an *arad* mini-trawl net, banned from the inshore zone Ia. The crews on the Surodadi boats threatened the Muarareja fishers, telling them to stop using the *arad* net and asked for the nets. This created more tension. They then forced the Muarareja crew to leave their boat and come onboard the Surodadi boats. Suddenly the Muarareja boat (owned by a trader) caught fire. The navy sent a fast patrol boat to calm the situation down, but managed only to save the burnt boat's motor.[28]

Significantly, the press described the conflict primarily as a law and order issue, and a protest about compensation for damaged property, rather than a conflict between two communities over access to the fishery in the context of declining livelihoods. Stories focused on how successful the navy and the police had been in quelling a violent clash between angry fishers from two different communities. Burning a boat was a violent criminal act and village leaders were ordered to 'calm the anger of the community'. Press sympathy was with the *bakul-pengijon* boat owner, despite the fact that the boat had been using banned nets illegally in the inshore zone Ia.[29] The press reports made no mention of the underlying cause of the conflict, namely that boats using *arad* mini-trawl nets from Muarareja, illegally operating within the three-mile zone, often collided with *kejer* nets owned by Surodadi boats, resulting in increased expenses due to net repairs and loss of livelihood (Mardiyono 2006: 6–7). A prominent spokesperson for the Muarareja community did admit that 'the community was wrong to use the *arad* nets', but 'it wasn't necessary to burn our boat', this was 'taking the law into one's own hands'. Predictably the ex-navy Mayor of Tegal said he wanted the authorities to solve the problem as quickly a possible, 'so the excessive behaviour would not spread'.[30]

The Semarang provincial daily *Suara Merdeka* ran a three-quarter page photo with the key leaders on both sides of the dispute embracing each other in a 'spirit of peace' after an agreement was eventually reached.[31] Maintaining social harmony and order (a legacy of Suharto's New Order) was as important as negotiating competing claims and protecting marine resources. The Tegal chief of police intelligence was the only official who publicly acknowledged that both sides were breaking the law:

> Burning a boat can get a harsh penalty. So can the use of trawl nets which the government has banned for a long time: 'If both sides cannot find a peaceful way out of the dispute then the rule of law will decide the conflict. I hope that everyone realizes that the main issue is how fishers can make a livelihood at sea in a safe and peaceful way. Having food security (*urusan perut*) is the main issue. Not obstinately accusing each other of being in the wrong'.[32]

Since the beginning of the *reformasi* period, government officials have been saying that law-making and law enforcement are at the heart of the problem of balancing economic viability and resource security in the fisheries industry. Mini-trawl nets are three times as profitable as traditional *kejer* nets,[33] but involve a clear trade-off between immediate and long-term needs. An effective resource management regime would need to find some way to ensure that both short-term needs and long-term sustainability can be met. According to the head of the municipal Fisheries Department, restricting mini-trawl nets impacts most on small fishers, whose livelihoods are all under threat. This is why he says, 'With *arad* I close one eye, if not there is continual conflict … In Indonesia trawl nets are considered as a solution to social conflict rather than [a problem for] protecting natural resources'.[34]

It is not surprising that mini-trawl nets are seen by officials as part of the solution as well as part of the problem. Ironically, since this incident in 2000, many Surodadi fishers have been forced to adopt the nets they were protesting about to make a livelihood for themselves. Thus in June 2005, 50 per cent of fishers (30 boats) in Surodadi TPI no. 1 were found to be 'active *arad* users' within the Ia zone during two months of the west monsoon (December–January) when they cannot fish with other nets.[35] The matter thereafter disappeared from the public domain, although the legal and long-term environmental issues remain.

In a subsequent investigation into the conflict, Co-FISH also found that modified nets were in widespread use: beach seine nets (*pukat pantai*) had been modified by attaching otter boards to widen the net, making it possible for one boat to trawl 14.3 ha of seabed in the Java Sea in one day; boats with *arad*[36] trawl nets were operating inside the three-mile limit (which would automatically bring them into conflict with smaller boats using set nets); and that the number of *arad* nets in Muarareja had more than doubled in four years to 2000.[37]

The Muarareja–Surodadi conflict in 2000 shows how fishers use direct action to protect their livelihoods in the absence of government enforcement of fishing zones, but also suggests the futility of such actions in the absence of law enforcement on a wider scale. From the local Fisheries Department viewpoint, the central government Fisheries Instruction 340/1997 muddied the waters,[38] making the legal situation unclear and inconsistent. The regulatory regime being applied to modified trawl nets needs reforming before proper law enforcement can occur.

Following another highly publicized series of incidents of open conflict in 2002 between Pekalongan and armed Masalembo fishers who captured north coast boats fishing within the 12-mile zone,[39] a delegation of fishers from Tegal went to Jakarta with five demands for Sarwono Kusumaatmadja, then Minister of Fisheries and Maritime Affairs. These demands were: (1) that the police take 'strong action' against the boat burners; (2) that purse seine boats banned from the Java Sea in July 2000 be allowed back in, but only to operate in the Exclusive Economic Zone beyond the 12-mile territorial waters limit;[40] (3) that the Regional Autonomy Act needed an implementing regulation (to clarify the meaning of 'authority' for marine affairs in the context of regional autonomy); (4) that foreign boats captured 'stealing fish in Indonesian waters' should be processed by Indonesian courts; and (5) if the boats were found to have been 'acting illegally,' the government should destroy them. Minister Sarwono agreed to all five HNSI demands, although they have not been actively implemented by his successors.

Profit sharing and moneylending

When Tegal was preparing for the visit of President Megawati to open the new Jongor boat harbour and auction complex (built at a cost of Rp 73 billion) in July 2004, a group of fishers announced they would present her with two demands. First, that no further foreign company fishing licences be issued granting access to local fish stocks; and secondly, that the government must provide credit because 'small fishers are trapped by money lenders' while costs of fishing are skyrocketing.[41] Apart from having to use trader-moneylenders instead of banks for credit, and sharply rising fuel costs, income-sharing arrangements between boat owners and their crews also keep wages low. Income-sharing relations between boat owners (*juragan*) their assistants (*pengurus*),[42] captains (*nahkoda*) and deckhands (*ABK*),[43] are not regulated. Shares vary between 50–50 and 60–40 in the boat owner's favour. Exploitative income-sharing arrangements are commented upon in the Co-FISH socio-economic assessments of Tegal region fisheries, and by local NGO workers, who have observed that in some places an owner will take his/her share *before* deducting expenses (as in the case below). Because shares are now so low, many deckhands are leaving their communities to find work either outside the industry or on larger foreign boats, often staying away for three to six months. This labour exodus has not improved livelihoods for local crew, however:

> Fisher livelihoods are not improving much at present. Especially if the owners are local *juragan* ..., the shares system is very unjust. Usually the boat owner takes 50%, then 25% is deducted for expenses for the next trip; the remainder is divided between the navigator/captain (*jurumudi/nahkoda*) and the crew (ABK). The crew say that they don't understand the share system and are never told the costs of supplies (rice, cooking oil, spices, ice, fuel), which are decided directly by the boat owner, who usually has a shop selling supplies as well.[44]

Another issue is the depreciation deduction of 25 per cent on every catch by owners of larger boats. The practice of assigning extra shares to skilled crew further reduces the shares obtained by unskilled deckhands.[45]

The role of trader-moneylenders on Java's north coast has existed since the mid-nineteenth century (Butcher 2004: 50–1). Their multiple roles make indebtedness a complex social issue. Traders in Tegal have many functions. As brokers, they handle all kinds of fish, buying at auction and selling to the big city markets of Java. They must know the markets they are buying from and selling to each day, so that they can bid competitively to make a profit. Those traders that also lend money for boats and supplies, referred to as *bakul-pengijon* in Tegal,[46] buy fish at a cheaper price from indebted boat owners. Many act as patrons, contributing cash whenever their 'client' fisher families need assistance for marriages, funerals and even for routine living expenses (Satria 2002).[47]

The function of providing credit to fishers is important, because institutional credit is not available, despite the government's concern to get banks to lend money to fishers.[48] For big and middle-size purse seine boats, selling a small percentage of the catch to *bakul-pengijon* before the bulk is sold at TPI, is not an issue. But medium and small boat owners (under 10 GT) have no choice but to borrow from *bakul-pengijon* for boat and equipment repairs 'with the condition that all their catch has to be sold to the trader, which means that the sale price is determined by the creditors in a one-sided way'.[49]

There are three views on the role of trader-moneylenders. The NGO view, often reflected in the local media, is that fishers would be better off if the government paid off their debts to *bakul-pengijon* so that all of their catches could then be sold through the auction system, where prices are usually 10–20 per cent higher, depending on size of the catch, quality and type of fish. At present in the Tegal region up to 40 per cent of the catch is sold outside the auction system (60–70 per cent in Muarareja) because of the debt relation with *bakul-pengijon*. Many TPI have ceased to function for this reason. But selling at auction should mean greater returns to fishers; and the 5 per cent levy on catches at TPIs would produce more welfare funds for supporting fisher livelihoods as well. From this perspective, *bakul-pengijon* are a major cause of poverty in the fishing industry.[50]

The following case supports the NGO view. In 2004, a group of 50 mini-trawl (*arad*) fishers in Muarareja began selling off boats built and financed by a well-known *bakul pengijon*. They were fed up with having to sell their shrimp catch to her at a price 30 per cent lower than they could get at the TPI. The *bakul-pengijon* who had supplied the fishers' boats (worth Rp15–20 million each) reported them to the police, saying they were selling boats that didn't belong to them. The police summoned the fishers to the police station to explain their actions. They refused to attend, saying they had done nothing wrong, because the boats in question now belonged to them. They argued that, because they had been selling shrimp to the *bakul-pengijon* at a price substantially lower than the price they would have received at the fish auction for the past seven years, their calculations showed they had effectively paid off the original cost of the boats. In the fishers' view, 'it was only fair that the debt be considered paid' and that they should have no further financial obligation to the trader-moneylenders. There the matter ended without going to court.[51]

This case highlights the ambiguous nature of the fisher/trader-moneylender financial relationship. Trader-moneylenders finance the building and operation of boats on the condition fishers sell all of their catches to them. They also provide money to buy nets and supplies needed for fishing trips, as well as repairs to boats and equipment. Fishers are not normally required to pay back this assistance, a situation which leaves them in a state of permanent obligation to the trader-moneylenders.

A second perspective focuses on the essential contribution women traders, often the wives of fishers, make to the fish trading networks. This role in the networks of fish distribution (supplying regional and Jakarta markets) as well as the fish processing home industry was clear from a study of Juwana harbour in East Java. While the moneylender roles of traders are mentioned in passing in this study, indebtedness and the impact of traders buying large quantities of fish outside the public auction system are not mentioned as an issue in Juwana in the early 1990s (Antunes 1998a: 253–7).

Trader-moneylenders have other functions. They buy all the fish caught individually by deckhands on medium and large purse seiners. Many small fishers prefer to sell to *bakul-pengijon* because they get their money quickly compared with sale through TPI auction. TPI usually have only one auctioneer, and sales can take hours. Fishers are afraid that prices will have fallen drastically (*harga ambrol*) by then.[52] Further, banks will not lend money to fishers to buy boats or to finance their fishing trips because they do not accept boats as collateral, and very few fishers have certificates of land ownership, which can be mortgaged.

Over time traders become moneylenders and then buy boats themselves. One female *bakul-pengijon* in Sawojajar (Brebes district) finances 100 of the 140 *sopek* boats in this fishing community, and loans out an average of Rp 5 million to each fisher for the purchase of boats, gear and supplies. Not surprisingly, the Sawojajar TPI is not functioning because the catch is sold through one or two *bakul-pengijon*.[53]

The third view on *bakul-pengijon* emphasizes the patron–client as well as economic roles of traders (Satria 2002: 3–6), which are important for the survival of small boat owners. The Sawojajar trader mentioned above said that she gave assistance for fishers to repair nets, motors, and damage to boats that occurred while at sea. Also 'if fishers don't have enough money for daily needs, I help them. I have to give them what they need, because they are my clients (*anak buah*)'.[54] As we have already noted, these forms of personal assistance are not treated as loans, are not paid off, and no interest is charged. Instead, clients (small fishers in boats of less than five GT) sell their entire catch to the trader. In communities where there is no TPI, or it is non-functioning, this marketing role is essential.

Because of the impacts of overfishing already discussed, trader-moneylenders are also starting to feel the pinch. Boat owners they have sponsored are unable to pay back the loans. Thus one of Tegal's biggest *bakul-pengijon* owns four large purse seiners and has loans of Rp 5–20 million out to another 25. She has had to take back 10 boats, after no repayments were made for five years. These boats were sold at a loss on the original cost.[55] In 2002–2003 Pak S. (a large trader-moneylender) says that he ran out of working capital because 'there were too many fishers and too many boats. Yields on the big purse seine boats declined by 50 per cent. But boats with *cantrang*[56] nets declined by 80 per cent. I had to sell a car for Rp 100 million in order to pay for my boats to go to sea.' The four main reasons for the decline in income, according to Pak S, were: foreign boats 'stealing' fish in Indonesian waters, too many Indonesian fishing boats, the increasing proportion of undersized fish being caught (indicating current fishing practices are unsustainable), and the rising cost of supplies, meaning returns from fishing no longer cover operating costs. 'To send a purse seine boat to sea now costs Rp 20 million; it used to cost Rp 11–12 million a few years ago. If the price of diesel goes up, the price of everything else goes up. How can fishers improve [their livelihood] when they are beaten into the ground, and powerless (*mati kutu*)?'[57]

Programmes for improving livelihoods

Government and NGO programmes have to work in the context of a declining sense of common interest amongst fishers. This is reflected in a weakening of collective responsibility that underpins various rituals performed by fishing communities (such as the *sekedah laut selamatan* or *nyadran*) that traditionally ensured the safety of boats at sea, but which nowadays are performed more for their entertainment value.[58] But beyond the commercialization of rituals everywhere apparent, is the declining sense of solidarity that they once expressed. Iwan Nurdin explains:

> In the past fishers in Batang [east of Pekalongan] knew all the kampung people from one end to the other, all the names of the families, who the parents were, and their character (*tabiat*). This is no longer the case. In the past fishers went out on day trips, so they had plenty of time to socialize. Nowadays a fisher goes away for one day to six months at sea. I think this is the reason that tradition and sense of community (*rasa kebersamaan*) amongst fishers is declining.[59]

According to an NGO worker with the Co-FISH project in Tegal from 2000–2004, community ties are being eroded by an attitude of 'do as you like; forget about others' (*masa bodoh dengan orang lain*), with social jealousy, suspicion and unwillingness to cooperate between individuals, groups and communities. When the project invited fishers to discuss the degradation of the local Karang Jeruk reef, each community blamed the other for the damage.[60] In this context, the ADB funded Co-FISH Project has attempted to deal with the decline of this significant common pool resource. Aiming to promote community-based sustainable management of artisanal fisheries, and reduce poverty in coastal areas (Co-FISH 2004: 3),[61] the project has funded a new Rp 73 billion fishing harbour and port infrastructure, including a breakwater, fish auction centre, trader kiosks and offices. There is a poverty eradication programme targeting small fishers (Departemen Eksplorasi Laut *et al.* 2004) as well as the programme to reduce conflict caused by *arad* nets already discussed. Communication, information and education (KIE) programmes aim to raise consciousness in order to protect fish resources for the future. To this end Co-Fish has created the Karang Jeruk Fish Sanctuary, 4 hectares in area, six kilometers off the coast of Tegal, as a marine protected area. Recognizing its importance as a fish breeding ground to local communities, it also created artificial coral reefs (TKB, *terumbu karang buatan*) made of rubber tyres tied to bamboo frames and weighed down with concrete blocks to create buffer areas around a nuclear zone.

Buffer areas were needed because the Karang Jeruk reef was being over fished and undergoing environmental damage from boat anchors from more than 200 boats (two to five GT) mainly from two communities.[62] Under the auspices of Co-FISH, meetings were held throughout 2001 with communities using the reef, who agreed on the problem, but not the solution. The problem was habitat degradation (destruction of mangroves, sea grass, and coral), but fishers could not agree on how much of the reef should be totally protected as a nuclear zone.[63]

A working group of stakeholders,[64] the Karang Jeruk Fish Sanctuary Group

(KFSKJ), was also formed to manage the reef and to explain and promote the agreement in their respective communities. As in the Lindu case study (Acciaioli, chapter 4) a transnational actor intervened to try to create a resource-management regime due to state failure. There was no legal basis for enforcement in the form of a district law (Perda, *Peraturan Daerah*) because the Tegal municipal assembly (DPRD) was not interested in the issue, since it was not a lucrative investment or significant contributor to the regional budget. Nor were there any agreed sanctions for fishers who broke the agreement (except for bombing and mining of the coral).[65] Members of the KFSKJ explained to fishers that the inner zone was a protected area where fishing was not allowed, but could not enforce the agreement 'because they were faced with an undercurrent of [disaffection from] fishers who had to make a livelihood'.[66] These governance failures reflect the continued difficulties of imposing exclusions on common-pool resource regimes even when significant funding and efforts to engage local participation are deployed.

The environmental benefits of the programme were nevertheless demonstrable. The artificial reefs were quickly covered with sponges, mollusks, and other sea plants (*Montipora, Helliopora*), showing that environmental regeneration was possible.[67] Fishers no longer put their nets near the TKB for fear of catching and tearing them on the reef, which meant added protection for the re-established breeding grounds. In addition to the artificial reef around the nuclear zone for pelagic species laid down in 2001–2004, the project put in place a group of artificial reefs for demersal species. Fishers began to see the importance of these constructed reefs as spawning and nursery grounds for both pelagic and demersal species (Ministry of Marine Affairs *et al*. 2004: Table V-1), and volunteered to help the Fisheries Department construct them; but when placed in the sea to demarcate the boundaries of the proposed MPA, local fishers stole the floating markers for their own use within a week. They were not replaced for a year. One fisher, who knew the location of the now unmarked reef, was reputed to have sold one catch from there for Rp 13 million. Rumours quickly spread, but the fishers who knew its location refused to share this information with others, or even with Co-FISH NGO workers who did not know the location of the reefs once the buoys were stolen.[68]

This highlights the continued competition among fishers, which represented an attempt by a few to monopolize the resource, keeping their strategic knowledge a closely guarded secret for private interest. The common-pool resource has not been regulated in a way which gives fair access because the institutional framework for reef management is neither clear nor supported by legal mechanisms. As a result, corals are being taken by fishers (to sell for decoration and jewellery), and boats larger than five GT using modified nets (illegal in zone 1b where the reef is located) are operating around the reef.[69] The unclear legal status of the reef as a protected area is partly because there has been no regional government regulation (Perda, *Peraturan Daerah*) promulgated to support negotiated arrangements for its use and protection. There is also no regional budget for the reef's ongoing management,[70] despite the fact that Co-Fish spent Rp 330 million on surveys, workshops with fishers, and the construction of 42 artificial reefs during 2001–2002. Co-Fish, local government and fisher communities have failed to work out how to manage the reef sustainably, in the face of continual overexploitation by illegal fishing technologies, and subversion by local fishers with inside knowledge of where the artificial reefs were placed. Thus the creation of the Karang Jeruk fish sanctuary and construction of artificial reefs for fish breeding is a cautionary case study of an only partially successful intervention to solve environmental and resource rights issues in the face of increasing competition and resource decline.

During this period local governments also implemented a national programme to give financial assistance to fishers to offset the sharp rise in the cost of fuel since 2000. The PEMP (*Pemberdayaan Ekonomi Masyarakat Pesisir*, Coastal Communities Empowerment Programme) was implemented in 125 coastal districts/municipalities across 30 provinces, as part of the nation wide PPD-PSE (*Program Penanggulangan Dampak Pengurangan Subsidi Energi*) programme to offset the impact of the reduction in the national fuel subsidy, which has resulted in crippling rises in fuel prices. Fishers in the programme received direct loans of Rp 1–5 million. These were supposed to be repaid with 3 per cent concessional interest over a 20-month period into a revolving fund, which would then be lent to other fishers. The programme, which ran for two years, went broke because of the level of delinquent loans.[71] The state blamed the fishers for not repaying their loans,[72] but many fishers said they did not make enough to meet both daily needs and repayments.

In contrast, the Tegal district fisheries department office has twice provided Rp 2.5 million in individual revolving loans for nets and outboard motors to Munjungagung village for 110 fishers in 2002–2003. In two years 85 per cent of the loans had been repaid. This high repayment rate happened for a number of reasons. A condition of the loan was that fishers had to sell all of their catches at the local TPI. This meant that monthly repayments deducted from catches via a 10 per cent levy were made to the Fisheries Department. Other factors accounting for the high rate of loan repayments were the location of the new auction centre near where boats were moored, and the role of the KUD leadership, who at the time worked closely with the TPI in explaining the benefits of using the auction system. Repairs to boats and gear as well as welfare payments were paid out of a further 5 per cent levy, also deducted from the catch.[73] The Larangan fish auction system worked because local trader-moneylenders agreed to cooperate by buying fish at the TPI, and receiving an annual savings (*tabungan*) payment (0.25 per cent of the value of fish purchased). Because there were benefits for them, *bakul-pengijon* also agreed to deduct the 5 per cent levy on shrimp catches they buy direct from fishers, and pay this to the Larangan TPI themselves. In short the trader-moneylenders were important in making the Larangan fish auction system work.

To the west of Larangan, at Surodadi TPI no. 1, a small harbour of 70 boats, the KUD loaned 40 fishers (who actively sold their catch through the TPI) Rp 2.6 million each to purchase new locally produced Chinese outboard motors. In a similar scheme, the loans are being repaid by deducting 10 per cent of the value of each auctioned catch over two years. In September 2004 repayments had been continuing for eight months, and some fishers have nearly completed their loan repayments, suggesting the Larangan experiment is replicable.[74]

An employee of the local Surodadi fish auction centre (TPI no. 1) believes this programme can be a model for repayment of loans to *bakul-pengijon*. 'If I were a member of the local KUD board today I would borrow money from the bank (with the recommendation of the bupati [district head] of Tegal), and pay off the debts of each fisher', who would then have to sell all their catch through the TPI. The KUD would collect a levy of 10 per cent deducted from the catch by the TPI, to repay the bank loan and to give loans to fishers for supplies, which was the original function of KUDs in the 1970s before they went bankrupt. *Bakul-pengijon* would also have to buy their fish from the fish auction centre. The total debt of fishers in Surodadi (70 boats × Rp 3 million per boat) is about Rp 210 million. The KUD could pay that loan off in one to two years if all fishers auctioned all of their catch in the TPI.[75]

Could the KUD replace the *bakul-pengijon* as a patron? This would depend on how the KUD board, which at present is mistrusted because of corruption, can manage the scheme. Most of the fishing cooperatives set up in the 1970s by the New Order government to provide credit to fishers in the Tegal region are now bankrupt because of bad debts, corruption and mismanagement. According to local fishers, only those close to KUD board members – usually large operators – got loans.[76] There is also a broader problem of official oversight and regulation of the KUD system that has to be considered here. This involves issues of responsibility for the management of cooperatives – fishers have limited effective power to monitor activities of KUD board members. Neither does the fisheries department that provides credit to KUDs. It is the district office of the Cooperatives Department that is supposed to monitor KUDs, but supervision by local officials is limited to attending KUD annual general meetings. These cooperatives could be supervised more effectively via a monthly reporting system and more formal involvement of the membership in monitoring.[77]

State policy towards fisheries – where to from here?

Given the Minister for Marine Affairs and Fisheries' commitment to 'raise economic welfare and equalize disparities' (quoted above) through a decentralized system, it is relevant to ask here, what has the state been doing to limit access of large-scale companies (including foreign fishing fleets), exploiting Indonesia's marine resources? So far there do not appear to be any lasting policy changes or greater law enforcement to support the then Minister's policy claims. In Tegal municipality the implementation of a Co-FISH sponsored community surveillance system (SISWASMAS) has identified 'stealing' (i.e. illegal fishing) of fish by unlicensed foreign boats, and pressure on traditional fishing grounds from licensed foreign boats among the key issues concerning fishers,[78] but the project so far has failed to make an impact on these problems.

Despite the policies and programmes introduced since 1998, marine resources in the Java Sea are under severe threat of collapse from overfishing. After trawl boats were banned in 1980 purse seiners began fishing pelagic species at rates that were unsustainable by the end of the decade. Resource decline has been compounded by the introduction of mini-trawls, partly in response to rapidly declining catches. According to the Semarang fisheries department, there are 520,000 fishers trying to make a livelihood in the Java Sea, while the optimum is estimated at less than half that number (249,000). As the provincial fisheries department official who provided these figures remarked 'Where do the rest go? You can pay off their debts, but where do the fishers go?'[79]

For boats under seven GT (*sopek* and *compreng*), rising costs, declining catches, seasonal variations and contracting fishing areas mean incomes per day for ordinary crew often go below Rp 10,000 (US $1.20), so boats do not go to sea. Because returns are so low, Tegal boats are now finding it difficult to recruit deckhands.[80] Many boats have to recruit crews from outside the municipality. This has not had any noticeable effect on wages, which are based on the size and value of catches that continue to decline.

The decentralized fisheries management and the reintroduced fishing zones were attempts by the state to protect the marine resource. Lack of enforcement capacity by both the state and by local fishers, however, has left fishers unable to confront the problem of mini-trawl nets. The role of the ADB in trying to create a better fisheries regime through the Co-FISH programme is important. While its efforts to raise fisher incomes in the short term have been generally unsuccessful, the idea of creating a marine protected area, building artificial coral reefs, and negotiating agreement from fisher communities for a strategy to protect the Karang Jeruk breeding grounds,[81] does represent a step forward in efforts to balance economic needs and environmental protection. It remains to be seen whether long-term arrangements can be put in place to manage the reef sustainably. To do this the Tegal fisheries resource needs a fair, well-considered and enforceable regulatory regime, which involves fisher communities as well as local fisheries departments and the naval and police units based in Tegal harbour. At present the state does not have the capacity or the political will to police Karang Jeruk Fish Sanctuary, let alone coastal fishing zones.

Both equity and income security questions are implicated in any resolution of the resource protection question. Profits from mini-trawl nets and high overall indebtedness of fishers are currently driving overexploitation. If the regulatory and credit regimes were fairer, some fishers would be able to gain a better livelihood from these fisheries. But many would still have to leave the industry to make it more sustainable.

With decentralization we have seen attempts to territorialize fisheries, that is, to institutionalize property regimes under which local people would have exclusive rights to local fishing grounds. The Indonesian government and the Asian Development Bank seem to have encouraged the creation of new forms of territorialization under district government projects in response to fishers' direct action. However, these are dependent on local enforcement, and fisheries are in a weak position politically with local governments because they are not significant revenue raisers for their budgets.[82] On the other hand, fisheries employ a quarter of the local workforce, whose needs could translate into political pressure with better organization and a sense of common interest.

To date there has been a trade-off between ecological and livelihood considerations. But the continuing unsustainable exploitation of the fishery will lead ultimately to the collapse of fishers' livelihoods. In order to create a sustainable fisheries regime the enforcement issue is central. Community-based enforcement of fisher zones should include encouragement to report sightings of large boats inside the three-mile fishing zone to the police, who would have the patrol boats to intercept and fine these boats. Fishers are unlikely to do this if no action is taken. Districts must play a role in managing the three-mile zone including restricted and conditional issuing of licences to boats under 30 GT.[83] Fishing licences will have to be used to control the number of boats, and licences must be revoked if boats infringe the zone regulations. This would help relieve pressure on the environment, although only alternative economic options will solve the livelihood problem.[84] While police still accept bribes from boats, however, strict and fair enforcement of fishing zones will not occur.

What accounts for the relative success or failure of some of the interventions discussed in this study? Outcomes from several of the interventions considered here suggest that there are serious prospects for improving both fish stocks and livelihoods under certain conditions. The TPI–KUD auction/loan/social security system was able to ensure that fishers obtained credit, insurance, fair prices, and could repay loans in a well-managed system. A rejuvenated auction and cooperative system with trader involvement offers the best chance of reducing onerous debt faced by fishers. Restricted licensing and enforcement of zoning regulations must be accompanied by loans and training to support alternative livelihoods. The creation of the Karang Jeruk reef as a marine protected area is generally supported by local fishers, although the Tegal local government has not yet given the reef legal protection with a district decree (Perda). The marine protected area offers the prospect of improving fish stocks and long-term livelihoods if agreement on its regulation can be reached among local fishing groups and if provided with legal protection and implementation through local government. But when talking about how to preserve the reef, local fishers say, 'Don't just tell us small operators to be aware [of the problems], tell the big fishers to be aware as well, because they always hurt the small fishers. Our protests always hit the wall of powerful interests who always side with the big boat owners.'[85] Alongside law enforcement, inequality remains a critical problem for sustainable management.

Acknowledgements

The research for this chapter was carried out under an Australian Research Council grant. I would like to thank John Butcher, Jim Schiller and Carol Warren for their comments, as well as Duto Nugroho, Brian Fegan and Mardiyono for assistance with research materials.

Notes

1 *'Berjuang diatas perahu'* ('struggling on the boats') is a line from the song 'Nelayan' (fishers) by the popular singer/songwriter Iwan Fals, from his CD *'Salam Reformasi 2'*.

2 Rokhmin Dahuri, Minister for Fisheries and Marine Affairs, *'Kata Sambutan'* (Preface) in Satria *et al.* 2002: xv (author's translation). Unless otherwise noted, all text and interview translations are by the author from the Indonesian language.

3 Tegal region refers to the coastal communities of Brebes and Tegal districts in north Central Java. Fishers interviewed in this study are located at Kluwut and Sawojajar (Brebes district), Kalibacin, Muarareja and Pelabuhan (Tegal municipality), Munjungagung and Surodadi (Tegal district).

4 Catch per unit of effort (CPUE) is the catch in tons per number of (operational) days at sea. Fishing effort is the estimated number of days at sea derived from the number of trips multiplied by the averaged days at sea per trip. Tegal municipal fisheries data only give total fish catches and their values auctioned at TPI, according to type of fish and type of technology (Pemerintah Kota Tegal 2004).

5 Co-FISH (Coastal Community Development and Fisheries Resource Management Project) is a project of the Asian Development Bank and the Department of Marine Affairs and Fisheries, initially running from 1999 to 2005 at four sites, now extended to 2008. Two of the project's aims relating to sustainability were to reduce the fishing fleet of smaller boats (less than 10 GT) by 40 per cent, which meant providing alternative employment for fishers; rehabilitating habitats by planting mangroves; construction of artificial coral reefs, and creating MPAs (interview with Gde Wiadnya, former Trenggalek Co-FISH advisor, Denpasar, 11 July 2005).

6 Nevertheless patronage relations often inhibit the enforcement of existing regulations. One fishing village well connected to the local political elite is Muarareja, whose fishers continue to use modified mini-trawl nets, which are technically banned. Muarareja's leading spokesperson, Tambari Gustam (married to a wealthy *bakul-pengijon*), used to run a small local paper (the *Muara Pos*), and established an NGO (Cordova) to build patronage networks. When a larger daily, *Radar Tegal*, reported that the local branch of the Indonesian Fishers' Association (HNSI, Himpunan Nelayan Seluruh Indonesia) had protested about trader-moneylenders keeping fish prices down in 2003, Tambari is said to have initiated a demonstration and forced *Radar Tegal* to issue an apology.

7 Data from the Tegal Municipal Fisheries Department states that there were 593 boats registered in Tegal municipality in 2004; 87 per cent were less than five GT and powered by outboard motors. Of the 716 fishing gears registered, 45 per cent were *cantrang* (modified mini-trawls which damage the marine environment in the same way as *arad*). The rest were trammel nets (12 per cent); *prawe* lines (11 per cent); *klitik* nets (10 per cent). Tramel nets (*jaring kantong*) have been modified with a mesh of 1.7 inches and length of 200 m and trawled in zone Ia like *arad* to catch both demersial and pelagic species (*udang, sontong* and *blekutak*). *Prawe* are fishing lines (*pancing*) originally used on small boats (*sopek* of five GT) in Zone Ia; as many as 700 *prawe* lines (20 m to 50 m long) are now used on larger boats (10 GT) to catch *manyung* and *remang* in an unsustainable way in zone Ib. *Klitik* are set nets 200 m in length with a mesh width of 1.75 inches to catch *kembung* and *tetek*, purse seine (9 per cent) and gill nets (9 per cent). Although no *arad* are mentioned in the municipal Fisheries Department statistics because they are banned, fishers told my informant (a Co-FISH NGO worker) that there were 300 *arad* nets in the subdistrict of West Tegal (which includes Muarareja) in 2000, double the number in 1996 (Mardiyono, personal communication, 31 May 2005).

8 Recently a modified net has appeared, known as the *jaring teri pursin*, with a net mesh width of only 0.75 inches, which catches small fish and shrimp, supposedly in zone Ib. When catches are small, these nets are being used in zone Ia on motorized boats, which is illegal.

9 Local preference for particular types of nets is due to the demonstration effect. One

fisher tries a net modification which results in a larger catch, so other fishers from the same community follow suit in the expectation of also obtaining bigger catches (Mardiyono, personal communication, 03 May 2007).

10 This study found that the annual average number of trips per boat declined from 9.1 to 5.2 between 1986 and 2002, while the number of operational days at sea increased from 377 to 401 in the same period (Suherman and Nugroho 2004).

11 One fisher will lay 150–500 *badong* traps along the sea bed, seven metres apart, taking up an area of 1,000–3,500 metres.

12 A Dutch colonial survey in two coastal East Java districts (Rembang and Tuban) and in Madura in 1910 found that 'Each village has ... its own fishing ground and will not leave it, even if more fish are being caught in neighbouring areas' (Butcher 2004: 100 and Map 4.4).

13 The total annual fish catch auctioned in Tegalsari TPI fell from 2,844,886 kg (valued at Rp 1.9 billion) in 1999 to 923,445 kg (worth Rp 911 million) in 2003. Villagers whose livelihoods depended mainly on fishing declined from 4,820 (1999) to 1,254 (2002) (Departemen Kelautan dan Perikanan *et al*. 2004a: Tables II-67 and II-14). There was a greater decline in fish catches sold at auction in neighbouring Muarareja (from 5,448 kg, valued at Rp 58 million, to 783 kg, valued at Rp 15.6 million) in the same period, while the number of villagers whose livelihood depended on fisheries increased marginally from 1,944 to 1,990 (Departemen Kelautan dan Perikanan *et al*. 2004a: Tables II-67 and II-5).

14 See Departemen Kelautan dan Perikanan *et al*. (2004a): VI-4 for a list of these species. For identification of these species I am grateful to Duto Nugroho; see also Subani 2002.

15 Interview in Kluwut, 14 September 2005. Boats under five GT now have to go beyond the six mile (*Jalur* Ia/Ib) limit, and stay longer at sea (for three days instead of overnight), which adds to operating costs and reduces the profit shares of crews (ABK).

16 In Surodadi, diesel rose from Rp 900/litre in October 2001 to Rp 4, 750 per litre in September 2006, while prices of *teri nasi* (the main catch) fell from Rp 25,000 per kg to Rp 18,000 per kg in the same period (these figures do not show the large annual fluctuations in fish prices). Fishers say that with no cold storage facilities (only Pekalongan has this facility), traders are able to manipulate prices (Mardiyono, personal communication, 15 December 2005).

17 This was not a new phenomenon (see n. 12). Fishing zones were first introduced in 1976, but because of the government's inability to enforce these zones, conflicts between trawlers and small fishers continued to increase until in 1980–1983 trawlers were banned in all Indonesian waters except the Arafura Sea (Semedi 2001: 1). These zones were reintroduced in revised form under a new 1999 Ministerial Regulation (see n. 19).

18 The issue of surveillance has led to the creation of SISWASMAS (*Sistim Pengawasan Masyarakat*-Community Surveillance System), a joint security patrol programme between the Provincial Fisheries office, the Tegal Water and Air Police (PolAirud) and the Navy (DANAL), which is hampered by lack of coordination and lack of patrol boats. Only one provincial fisheries department boat patrols the fishing zones once a month because of lack of funds. Tegal PolAirud has one speed boat, one rubber dinghy and one remote control radio.

19 According to Minister of Agriculture Decree (SK, Surat Keterangan) No. 392/1999, legal fishing gear refers to set nets, and unmodified moveable nets (*Jalur* Ia 0–3 miles) used by unmotorized boats under 10 m length; 150 metre purse seine nets and 1,000 metre drift gill nets used by motorized boats no larger than five GT (*Jalur* Ib 3–6 miles); 6,000 metre purse seine nets, tuna long lines and 2,500 metre drift gill nets used by boats no more than 60 GT (*Jalur* II 6–12 miles). *Jalur* III (12–200miles) is the Exclusive Economic Zone for purse seine boats above 200 GT (on this zone, see Butcher 2004: 242–6). Importantly, *Jalur* Ib has a category 'modified nets which are not stationary' which refers to modified trawl nets. One study of decentralization of marine resources argues that in the decade following the banning of *puket harimau* trawl nets

in 1980, many fishers started using *arad*, *cotok* and *cantrang* as 'smaller and cheaper versions of trawl nets' (Satria *et al*. 2002: 54). Initially this caused conflicts between those fishers who could afford such nets and fishers using more traditional nets; but now that their use is very widespread, the study suggests there is little conflict, a claim which is not verified by this case study. On conflicts over *cotok* nets see Chapter 2. The difference between *arad* and *cantrang nets* (apart from the trawling issue) is that *arad* have 'otter boards' (*siwakan*) 1x 0.5 metres attached to each side of net lines close to the boat which keep the net open to a width of 10 metres. *Cantrang* nets have much smaller otter boards and therefore they do less environmental damage than *arad* (Kusnandar 2000).

20 The Tegal harbour master says *sopek/compreng* boats under seven GT are supposed to be managed by districts/municipalities, but the Department of Marine Affairs and Fisheries in Jakarta will not give up its authority to issue licences because it makes money for officials. Ministerial Decision No. 45/2000 on licensing of fishing enterprises defines only what records and licences boats must carry. In theory, under Governmental Regulation PP 141/2000, provincial governors can delegate to district officials the issuing of fish enterprise licences (IUP) and fishing licences (SPI) for boats less than 30 GT with engines less than 90 hp which are not using foreign capital or foreign crews (Direktorat and Co-Fish 2002: 13). This has not been implemented in Central Java (interview with Tegal harbourmaster, 23 September/2003).

21 Interview with bankrupt boat owner in Tegal on 13 September 2004; *Muara Pos*, May 2004.

22 In 2001, the Tegal Municipal Fisheries Department proposed that the local government buy three large (40 GT) purse seine boats, each worth Rp 800 million, financed from the municipal budget. The DPRD supported the plan, as it would supplement Tegal's regional income and also provide jobs for 40 deckhands. Small fishers protested, saying the capital was given to boat owners (*juragan*) instead of small fishers who would have benefited more. The chair of the local HNSI said small fishers could not finance these purse seine boats (Rp 29 million per trip) to go to sea (*Suara Merdeka*, 30 November 2001). The boats contributed Rp 30 million to regional income in 2003, but only Rp 5 million the following year, the reason being that 'returns only covered costs of running the boats' (Mardiyono, personal communication, 30 November 2005).

23 Interview with Tegal harbourmaster, 23 September 2005.

24 According to a development sociologist working on the management and conservation of Indonesian fisheries, 'it is not clear what national purpose is served by the licencing system. There has been no attempt to use it to manage any stock by limiting [fishing] effort. It is not clear whether the revenues from licences cover the cost of their collection' (Fegan 2001: 49, quoted with permission).

25 *Kompas*, 12 August 2004.

26 The 5 per cent levy is then divided between provincial government (1.90 per cent), municipal government (0.95 per cent), and two cooperatives, the local KUD Mina (1.45 per cent) and the provincial level PUSKUD Mina Baruna (0.70 per cent): interview with chief fish auctioneer TPI Pelabuhan (Tegal harbour), and data collected from Tegal HNSI office, 05 December 2004.

27 Fisheries (including aquaculture) contribute 4.11 per cent of Tegal municipality's total regional gross domestic product (calculated from Lampiran Peraturan Daerah Kota Tegal, Table 3.8 'Produk Domestik Regional Bruto Atas Dasar Harga Konstan Kota Tegal 1996–2001', III-9). According to the former chair of the Economics Commission (Komisi B) of the Tegal municipal DPRD, Tegal fisheries contributed roughly 5 per cent (or Rp 20 billion) of total municipal tax revenue (*retribusi*): interviews with Abdullah Sungkar in Tegal, 7 and 30 September 2005.

28 *Suara Merdeka*, 9 March 2000.

29 *Suara Merdeka*, 9 March 2000. The boat was variously reported to be worth between Rp 11 and Rp 25 million. Even so, the description of the female owner as being 'in

deep shock and hit hard' by what happened suggests that Muarareja leaders were buying favourable press reports, a serious corruption issue in Reform Era Indonesia.

30 See e.g. *Suara Merdeka*, 9 and 10 March 2000; *Wawasan*, 10 March 2000.

31 *Wawasan*, 13 March 2000. Surodadi fishers collectively paid Rp 7.5 million as compensation (*uang santunan*) to the Muarareja boat owner (*Suara Merdeka*, 9 May 2000). Muareja fishers for their part agreed not to use *arad* nets inside the Ia zone in the Surodadi region (but note not elsewhere).

32 *Suara Merdeka*, 9 May 2000.

33 An *arad* net costs Rp 300,000–500,000 while *kejer* nets cost twice that price (Rp 900,000–1,500,000). *Arad* nets during the fishing season will yield a gross catch on a good three-day trip worth Rp 2 million, which after deducting fishing costs (*perbekalan*) yields a net profit of Rp 1,150,000. Divided 50/50 between the owner (*juragan*) and the three crew, each crew member is paid Rp 64,000 (or A\$9.40) per day. Compare this with a good *kejer* net that will yield a gross return of Rp 200,000. After deducting costs, the net profit is Rp 153,000, shared 50/50, which leaves each crew member Rp 25,000 (or A\$3.80) per day (Mardiyono 2006: 5).

34 Interview with head of Tegal Municipal Fisheries Department, 29 September 2004.

35 Mardiyono, personal communication, 14 June 2005. Because *arad* nets are classified as modified trawl nets and are therefore illegal in zone Ia but not zone Ib if used by boats of less than five GT, no statistics are collected on their use (Departemen Explorasi Laut *et al.* 2002: 1–4). A lecturer in the Faculty of Fisheries at Pancasakti University in Tegal said he was certain that the main spokesperson for the Muarareja community and his family used *arad* trawl nets during the economic crisis (Krismon) of 1997–1998 (interview, 8 September 2004).

36 According to Co-FISH, an *arad* net with a width of 6.61 m being towed by a boat with a 16 hp motor at two knots per hour could trawl 3.57 ha of seabed in one trip of 1.5 hours (Departemen Explorasi Kelautan *et al.* 2002; Lampiran n.d.: 1–3).

37 From approximately 150 units in 1996 to 350 units in 2000 (Departemen Explorasi Kelautan 2002: III–3). The total number of *arad* nets in Tegal in 2003 was 339, the second largest counted after *cantrang* (347), while purse seine (197) was third (from a total of 1066 nets). Since the violence over unfair fishing practices occurred in March 2000, the number of reported *arad* mini trawl nets in Tegal municipality has declined from 402 nets (2001) to 339 nets (2003) (Pemerintah Kota Tegal 2004: 9, Table 1.9), possibly reflecting a decline in the number of boats of 5–10 GT.

38 Confusion over which technologies are allowed to operate inside the three-mile limit arose because an earlier Fisheries Directorate General Decree (SK No. 340/1997) allowed small-scale fishers with motorized boats (*sopek* and *jukung*) under five GT/15 hp motors to use modified trawl gear including *arad* in that zone (interview with head of Tegal Municipal Fisheries Department, 29 September 2004). See n. 19 and Table 3.1 for which fishing technologies are currently allowed to operate in which fishing zones under Minister of Agriculture Decree No. 392/1999 (which should have superseded the earlier Fisheries regulation). The situation is further confused by the constant modification and development of new nets used by motorized boats in *jalur* Ia reserved for set nets and unmotorized boats under the existing legislation, as well as the use of boats over five GT in *jalur* Ib (see n. 35) to try and maintain livelihoods in the face of declining fish stocks.

39 They confiscated equipment, and demanded ransom (initially Rp 1.5 million rising to 400 million) for boats and fishers seized during several violent incidents. *Suara Merdeka*, 8 November 2000.

40 This meant revoking an instruction issued five months previously to ban all large purse seine boats over 30 GT from operating anywhere in the Java Sea, an important policy measure to address the overcapacity in the industry (*Suara Merdeka*, 30 September 2000. See discussion on overcapacity below).

41 *Radar Tegal*, 3 July 2004. In the event, local fishers were unable to personally present

their demands to President Megawati, who said in her speech opening the new fishing harbour that 'fishers are consciously protecting our seas from rampant smuggling', and that the new harbour would shorten fishing trips. The president ended her speech by advising fishers that if the weather was too rough to go to sea they should have fish farms (*Radar Tegal*, 6 July 2004).

42 A former boat owner said that a *pengurus* 'could never make a loss (*untung melulu*)'. The *pengurus* buys supplies, and gets a commission on the diesel fuel purchased for the boat. If the catch is good, he gets a percentage, if the boat is lost at sea, the *pengurus* already has made his cut on the supplies (interview in Tegal, September 2004).

43 ABK, *anak buah kapal*; also referred to as *pandega*, or *buruh nelayan*.

44 Personal communication with a leader of FPPN-BP (Forum Perjuangan Petani dan Nelayan Batang Pekalongan, the Batang Pekalongan Farmers and Fishers Struggle Forum), 23 May 2005. ABK know the practice of inflating prices by *juragan* is widespread.

45 For example, a large purse seine boat of 80 GT with a crew of 35 ABK, on an average fishing trip of 45 days, obtained a net profit of Rp 31 million. Income derived from the complicated share system saw 20 unskilled deckhands paid Rp 583,600 for a 45-day trip, or Rp 13,000 per day, which is roughly the same as ABK on boats of less than seven GT if the catch is a good one. Skilled crew (the captain, mechanic, winch hand, fish lure operator, fish preserver, cook, etc.) get paid additional shares, as does the *pengurus* (buyer of supplies) and the spiritual counsellor (*suhu* or *dukun*) who provides protective talisman for the boat and crew (interview with Sumito in Tegal, 22 September 2005).

46 Trader-moneylenders are called *langgang warung* in Brebes, *pelele* in Pemalang, or more generally *tengkulak* (middlemen).

47 The Tegal fisheries office lists 40 large, 166 medium, and 83 small traders, many of them moneylenders, operating in 2003 (Pemerintah Kota Tegal 2004: 13, Table 2.5). My observation is that while large *bakul* are men (with one or two notable exceptions), medium and small trader moneylenders are usually women.

48 Individual loans are impossible to obtain because boats are not accepted as collateral and most fishers do not have land certificates they can mortgage. Bank Bukopin financed the fisheries development (PEMP) programme (see below) for Central Java with Rp 6 billion in 2001. Fishers complained that interest rates were too high. The Fisheries Department said only 38 per cent of loans were repaid.

49 *Kompas*, 15 November 1999.

50 In the Riau archipelago, the government considers trader-moneylenders, called *tauke* (usually Chinese Indonesian) 'as one of the most import factors in the underdevelopment of fishing households' because of a similar debt loan (*hutan pinjam*) system to that on Java's north coast (Osseweijer 2005: 177).

51 Mardiyono, personal communication, 20 July 2006.

52 Interviews in Muarareja Tegal, 05 September 2004.

53 See *Suara Merdeka*, 10 July 2001.

54 Interview in Sawojajar, 16 September 2004.

55 Interestingly, this trader cited the use of trawl nets, as well as the price of diesel, and the small size of fish as the reasons why *juragan* cannot make repayments to her. She replaces her own boat captains if they make a loss for four consecutive trips. In five years she has dismissed two captains (interview with Carimah, 19 September 2004). Others in the industry believe boats financed by this trader used trawl nets, at least during the economic crisis of 1997–1999, and probably still do.

56 *Cantrang* trawl nets are similar to the banned *arad* mini-trawl nets, but they do not open as wide when being trawled, and therefore do less damage to the marine biomass.

57 Interview in Tegal, 05 December 2004. Another former boat owner, decided to leave the industry (or went bankrupt) because of the price of fuel, reduction in in-shore pelagic fish stocks, and lack of commercial cold storage facilities in Tegal municipality's three TPI (interview in Tegal, 05 September 2005).

58 *Sedekah laut* is a ritual performed at the start of the month of Suro, the Javanese New Year. Fishing fleets go to sea to cast offerings to 'the lord of the sea' in order that boats will be safe and the catch will be big (*along*). (Mardiyono, personal communication, 23 May 2005; *Tegal Tegal*, 17 July 1099, 20 July 1999, 5 August 1999).

59 Iwan Nurdin, personal communication, 20 July 2005.

60 Karang Jeruk is the main breeding ground for shrimp and other demersal species in this part of the Java Sea.

61 Co-FISH has four components, namely coastal fisheries resource management, community development and poverty reduction, fish landing centres and institutional strengthening (Co-FISH 2004: 5). In financial terms, NGOs and fishers believe a disproportionate amount of its funds has been spent on infrastructure (the Jongor fish landing and auction centre).

62 From Kalibacin (West Tegal subdistrict) 90 single crewed *compreng* boats (2–3 GT) using hand held lines to catch *ikan cracas* used the reef, while another 120 boats (with crews of 5–7) from Munjungagung village (Tegal district) fished the reef for *ikan teri*. More than 720 families depend for their livelihood on the waters surrounding this reef.

63 In the end an inner area of only 100 m × 100 m (later extended to 150 m) was declared the nuclear zone in which no fishing was to be permitted. From 2001–2004 TKB were laid to protect the reef nuclear zone. However, half the remaining coral reef was still unprotected.

64 The stakeholders finally identified were three fishing communities, Muarareja, Kalibacin (Tegalsari), Surodadi, the navy base, the water and air police, the Tegal Municipal Fisheries Department, and the local branch of the HNSI, which is supposed to represent fishers' interests, but seems to be ineffective in the Tegal region.

65 Even these were difficult to enforce. In February–March 2002 fishers from outside the Tegal region mined coral from the reef. Local fishers did not dare catch them, and were afraid of being threatened if they reported the boats that damaged the reef (Mardiyono, personal communication, 11 June 2005).

66 Mardiyono, personal communication, 11 June 2005.

67 For the increased number of fish and plant species identified during monitoring of the artificial reef from 2001–2003 see Ministry of Marine Affairs and Fisheries *et al*. 2004, ch. V-1 compared with Ministry of Marine Affairs and Fisheries *et al*. 2001, Table IV-4.

68 When an NGO worker asked the location of the artificial reef that Co-FISH had built, the fisher said he had 'forgotten' where it was (Mardiyono, personal communication, 11 June 2005).

69 Tim Perumus (2004).

70 Co-FISH is located and works with the Tegal Municipal Fisheries Department, while official responsibility for the reef lies with the Tegal district fisheries department. This administrative split contributes to the lack of effective governance of the reef, although neither department has patrol boats to enforce regulations.

71 In Tegal municipality only purse seine boats more than 30 GT received grants under this programme, not the most needy small fishers (interview with head of Municipal Fisheries Department, 29/09/04). According to this official 'There is no concrete solution on how to help small fishers' (English original).

72 In Tegal, individual loans of Rp 1 million were given to 67 boat captains in 2002; by mid-2004, 40 had repaid their loans. A government official said that fishers 'lived as they pleased (*ndableg*).' (*Radar Tegal*, 20 July 2004).

73 Apart from assistance to repair boats and gear, fishers rely on this 5 per cent levy (paid since the early 1980s) for rice to meet daily needs in times of low catches and to pay for medical expenses incurred from accidents on fishing boats (Mardiyono, personal communication, 17 April 2007).

74 In the 1970s the entire catch was sold via the local auction centre. Because the government cooperative (KUD) was able to give loans to fishers for buying supplies to

84   *A. Lucas*

go to sea, fewer fishers were dependent on *bakul-pengijon*. This was when the KUDs were still new and still had money. Repayments were deducted each time fishers auctioned their catch. At first all the Tegal harbours were under one central district level KUD. Then in 1987 the Tegal KUD went bankrupt, so the local fishing community was told to form their own KUD. Because of corruption and poor management, most of these are now bankrupt too (interview in Surodadi, 24 September 2004). How two local KUDs were able to operate in an efficient and accountable way while the rest went bankrupt needs further research.

75 Interviews at Surodadi TPI, 24 September 2004.
76 The Co-FISH programme repeated the same mistake as the KUDs, giving capital loans to larger fishers, not to those who needed them.
77 Mardiyono, personal communication, 02 October 2007.
78 The others are the 'illegal' sale of fuel (by non-licensed agents), sale of catches to offshore foreign boats, overcrowding of fishing grounds, destruction of coral reefs and fish populations, and limited ability of local government to monitor fisheries (Tim Perumus 2004).
79 Interview with Ir. Hari Purnomo,Tegal, 29 September 2004.
80 One fisher informant in Kluwut (Brebes district), who can no longer afford to go to sea, said deckhands find work as unskilled labourers (*kuli kasar*), as petty traders (*asongan*), as pedicab drivers, or as fish traders in local markets (interview, 14 September 2004).
81 Fish sanctuaries are essential if ecosystems are to rejuvenate, 'by giving the spawning biomass of many species the opportunity to rebuild, such sanctuaries supply fish to adjacent waters and in this way increase catches outside their boundaries' (Butcher 2004: 290–1).
82 Although recognizing Tegal as a 'maritime city', fisheries is a sub-sector (with food crops, and livestock) of Agriculture and Forestry in the budget, and received no project aid during the current mayor's first term in office (1999–2004) beyond the purchase of large fishing boats mentioned earlier (Pemerintah Kota Tegal 2004: 33–5).
83 The mayor of Tegal announced that fishing licences (SPI) could be obtained locally from mid-2004 for boats under 30 GT as part of the decentralization of fisheries management, but this has not yet been implemented (*Suara Merdeka*, 28 May 2004). Currently larger boats detained for violations do not have their licences revoked.
84 The Co-FISH programme was also supposed to set up alternative livelihoods scheme for fishers but failed. Some government loans also went to large boat owners or aquaculture farmers instead of poor fishers.
85 '*Lagi-lagi protes mereka terbentur oleh tembok kekuasaan yang selalu berpihak pada juragan besar*'. Notes from a meeting of 15 Larangan fishers on 05 July 2005 organized by Co-FISH to explain the sustainable management of marine resources and biodiversity. Fishers who attended this meeting protested that large purse seiners had been seen operating illegally in the region of Karang Jeruk reef located 3.15 miles offshore, in the week previous to the meeting. These boats are never fined for breaking the law on fishing zones (Mardiyono, personal communication, 06 July 2005).

Bibliography

Adhuri, Dede Supriadi (2002) 'Does the sea divide or unite Indonesians? Ethnicity and regionalism from a maritime perspective', Working Paper No. 48, RMAP (Resource Management in Asia-Pacific Program). Canberra: Australian National University.
Antunes, I. (1998a) 'Ladies run a gold mine in Juwana's fish auction market', in J. Roch, S. Nurhakim, J. Widodo and A. Poernomo (eds) (1998) *SOSEKIMA: Proceedings of Socio Economics, Innovation and Management of the Java Sea Pelagic Fisheries Conference*, Bandungan 4–7 December 1995, European Union, Central Research Institute for Fisheries and ORSTROM (French Scientific Research Institute for Development through Cooperation): 243–57.

Antunes, I. (1998b) 'Setting the Seine: a matter of luck, knowledge and beliefs of purse seine captains in Juwana', in J. Roch, S. Nurhakim, J.Widodo and A. Poernomo (eds) (1998) *SOSEKIMA: Proceedings of Socio Economics, Innovation and Management of the Java Sea Pelagic Fisheries Conference*, Bandungan 4–7 December 1995, European Union Central Research Institute for Fisheries and ORSTROM (French Scientific Research Institute for Development through Cooperation): 161–174.

Atmaja, S. B. and Nugroho, D. (2004) 'Indikator penyusutan sumber daya ikan pelagis kecil di Laut Jawa dan sekitarnya', paper presented at the Seminar Nasional Perikanan Indonesia, Jakarta.

Berkes, Fikret (2004) 'Commons theory for marine resource management in a complex world', in Nobuhiro Kishigami and James M. Savelle, *Indigenous Use and Management of Marine Resources*, Senri Ethnological Studies 67.

Bush, Simon R. and Hirsch, Philip (2005) 'Framing fishery decline', *Aquatic Resources, Culture and Development* 1 (2): 79–90.

Butcher, John G. (2004) *The Closing of the Frontier: A History of the Marine Fisheries of Southeast Asia c.1850–2000*. Singapore: Institute for Southeast Asian Studies.

Butcher, John G. (2005) 'Bringing the state into explanations of fisheries depletions in Indonesia', paper presented to 'People and the Sea III: New Directions in Coastal and Maritime Studies', Conference, Amsterdam, 7–9 July.

Co-FISH (2004) Profil bagian proyek pembangunan masyarakat pantai dan pengelolaan sumberdaya perikanan Jawa Tengah Co-FISH PIU Tegal, Co-FISH PIU Tegal Jawa Tengah.

Departemen Eksplorasi Laut dan Perikanan Direktorat Jenderal Perikanan and Co-FISH (Coastal Community Development and Fisheries Resources Management Project) (2000) *Laporan Akhir: Rencana Pengelolaan untuk Mengurangi Konflik antara Pengguna Alat Tangkap Arad dan Pengguna Alat Tangkap Lain*. Kendal: LP2LK (Lembaga Pengkajian dan Pemberdayaan Lingkungan dan Konservasi Alam).

Departemen Kelautan dan Perikanan, Direktorat Jenderal Perikanan Tangkap and Co-FISH (2001) *Penataan Fish Sanctuary*. Kendal: Lembaga Studi Pembangunan Daerah (LSPD).

Departemen Kelautan dan Perikanan, Direktorat Jenderal Perikanan Tangkap and Co-FISH (2004a) *Socio Economic Assessment II Laporan Akhir*. Semarang: PT Swarna Dasakarya Konsultan Engineering Development and Management.

Departemen Kelautan dan Perikanan, Direktorat General Perikanan Tangkap and Co-FISH (2004b) *Laporan Pendahuluan Pendampingan dan Pemberian Paket Bantuan Pengelolaan Usaha Perikanan Skala Kecil*. Jepara: Al Hikmah.

Direktorat Jenderal Perikanan Tangkap and Co-FISH (2002) *Peraturan dan Tata Cara Permohonan Perijinan Usaha Penangkapan Ikan*. N.p.: Proyek Pembangunan Masyarakat Pantai dan Pengelolaan Sumberdayaan Perikanan Jawa Tengah.

Fegan, B (2001) 'Field report on April 2001 socioeconomic research on management and conservation of the *Terubuk* fishery in Bengkalis, Riau, Indonesia', Project No. FIS/2000/128, Community-based management of the *Terubuk* fishery in Riau, Indonesia.

George, Susan (2004) *Another World is Possible if....* London: Verso.

Kusnandar (2002) 'Perikanan cantrang di Tegal dan kemungkinan pengembangannya', Bogor, Program Pascasarjana IPB (Institut Pertanian Bogor), unpublished S2 thesis.

Lampiran Peraturan Daerah Kota Tegal Tentang Rencana Tata Ruang Wilayah Kota Tegal Tahun 2004–2014. N.p., n.d.

McElroy, J. K. (1991) 'The Java Sea purse seine fishery: a modern-day tragedy of the commons' *Marine Policy*, July: 255–71.

Mardiyono (2006) 'Senja kala perikanan: konflik perikanan di Tegal dalam tinjauan antropologi', unpublished paper.

Ministry of Marine Affairs and Fisheries, Directorate General of Capture Fisheries and

Co-FISH (2001) 'Fish Sanctuary Arrangement Final Report'. Kendal: LSPD (Lembaga Studi Pembangunan Daerah).

Ministry of Marine Affairs and Fisheries, Directorate General of Capture Fisheries and Co-FISH (2004) 'Assistance and Establishment of Artificial Reef Final Report'. Kendal: LSPD (Lembaga Studi Pembangunan Daerah).

Mous, P. J., Pet, J. S. *et al.* (2005) 'Policy needs to improve marine capture fisheries management and to define a role for marine protected areas in Indonesia', *Fisheries Management and Ecology*, 12: 259–68.

Nugroho, D., Atmaja, S. B. and Natsir, M. (2004) 'Pengelompokan alat tangkap yang beroperasi di pesisir utara Pulau Jawa', paper presented at the Seminar Nasional Perikanan Indonesia, Jakarta.

Osseweijer, M. (2005) 'The future lies in the sea: fisheries development programs in island Riau', in P. Boomgaard, D. Henley and M. Osseweijer (eds) *Muddied Waters: Historical and Contemporary Perspectives on Management of Forests and Fisheries in Island Southeast Asia*. Leiden: KITLV Press.

Pemerintah Kota Tegal (2004) *Penanggungjawaban Ahkir Masa Jabatan Walikota Tegal Tahun 1999–2004 Kepada Dewan Perwakilan Rakyat Daerah Kota Tegal Buku I.*

Pemerintahan Kota Tegal Dinas Pertanian dan Kelautan (2004) *Perikanan Kota Tegal Dalam Angka.*

Perumus, Tim (2004) 'Kesepakatan dari Hasil Diskusi Peserta Pelatihan Sistem Pengawasan Berbasis Masyarakat (SISWASMAS), 17 December.

Potier, M., Boley, T. and Nurhakim, S. (1990) 'Study on the big purse seiners fishery in the Java Sea: V. Estimation of the effort', *Jurnal Penelitian Perikanan Laut* 54: 79–85, in *Collected Reprints on the Big Purse Seiners Fishery in the Java Sea*, vol. 3: *Years 1989–90*, Scientific and Technical Document No. 1, July 1991 Agency for Agricultural Research and Development, Research Institute for Marine Fisheries and ORSTROM, Commission of the European Communities Java Sea Pelagic Fishery Assessment Project (ALA/INS/87/17).

Potier, M. and Sadhotomo, B. (1995) 'Exploitation of the large and medium seiners fisheries', in M. Potier and S. Nurhakim (eds) *BIODYNEX: Biology, Dynamics, Exploitation of the Small Pelagic Fishes in the Java Sea*, Agency for Agricultural Research and Development, Ministry of Agriculture and ORSTROM, pp. 195–214.

Roch, J. and Clignet, R. (1998) 'Income uncertainties management in the Java purse seiners' fishermen', in J. Roch, S. Nurhakim, J. Widodo and A. Poernomo (eds) *SOSEKIMA: Proceedings of Socio Economics, Innovation and Management of the Java Sea Pelagic Fisheries*, Bandungan 4–7 December 1995, European Union, Central Research Institute for Fisheries and ORSTROM: 99–110.

Satria, Arif (2002) *Pengantar Sosiologi Masyarakat Pesisir*, Jakarta: PT Pustaka Cidesindo.

Satria, Arif *et al.* (eds). (2002) *Menuju Desentralisasi Kelautan*, Bogor, Pusat Kajian Agararia, Institut Pertanian Bogor and Partnership for Governance Reform in Indonesia.

Semedi, P. (2001) 'Otonomi Daerah di Sektor Penangkapan Ikan', paper presented at the Second International Seminar the Dynamics of Local Politics in Indonesia: the Politics of Empowerment, Yayasan PERCIK and Riau Mandiri, Pekanbaru, 13–16 August.

Subani, W. (2002) 'Jenis-jenis ikan laut ekonomis penting di Indonesia: economically important marine fisheries in Indonesia' (wall chart), Jakarta: BRPL (Balai Riset Perikanan Laut), Pusat Riset Perikanan Tangkap Departemen Kelautan dan Perikanan.

Suherman, B. A. and Nugroho, D. (2004) 'Indikator penyusutan sumber daya ikan pelagis kecil di Laut Jawa dan Sekitarnya', unpublished paper presented at the Seminar Nasional Perikanan Indonesia, Jakarta.

Wudianto and Linting, M. L. (1986) 'Telaah Perikan Pukat Cincin (Purse Seine) di Daerah Tegal, *Jurnal Penelitian Perikanan Laut* 34: 57–68, in *Collected Reprints on the Big Purse Seiners Fishery in the Java Sea*, vol 1: *Years 1983–1987*, Scientific and Technical Document No. 1 July 1991, Research Institute for Marine Fisheries in the [Indonesian]

Agency for Agricultural Research and Development (AARD) in conjunction with ORSTROM, Commission of the European Communities Java Sea Pelagic Fishery Assessment Project (ALA/INS/87/17).

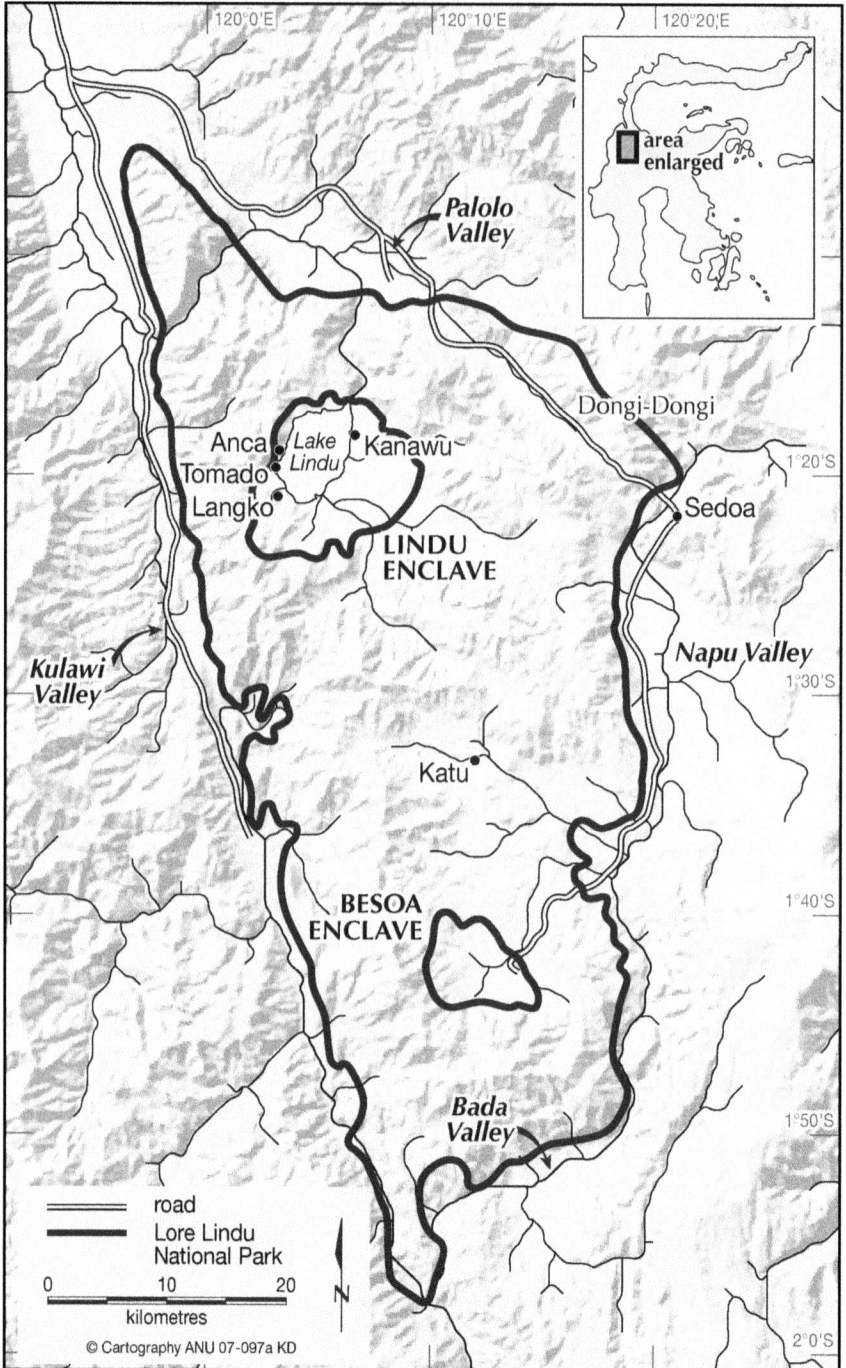

Map 4 Lore Lindu National Park, Central Sulawesi Province.

4 Conservation and community in the Lore Lindu National Park (Sulawesi)

Customary custodianship, multi-ethnic participation, and resource entitlement

Greg Acciaioli

Introduction: interrogating community in conservation contexts

In their report *Whose Common Natural Resources? Whose Common Good?* Lynch and Harwell (2002: xxvi) seek to 'articulate a new paradigm that emphasizes local community well-being as an integral and important part of the national interest'. As in many other studies that consider community participation as the key to enhancing sustainability and achieving environmental justice, their exploration of the possibilities of gaining recognition for the community-based property rights of local peoples takes for granted the analytic viability of the notion of 'community' as a locus of what we in this volume have labelled the 'commonweal' (Warren and McCarthy 2002). Others, however, have disputed the theoretical utility of dependence upon this notion. In their classic article 'Enchantment and disenchantment: the role of community in natural resource conservation', Agrawal and Gibson (1999) analyse the problematic character of this focal concept, noting how many of the distinctive features used to construct this concept are often belied by social arrangements on the ground. They demonstrate how the criteria of small-scale, fixed territorial habitation, homogeneity of social structure, commonality of interest and shared allegiance to norms actually mask not only the divisions within communities, but also the intra-community conflicts that catalyse contestation of resource entitlements rather than conservation of scarce resources (Agrawal and Gibson 2001: 7–12).[1] In their view, cleavages along ethnic, religious, linguistic and other lines often disrupt conservation efforts. The lack of shared norms among different segments, and even individuals, of a purported community requires considering the actual practices of local actors: 'A more acute understanding of community in conservation can be founded only by understanding that actors within communities seek their own interests in conservation programs, and that these interests may change as new opportunities arise' (Agrawal and Gibson 2001: 13).

The co-management of national parks provides one context for investigating how institutional arrangements and the practices of local actors construct community in the larger context of impinging national and international forces that set the parameters of conservation practice. This study takes as its focus the relationships of peoples to each other and to their environment in one area of what has now become the Lore Lindu National Park (Taman Nasional Lore Lindu [TNLL]) in Central Sulawesi. NGO publications (e.g. Sangaji *et al.* 2004) have depicted the inhabitants of the highland plain surrounding Lake Lindu, now an enclave within TNLL, as a homogenous community, constituting a *masyarakat adat* (customary community or indigenous people). Such a depiction is part of the general strategy of portraying local people whose livelihoods are now constrained by the presence of the park as long-time custodians of the region and whose own cultural norms have conserved the region more comprehensively than the state imposition of park status on their homeland. Although the promulgation of such images has not achieved the abolition of the national park for which some activists have hoped, the strategy has been successful in bringing about the introduction of some elements of community-based conservation and co-management by park authorities, in this case the national park management office (Balai Taman Nasional Lore Lindu [BTNLL]) and its partner, The Nature Conservancy (TNC). However, the assumption of the homogeneous community that underlay some of these arrangements, as evident in conservation agreements made with communities around the park, masks the complexities and tensions of local involvement in resource management and the continuing marginalization of long-resident migrants in the region. The revitalization of local *adat* (framework of local custom) that has accompanied the general devolution of governmental authority with the implementation of regional autonomy (Acciaioli 2002) has not only provided a basis for the institutionalization of co-management efforts for conservation in TNLL and other parks, but it has also increased the potential for neglecting rights and interests of spontaneous migrants and local resettlers. These latter risk becoming invisible members of the villages in which they have resided for decades, now conceptualized primarily as customary communities in many conservation agreements. Although they have subsequently been included in some of the participatory mechanisms that have been instituted, such as the village conservation organizations (*lembaga konservasi desa* [LKD]) established in Lindu and other communities surrounding TNLL, the dependence of these organizations upon the authority of *adat* functionaries reveals how such organizations can be co-opted to the project of asserting indigeneity. Migrants may thereby be subordinated or excluded from the wider reformulation of the commonweal towards which these institutions aim.

Communities at Lindu before TNLL: divergent livelihood strategies and resource depletion

Since at least the 1950s, the Lindu plain has no longer been occupied by a single ethnic group. Kaudern (1925: 8) opines that the people who now identify as indigenous To Lindu had first settled this highland plain by the seventeenth century, having arrived from the Lore highlands to the east. The missionary ethnographer A.C. Kruyt reports that the To Lindu in the nineteenth century stressed their autochthony, claiming to have 'descended from a heavenly being, who appeared from within a *kole* tree' (Kruyt 1938: 142).[2] While commoners largely intermarried among themselves in the Lindu plain, nobles intermarried extensively with fellow aristocrats of Kulawi (Kaudern 1940), the highland valley to the west of Lindu. Dutch colonial penetration of the western highlands of Central Sulawesi at the beginning of the twentieth century did not initially change Lindu relations with domains outside the plain, but by the 1920s the Dutch had effectively resettled the To Lindu from numerous smaller settlements in the hills surrounding the plain to the three villages on the lake's shore – Anca, Tomado and Langko. These villages continue as the primary sites of habitation of the To Lindu today. Effectively, the To Lindu had become the first resettlers within their own land. Although the Dutch initially improved the trails to the Lindu plain, as well as irrigation works for wet-rice cultivation, after the discovery of schistosomiasis in the plain surrounding the lake, colonial authorities stopped the policy of improving access to the Palu Valley in order to minimize the spread of the snails that carried the disease.

However, even in the late nineteenth and early twentieth centuries Lindu was certainly not isolated. Politically and socially embedded in the larger region of the Palu Valley and the surrounding mountains by its tributary relation to Sigi and marital alliance relations with Kulawi, it was also economically tied to the region by the provisioning of fish from Lake Lindu to surrounding domains. In fact, when the missionaries Adriani and Kruyt first ventured to Lindu in the 1890s before pacification, their first encounter with the To Lindu was a meeting with a group of Lindu men resting in a temple from carrying smoked fish down from the lake (Kruyt n.d.: 82). The entry 'Lindoe-meer' in the *Encyclopædië van Nederlandsch-Indië* (Graaf and Stibbe 1918: 583) noted both the abundance of fish in the lake and the rudimentary technology of the To Lindu for harvesting it. More intensive harvesting of fish from the lake would come only with the arrival of migrants from South Sulawesi during the turbulent time of regional rebellions in the first years of independent Indonesia.

The first migrants to the Lindu plain were internally displaced persons of Bugis and Arab descent fleeing the demands of both the Muslim guerillas of Kahar Muzakkar's regional rebellion (1950–1965), later linked to the wider Darul Islam rebellion, and the national army, perceived as dominated by Javanese (Harvey 1974). Besides the push factor of the depredations of the *gerombolan*, as this unsettled time was labelled, there was the undoubted pull factor of the resources of Lake Lindu, the surrounding plain with its potential for wet-rice cultivation, and the resource-rich forests on its periphery. By the late 1960s a process of chain migration was bringing a steady flow of Bugis relatives of the first internally displaced persons from Donggala and Wajo' to the Lindu plain, as well as non-related Bugis migrants from such areas as Sidrap, Pangkep and Bone, who soon recruited their own kin to the plain (Acciaioli 1999, 2000).

While many of these Bugis engaged in land transactions with To Lindu, acquiring land for wet-rice fields on the basis of what was labelled 'compensation of loss' (*ganti-rugi*),[3] fishing in the lake was the predominant mode of livelihood for many of the Bugis migrants, especially in the early years after arriving. This livelihood choice differentiated them from the indigenous To Lindu, who remained wet-rice farmers, fishing largely for domestic consumption. Intensive fishing had been made possible only by the earlier stocking of Lake Lindu by the Fisheries Department with several varieties of fish, beginning in 1951, most significantly common tilapia (*mujair*),[4] although at the cost of the extinction of many of the endemic fish of the lake (Whitten *et al.* 2002: 52–3, 298).[5] Originally, this 'seeding' of the lake had been intended to foster livelihoods and increase incomes for the To Lindu themselves, as well as providing a source of protein for them and for the other highland peoples to whom the fish would be marketed. But in the absence of provisioning with relevant technology, the Lindu people had not been able to take advantage of this augmented resource. It was the first Bugis migrants who brought nylon gill nets to Lake Lindu, providing their incoming relatives with these in return for assured supplies of fish that they could market. In addition, some of these first gill nets were sold to the To Lindu at 40 times their cost in South Sulawesi.

By the early 1980s, when I first began fieldwork at Lindu, the Bugis and the indigenous To Lindu were clearly differentiated in the village of Tomado, where most of the migrants from South Sulawesi continued to live. Having settled on the western shore of the lake near the Tomado village centre and on the eastern shore in the hamlet of Kanawu, the Bugis were referred to as the 'shore people' (*orang pantai* in Indonesian, or *to sovea* in Tado, the language of the To Lindu), while the To Lindu themselves were labelled as the 'village people' (*orang kampung* in Indonesian, or *tumpu ngata* in Tado). This labelling encoded the difference in predominant livelihood orientation: the Bugis migrants to fishing and the To Lindu to wet-rice cultivation.[6] Although inhabitants of the same village, they were in most respects separate communities; Lindu resembled more Furnivall's (1944) model of a 'plural society' than an integrated community. At that time I recorded only one 'legal' marriage and one de facto relationship between Bugis men and To Lindu women, and no marriages between To Lindu men and Bugis women. Religiously, the two remained strictly differentiated, with the Muslim Bugis worshipping at the one mosque located in the shore community, and the To Lindu in the Salvation Army houses of worship found in each of their three villages.[7]

Although all were subject to the official village (*desa dinas*) administration that had become uniform throughout Indonesia with the imposition of the 1979 national law on village government[8], the Bugis tended to heed only their own 'shore' hamlet headman, himself a Bugis migrant from South Sulawesi. Even more importantly, the Bugis did not consider themselves as bound by Lindu custom (*adat*). Indeed, the village customary councils (*lembaga hadat*) operated only intermittently during that New Order period, as their participation in official regulation of local affairs had been abolished by the 1979 law. In addition, the scope of *adat* was largely curtailed by policies of the Department of Social Affairs and of the Education and Culture Ministry that sought to prescribe what constituted the legitimate content of *adat* among such 'isolated tribes' (*suku terasing*) as the To Lindu (Acciaioli 1985: 153).[9] Local *adat* councils (*lembaga adat*) still oversaw bridewealth negotiations among the To Lindu, but their jurisdiction over resource issues was no longer recognized. Disputes regarding land among the inhabitants were brought to the village headman, who attempted resolution with his administrative apparatus, assisted by such local respected 'social figures' (*tokoh masyarakat*) as schoolteachers and also individual members of the *adat* councils acting in their capacity as respected 'elders' (*totua* in Tado) rather than specifically as members of the village *adat* councils.

Despite the overarching administrative apparatus imposed by the state, the Bugis 'shore people' tended to regulate their own affairs. When faced with such problems as fish poaching, they resorted to their own notions of community justice, usually involving beatings and, as the ultimate sanction, exile from the Lindu plain.[10] But the consequences of their relative autonomy from the exercise of authority by the state-imposed village government and the customary sanctions of the Lindu *adat* councils came to be most apparent in the lack of regulation of fishing practices in the lake.

In the fish marketing system established by the Bugis (Acciaioli 2000), intermediate marketers purchased fish with a set price for each 'skewer' (*tusuk*), a ring of rattan cord with four fish fastened. The price was fixed, no matter what the size of the fish on the skewer. Of course, this manner of reckoning led fishermen to maximize their income by providing ever greater numbers of ever smaller fish. Accordingly, over time they would use gill nets (*landak*) of ever smaller mesh size, eventually providing fish that were so diminutive that they would become rotten before they had arrived at the fish markets in Palu. In fact, Bugis fishermen were harvesting such immature fish that they had not had a chance to reproduce. Alerted by the size (and sometime rotten state) of the tilapia being marketed in Palu, the Fisheries Department twice sent up representatives to Lindu during 1981–1982 on raids (*razzia*) to seize the fishing nets of insufficient mesh size and attempt to educate the fishermen in public meetings about the eventual environmental consequences of their actions. For a few weeks after each of these raids, the fishermen desisted. Gradually, however, they would yield to the inexorable market logic of maximizing their incomes by again purchasing nets of smaller mesh size in order to increase the number of fish harvested. The spiral of official raid and reversion to smaller mesh sizes continued unabated through the 1980s, resulting in the complete depletion of the tilapia stocks in the lake by 1989. For the first time in living memory, there were no local fish even for household consumption.

The depletion of tilapia stocks at Lindu, a condition whose effects were exacerbated by the earlier loss of endemic species due to the depredations of the tilapia themselves, can be read as a variant of the 'tragedy of the commons' (Hardin 1968). This situation had been made possible by the erosion under the New Order of *adat* authority and institutions that had earlier regulated access to resources on the Lindu plain and the impotence of state institutions to monitor and control resource exploitation in their place. While the To Lindu were resentful of the extermination of these resources, the *adat* institutions of the Lindu populace were powerless to exercise sanctions against such practices. In many ways the depletion of lake resources revealed the very lack of a coherent community among the fellow residents – indigenes and migrants – of the Lindu villages, and their inability to sustain any sense of 'commonweal'.

The revitalization of *adat* as a vehicle for community resource management

In many ways the creation of the Lore Lindu National Park has catalysed incipient senses of community and revitalized *adat* as a mechanism for custodianship of resources, a process furthered by the context of state decentralization. Based on a park management plan produced the previous year by the World Wildlife Fund in association with the Directorate of Nature Conservation[11] (PPA), TNLL was first declared[12] a candidate national park in 1982 (Schweithelm *et al.* 1992), five years after declaration as a UNESCO biosphere reserve in 1977 (Sangaji *et al.* 2004: 17).[13] The Lindu plain was one of two enclaves established in the candidate park by the first draft plan (Watling and Mulyana 1981: 10). While still in candidate status, the park itself had little immediate effect on the residents of the Lindu enclave. They noted the occasional visits of the PPA officials, politely listening to their exhortations to desist from hunting deer, deer pig and other protected animals, as well as from opening up new gardens and wet-rice fields, and then continued to hunt these animals and open their fields once the PPA officials had left. On several occasions during my initial fieldwork in 1981–1982, police were sent up from Palu to seize rattan that had been collected from the forests of the candidate park by Bugis and Lindu men and were being assembled, primarily by Bugis marketers, for shipment to Chinese merchants based in Palu, for eventual use in manufacturing cane furniture and other items. As in the case of raids on prohibited fish nets, the halt on collecting rattan usually lasted only a week after the seizures.

Enclavement within a national park later became a significant factor affecting the outcome of a proposed dam at Lindu. In 1988 the Central Sulawesi governor announced a plan to build a dam at the outlet of Lake Lindu as part of a hydro-electric project to supply the electricity needs of the rapidly growing provincial capital Palu (Sangadji 1996). As the projected dam would have caused the water level of the lake to rise over seven metres, inundating all villages and rice-fields, this project would have entailed the resettlement of all the inhabitants of the Lindu plain, a prospect that the inhabitants were unwilling to accept (Sangaji 2000). Their movement to resist the imposition of the dam was spearheaded by the local NGO, the Foundation for a Free Land (Yayasan Tanah Merdeka [YTM]), which led a coalition of local and national NGOs,[14] in a successful campaign to block construction of the dam.[15]

In mobilizing local support for this campaign to block the dam and prevent resettlement, YTM was focused solely upon the indigenous To Lindu. While later acknowledging the presence of migrant groups settled in the Lindu plain, including Bugis (Sangadji 2000: 17),[16] YTM worked exclusively with To Lindu representatives of the three Lindu villages, most intensively with members of the To Lindu village *adat* councils and the umbrella *adat* council of the plain as a whole (Lembaga Hadat Sedataran Lindu). These were the representatives whom YTM brought to Jakarta to present their case to members of the national Parliament (DPR), the National Committee on Human Rights (Komnas HAM), the central office of the State Electric Company (PLN), and Ministers of Mines and Energy, and of Forestry. Among the arguments made to counter the project was its incompatibility with the requirements and objectives of the national park, an argument that began to foster a consciousness among the To Lindu that engaging with the administration of the national park could work to their advantage in gaining recognition of their customary rights.[17]

The YTM strategy based on the political line of indigeneity[18] has enabled the To Lindu to assume what Li (2000: 163–8) has labelled a recognizable 'tribal slot'. In keeping with connotations of 'traditional wisdom' (*kearifan tradisional*) that permeate discourses of and about 'indigenous peoples' or 'first nations' (Maybury-Lewis 1992), YTM publications (Sangadji 1994, 1996) have represented To Lindu *adat* as functioning as a community-based resource management system. For example, in one YTM publication the Palu-based lawyer and human rights activist Laudjeng (1994: 155–8) outlines how the Lindu plain is indigenously classified into a number of land-use domains (*panjuaka* in Tado, *suaka* in Indonesian) that function to promote the 'conservation of the region'. Outsiders who wish to make use of such land must seek permission from To Lindu custodians and even then may only put it to uses allowed by *adat* according to Laudjeng's account.

My own investigations (August 2000) revealed a division into *pancua* (a dialectally different, but cognate term) in the Lindu plain differing somewhat, but recognizable from Laudjeng's trichotomy of land-use domains. Land that may be subject to human use is classified into two main types, *pancua ntodea* and *pancua maradika* (*ntodea* and *maradika* refer to the commoner and noble stratum respectively). Whereas the former land is open to all To Lindu and even outsiders for cultivation or gathering firewood or rattan and similar tasks, its use does require permission of the local nobles. *Pancua ntodea*, what might be glossed as commons land whose jurisdictional rights of access (*hak ulayat*) are under the regulation of the *adat* council, can encompass several labelled usage types. Once a person has gained permission and opens primary forest (*pangale*) for gardens (i.e. of cash crops) or dry-crop fields, he has established a precedential right of continuing access. Laudjeng (1994: 156) uncritically labels this as a 'private right' (*hak perorangan (privat) individual*) equivalent to private ownership ('*tanah itu sudah merupakan milik si pembuka pangale tersebut*'). Indeed, such is the understanding of many To Lindu cultivators who have some awareness of Indonesian land law and hope ultimately to gain government-issued ownership certificates for their land. However, members of the Langko *adat* council declared to me that ultimate jurisdiction remained with the *adat* council, and that first opening of fields within *pancua ntodea* only gave primary usufruct rights. For example, dry fields that were fallow (*ngura*) could be cultivated again without having to ask permission, as was the case for the annual cultivation of irrigated wet-rice fields, but such land could not be alienated.[19]

Pancua maradika is traditionally reserved for the use of the nobles, for example, for grazing their livestock. Commoners, should they own large livestock, could also ask permission of the nobles of a village to graze their cattle on the *pancua maradika* and once again to remove the cattle from it after grazing. Failure to observe the restrictions for each land type would result in a fine (*ragiwu* in Tado) imposed by the local village *adat* council, the predominant form of sanction for most transgressions (Acciaioli 2002: 225–6). This system of monitoring land use was explicitly represented by *adat* elders as an instance of indigenous provisions for protecting the environment from degradation through overgrazing and other damaging practices. In fact, some informants – uncomfortable about the implications of persisting rank distinctions inherent in the term *pancua maradika* that were in conflict with their more egalitarian Salvation Army convictions – tended merely to refer to *pancua ngata*, subsuming the distinction of *pancua maradika* and *pancua ntodea* into an undifferentiated village commons, whose jurisdiction was ultimately under the *adat* council of the Lindu plain.

In contrast to land that could be subjected to human use, another type of land may not be subjected to any forms of exploitation, whether for opening gardens or chopping down trees to obtain wood for building canoes or for other construction purposes.[20] Such land (*tana viata*[21] in Tado) is conceptualized as the preserve of local guardian spirits (*viata*). Indeed, several such clearly delineated stands of primary forest remained in the midst of intensively cultivated wet-rice fields, with stories of how people had suffered various misfortunes after transgressing these areas. Even when their Salvation Army convictions lead them to doubt the status or very existence of *viata*, middle-aged and younger members of the To Lindu community now present such interdiction as evidence of the local ecological wisdom encoded in *adat* institutions. One elder equated many of these *tana viata* with focal areas infested with schistosomes, remarking how the To Lindu ancestors thus understood the malign health consequences of treading in them, expressing this understanding in terms of the attacks of guardian spirits (*viata*). As many of these indigenous sacred areas are located in the surrounding forest, these informants point to the function of *tana viata* and the encompassing system of *suaka* in which it finds its place, in preserving watershed areas[22] and thus inhibiting erosion and the sedimentation that has been decreasing the depth of Lake Lindu, as evidenced by its noticeably receding shoreline. Some To Lindu explicitly point to the neglect of these prohibitions by migrants to the area who cut down forests and open up coffee and cacao gardens heedless of To Lindu prohibitions as contributing to the increasing shallowness of the lake and the streams feeding into it.[23]

The representation of the 'ecological wisdom' encoded in Lindu *adat* has indeed had some practical consequences. Local developers arguing for the erection of a new, environmentally less destructive hydroelectric scheme and for a water cleansing installation to provide potable water for Palu have had to accommodate in their plans how *tana suaka* under the jurisdiction of the To Lindu *adat* councils would be preserved. At a meeting of the LKD (see discussion below) of the four villages in the Lindu plain in Kanawu in 2004, the TNC representative announced that the zonation model presented in the draft plan for the park would be fine-tuned to align park zones more closely with the boundaries of *suaka* delineated in the community participatory mapping facilitated by YTM (Acciaioli and Warren 2005). Such representations constitute what I label a 'refunctionalisation' of *adat*; that is, they constitute a foregrounding of what undoubtedly had been latent (ecological) functions of *adat* stipulations related to control of land and other resources whose manifest functions – before their discursive reworking in such contexts as NGO publications toward gaining recognition of indigenous rights in resource struggles – could be interpreted as representing and preserving the authority of local elites (e.g. *maradika* in Lindu and adjacent traditional rajahdoms). This discursive refunctionalization has been a major factor in the successes achieved by the To Lindu, in cooperation with their NGO allies such as YTM, in gaining recognition as an 'indigenous people'. Such recognition has fostered reassertion of what To Lindu view as their traditional rights, based on ancestral connection to their land and their ecological wisdom in preserving its resources (Li 2000: 164), against the imposition of development and conservation schemes. It has also enabled them to gain new rights, however limited, as co-managers of the encompassing Lore Lindu National Park.

Indigeneity and co-management in TNLL

NGO depictions of the To Lindu as a 'customary community' (*masyarakat adat*) whose 'traditional wisdom' (*kearifan tradisional*) had supported an indigenous conservationist ethic from before the imposition of the national park have tended to represent the To Lindu as the sole inhabitants of the plain (e.g. Laudjeng 1994). In many aspects such depictions corresponded closely with the model of 'mythic community' deconstructed by Agrawal and Gibson (2001), presenting the image of a community of small scale with an immemorial territorial affiliation, a homogenous social structure and continuing adherence to shared norms. Such a depiction was consistent with, and indeed derived from the burgeoning indigenous people's movement in Indonesia,[24] which first declared the correspondence of 'customary community' in Indonesia with the globally recognized category of 'indigenous people' in the early 1990s. In fact, coinciding with the height of the Lindu anti-dam campaign, a declaration issued by a workshop organized by the Network for the Defence of Customary Societies (JapHama) in 1993 in Tana Toraja first set forth the criteria for *masyarakat adat* in this new understanding as 'indigenous people':

> ... social groups that have ancestral origins (which have persisted for generations) in a specific geographical region, along with possessing a value system, ideology, economy, politics, culture, society and region [i.e. territory] of their own (KMAN 1999).

The To Lindu were among the early beneficiaries of this burgeoning movement. As YTM publications depicted them, they constituted an almost ideal-typical *masyarakat adat*. Such status provided the basis for campaigning for special recognition of their continuing rights to land and other resources. Although first asserted in the context of resistance to their resettlement by the dam project, indigenous status was also essential for the subsequent recognition of the To Lindu role in co-management of TNLL.[25]

The international organization directly involved in determining the form of co-management in TNLL has been the US-based environmental NGO, The Nature

Conservancy (TNC), which has entered into agreements with the governments of countries throughout the world, especially in the Global South, to jointly manage parks and reserves in the interest of protecting biodiversity. Although rejecting the notion of 'sustainable use', and hence full custodianship by local peoples, as insufficient to ensure biodiversity conservation, TNC has committed itself to working with local partners in order to achieve this goal, particularly in the formulation of its Parks in Peril (PiP) programme (Brandon et al. 1998). It has thus acknowledged the need for local participation, including formal agreements with indigenous communities within and around reserves, as well as appropriate development programmes toward building sustainable livelihoods for such communities. At the same time, it has maintained its position on the necessity of core zones from which humans are excluded.

In keeping with its 'eco-region' approach to biodiversity protection, TNC park management plans and evaluations depend upon a zonation model,[26] differently elaborated in different contexts, with some park areas subject to regulated human use, including the recognition of enclaves, while others are designated as out of bounds. TNC has functioned as a co-manager of TNLL,[27] working with the park's Management Authority (BTNLL) and the Directorate General of Forest Protection and Natural Conservation, since TNLL's transition to full national park status by ministerial decree in 1993.[28] TNC's draft management plan for TNLL[29] acknowledges that it has had to carry out the task of fostering co-management 'at a time of great change and upheaval' during which 'the rigid directives of central planning' have been replaced by 'the needs and aspirations of the Park's diverse stakeholders' (TNC *et al.* 2001 vol. 1: 2).

Compared to earlier policies, the emphasis of TNC upon a collaborative management strategy with local stakeholders has been a salutary advance. However, in line with shifting priorities of the World Conservation Union (IUCN), and like other conservation and development organizations involved in park management, its focus has been primarily on the 'indigenous' peoples living in the park enclaves and around its boundaries. This orientation has been most evident in the various co-management agreements that have been brokered for TNLL.[30]

A number of other organizations involved in designing conservation agreements in the region of TNLL have emphasized a range of trade-offs in return for recognition of local customary rights over some resources. One of the very first NGOs involved was the Palu-based NGO YTM, which brokered a conservation agreement between the Katu people and TNLL management as part of the arrangement granting enclave status to the Katu, allowing them to maintain their stable (i.e. nonexpansive) swidden regimen within the boundaries of TNLL (Mappatoba and Birner 2004: 26; Sangaji 2002). Following that model, YTM has also facilitated conservation agreements with two other villages surrounding the park. In each case the agreement has been specifically focused on the *masyarakat adat* of the village, whose rights to land and resources are explicitly recognized in return for the community carrying out such activities as patrolling for rattan theft within the boundaries of TNLL.

The international relief and development agency CARE focused on providing rural development, infrastructure and agricultural extension in return for agreement by customary communities to abide by and help implement park regulations. Given its focus on community development, CARE has tended to work with the official (*dinas*) village government, rather than following YTM's practice of dealing primarily with customary institutions, such as the *adat* council. After establishing some dozen such agreements, CARE ceased to be involved directly, instead providing funds to a Palu-based NGO, Yayasan Yambata, which began the process of overseeing contracts in five villages where the protection of the Maleo bird constituted a major challenge. In contrast to CARE's focus, as a local NGO in regular contact with the other Palu-based NGOs, including YTM, Yayasan Yambata has followed a line more in keeping with these NGOs' orientation, recognizing the authority of customary institutions to oversee such contracts.

The Central Sulawesi Integrated Area Development and Conservation Project (CSIADCP), a long-term programme of rural development and conservation initiatives funded by the Asian Development Bank (ADB), has also pursued the drawing up of community conservation agreements.[31] Initially, under conditions set by the ADB, CSIADCP had supported plans to resettle indigenous groups out of conservation areas, but after the according of enclave status to the Katu community by BTNLL's director, it was forced to re-orient its policies. Between 2000 and 2004 CSIADCP began a lengthy process of arranging conservation agreements with 60 villages in the vicinity of TNLL. CSIADCP officials admit that these agreements constitute little more than an 'entry point', having been based only on consultations and workshops of one day's length in each village. As Li (2007: 131) has noted, CSIADCP-brokered agreements restrict the exercise of local resource rights in exchange for the 'livelihood improvements' that were part of the general mandate of the CSIADCP programme, including 'funding for village infrastructure such as minor roads, irrigation, and flood control'. Evident in the specification of *adat* sanctions and the 'local wisdom' underlying these stipulations is the predominant concern with harnessing the institutions of the local '*masyarakat adat*', notwithstanding the references to the official village government apparatus required by CSIADCP's status as a government development project. This focus aligns these agreements with those facilitated by YTM and Yayasan Yambata, all of which presume a relative homogeneity of the contracting community and the continuing authority of the customary council as adjudicator of transgressions.

TNC, however, has begun to take a somewhat different approach to involvement of other local stakeholders in drawing up its conservation agreements in the vicinity of TNLL. Beginning its efforts at about the same time as CSIADCP, by 2004 TNC had managed to initiate 14 conservation agreements, with five of them completed and approved by the TNLL management office (Mappatoba and Birner 2004: 18; Khaeruddin 2002). Working with both customary functionaries and administrative village officials, it has adopted a different strategy to deal with the issue of local-level monitoring and enforcement of conservation regulations, especially encroachment of gardens for cash crops such as coffee and cacao, as well as harvesting of timber and non-timber forest products such as rattan (Mappatoba and Birner 2004: 28). The emphasis in TNC's early agreements paralleled the strategies of CARE and CSIADCP in providing local-level development: for example, it promised to reward community commitment to observing conservation rules with provision of such services as drinking water and marketing assistance for organically grown coffee. More recently, in accordance with its interpretation of the 1999 Forest Law (TNC *et al.* 2001 vol. 1: 24) on community participation in forestry, TNC has linked community conservation commitments to recognition of customary rights, including allowing the right to access products from customary land (*tanah adat*) now incorporated into the national park.

As one example of this more recent approach, the TNC Conservation Agreement with the People of the Lindu Plain (Kesepakatan Konservasi Masyarakat Dataran Lindu, hereafter referred to as 'Kesepakatan Lindu' or Lindu Conservation Agreement), signed in March 2005, reveals a sophistication and range that far transcends such earlier agreements and opens the possibilities for negotiating an expanded notion of a 'community' of 'orang Lindu' embracing both indigenous To Lindu and migrants in the Lindu plain. It admits that the placing of TNLL's boundaries was a unilateral action taken without consultation, resulting not only in losses to local inhabitants, but also the failure of conservation programmes. It recognizes the prior existence of 'customary land/communal use/and living space for the societies of the area who have resided there continuously, long before the existence of the national park' (Kesepakatan Lindu 2005: 1). However, it states that such recognition must be balanced by measures for the preservation of biodiversity for the sake of sustainability in a way that is acceptable to all parties to the agreement. Sections of the agreement reference a wider range of laws and regulations, related not only to conservation but also to basic human rights and agrarian issues, than do other conservation agreements.

Balancing respect for the rights of the societies in the vicinity of the park with the control and management of natural resources is declared as the fundamental objective/purpose underlying the conservation agreement. Subsequent chapters seek to balance acknowledgement of customary institutions, such as the Adat Council of the Entire Lindu Plain and the *adat* councils of the enclave's four villages, and assertion of the authority of national park institutions. The document proclaims its commitment to a process of 'participatory management planning', but also insists on the park framework of zonation – core, wilderness, intensive use, traditional use, rehabilitation, as well as social historical, and tourist zones (TNC *et al.* 2001 vol. 2: 94–104). However, it admits the possibility of subsequently determining boundaries of zones on a participatory basis, taking account of both ecological and social considerations. Besides these basic principles and general stipulations, specific paragraphs set limits on felling trees (e.g. for house decorations, customary rituals, etc.), hunting, gathering damar, harvesting rattan and other natural resources – bamboo, enau sap (for palm toddy and sugar), roots and herbs for traditional medicines, stones and sand, honey, and so on – as well as opening up land for gardens, grazing livestock, and altering water courses in park land. In the final paragraph the agreement states as one of its aims to 'obtain acknowledgement of its [i.e. the local society's] management of natural resources in the customary territory that is located within the region of the Lore Lindu National Park'. This is a clear statement, at least on paper, that the notion of customary territory is to be respected, with practical consequences for management: the imposition of National Park status does not fully supersede *adat* territorial rights.[32]

Judicial implementation of the Lindu Conservation Agreement: reinforcing divisions among Lindu community stakeholders?

The roles allocated to various parties in the Lindu Conservation Agreement can be usefully analysed by reference to Agrawal's and Gibson's (1999: 11) dimensions of the exercise of authority in their institutional approach to community governance:

> Authority to manage resources effectively at the local level requires the exercise of authority and control by local actors over their critical domains … (i) making rules about the use, management, and conservation of resources; (ii) implementation of the rules that are created; and (iii) resolution of disputes that arise during the interpretation and application of rules.

The legislative role in these agreements, as Li (2007: 131) has also highlighted, is primarily lodged with TNC and the BTNLL, the authorities that have established park regulations, although consultation over such issues as the boundaries of zones is sought with local actors. However, judicial functions – adjudicating transgressions of park regulations and imposing sanctions upon guilty parties – have been shifted to the local *adat* councils,[33] whose deliberations are to be undertaken in the presence of park police, and, where possible, other witnesses, including park management staff, the village government apparatus, the village representative body (BPD), and the 'village conservation organizations'.[34]

The Adat Council of the Entire Lindu Plain began to exercise this judicial role in earnest on the basis of an earlier draft plan even before the final conservation agreement was formally ratified in 2005. On 31 January 2003 it convened a session to adjudicate the case of one 'rattan entrepreneur' from Kulawi who, according to the evidence presented, had not only been caught with 38 bundles of rattan, but over time had taken three tons of rattan from around the Lindu enclave, and approximately 20 tons from TNLL as a whole. The *adat* council judged him guilty and imposed the traditional fine, *sampole saongu*, consisting of three water buffalo (*bengka*), three traditional ikat cloths (*mbesa*), and 30 brass plates (*dula*), although the entrepreneur was allowed to pay a cash equivalent of Rp 800,000 in accordance with the contemporary practice of cash conversion for fines stated in a customary idiom.[35] In his comments as presiding elder, the head of the Langko *adat* council stated:

> It is certainly the case that you have heard that in 1982 [the year TNLL was officially declared a candidate national park by ministerial decree] the Lindu region was established as a region where it is prohibited to harvest forest products, and this has been repeatedly declared by our government, and we in

our capacity as the Adat Council have repeatedly appealed to the community not to take forest products for just any reason in the Lindu region, but this has evidently not been heard. And for all such trangressions there are Adat sanctions that are in force for the Lindu plain. (Lembaga Hadat Sedataran Lindu n.d.: 2)

This empowerment of the *adat* councils to adjudicate transgressions of park regulations must be seen in the context of the wider devolution of authority with regional autonomy, particularly in regard to the reintegration of *adat* institutions into village level governance made possible by the new regional government legislation.[36] This re-empowerment has extended to the exercise of *adat* sanctions over environmental transgressions within the Lindu enclave itself, an authorization bolstered by its assumption of judicial authority over local transgressions of TNLL regulations. This aspect of its operation has been most clearly evident in its resumption of authority over harvesting the resources of Lake Lindu, a role denied it under the New Order's policy of reserving authority over village matters to institutions established by the state (i.e. the official '*dinas*' village government apparatus), and over environmental management to state agencies (e.g. the Fisheries Department). As was noted above, the inability of these institutions to regulate fishing at Lake Lindu, especially the practices of the migrant Bugis fishermen, led to the depletion of tilapia stocks in the lake in 1989, a situation not rectified until the lake was restocked with tilapia in 2001 under CSIADCP auspices. Since that 'reseeding' the Adat Council has re-assumed the authority to declare *ombo*, a specified period of restriction on resource harvesting (in this case, fishing). Traditionally, *ombo* was declared on the death of the highest nobles (*maradika*), as a sign of respect for the passing of a high-ranking personage.

However, in keeping with the reconceptualization of the *adat* of indigenous peoples like the To Lindu as a community resource-management system, the function of *ombo* in maintaining the harmony of the environment has been foregrounded in NGO publications (e.g. Laudjeng 1994: 160–1), where it is presented in similar terms to *sasi* regulations in Maluku (Wahyono *et al*. 2000). An institution whose manifest function in the past was to display and reinforce the prerogatives of indigenous high rank is now invoked as an expression of indigenous 'environmental wisdom', a local resource-management stipulation to conserve threatened resources. The representation of *ombo* thus exemplifies the refunctionalization of *adat* institutions earlier discussed in relation to indigenous land-use domains (*suaka*).

However, such refunctionalization is not only a question of representation by NGO allies; it is now a matter of practical implementation in the contemporary context by *adat* elders as well. As Baudrillard (1981) would appreciate, representation has become reality. Compared to its relatively infrequent declaration in the past, *ombo* is now regularly proclaimed when members of the Adat Council of the Entire Lindu Plain or of one of the village *adat* councils have been informed by fishermen that tilapia and other fish are becoming smaller and more difficult to catch. For example, *ombo* was declared for one month at Lindu in March 2006 after such reports, and the To Lindu *adat* councils were active in the adjudication of transgressions of *ombo* during this period. One To Lindu man from Langko was caught 'stealing' fish from the lake during this declared *ombo* and was found guilty of the transgression in a meeting held by the *adat* council of Langko. As in the case of the rattan entrepreneur, the *adat* council imposed the *sampole saongu* fine – three water buffalo, three traditional *ikat* cloths, and 30 brass plates – which was equated this time by 'the art of the elders' (Acciaioli 2002) to Rp 1 million. Elders attributed this high cash conversion to the brazenness of his transgression and his unrepentant attitude. In contrast, a Bugis fish entrepreneur convicted for sending out his client fishermen on the very last evening of the *ombo* period had his *sampole saongu* fine converted to the lesser sum of Rp 800,000, since he claimed he had negotiated permission to do so.[37] It is significant that it was the Adat Council of the Entire Lindu Plain that adjudicated this case rather than the adat council of the village of Langko, in whose territory this fish entrepreneur collected catches of his client fishermen. Arguably, this *adat* council represented the entire indigenous To Lindu constituency acting against this migrant (*pendatang*). During the proceedings this Bugis fish entrepreneur was even threatened with exile from the Lindu plain as a sanction for his transgression, a threat not exercised against the indigenous To Lindu transgressor whose case had no extenuating circumstances.

Such differentiation in scope of the presiding customary council and in threatened sanction reveals the persistence of the social divisions apparent in the 1980s – indigenous To Lindu *orang kampung* acting as a separate community in concert against the newcomer *orang pantai*. Indeed, the conservation agreements negotiated with *adat* communities, re-empowering the *adat* councils as the judicial organ adjudicating transgressions of park regulations, have been explicitly intended as instruments to motivate local communities to keep the park resources from being raided by outsiders, who are regarded as the greatest threat to preservation of resources. By such means *adat* community members are made to feel that they are responsible parties with a stake in park protection. But the *adat* councils themselves have assumed this mandate in an even wider sense, extending their functions as adjudicators of transgressions of park regulations to the reassumption of environmental authority over resources within the enclave, as well as those located within the surrounding park. Such exercise of environmental authority on customary grounds also reveals differences from the situation of the 1980s, when the efficacy of *adat* institutions in resource matters had been suppressed by developmentalist and uniformitarian 'state simplifications' (Scott 1998) of the New Order. In contrast to earlier times, in 2006 the member of the Bugis community accepted the judgement and jurisdiction of the *adat* council. Indeed, his acknowledgement of his transgression before the *Adat* Council of the Lindu Plain was one factor in the cash equivalent of his *sampole saongu* fine being lower than that of the more brazen Lindu man. Acknowledgment of such indigenous *adat* authority by a migrant may be seen to constitute one sign of an incipient sense of a wider community membership whose parameters are at least partially dictated by local *adat* stipulations that have been re-acknowledged in the context of park governance.

Executive implementation of the Lindu Conservation Agreement: constituting a wider community?

Other aspects of the Lindu conservation agreement promote an altered sense of community that encompasses members beyond the indigenous To Lindu. Even in its specification of the *adat* councils as the judicial institutions concerned with transgression of park regulations, the agreement made no differentiation between the *adat* councils of Anca, Langko, and Tomado, villages all dominated by indigenous To Lindu, and the *adat* council of Puroo, a relatively recent village at the entrance to the Lindu plain exclusively inhabited by resettlers from elsewhere in Kulawi subdistrict. In contrast to the approaches of YTM, Yayasan Yaphama, and CSIADCP, which focused only on the indigenous groups who claim precedential status as the original inhabitants of the area, the TNC agreement is meant to encompass such settlers as participants and thus stakeholders as well.

Such a range of encompassment is made even clearer in the specification of the institutions that have been nominated to carry out many of the executive functions of the enforcement and policing of park regulations: the LKD. The LKD are labelled as 'the institutions that represent society in conservation efforts in TNLL at the village level', and function in an executive capacity to provide an umbrella for communication between local society and park authorities; inform and promote the Conservation Agreement to the local communities; carry out participatory planning with park management; and supervise, evaluate and report on evaluations of the Conservation Agreement to the village headman.

LKD members are selected 'on the basis of the Decision of the Village Head in accordance with the results of village consultations that have been attended by the Park Management of Lore Lindu National Park, the Village Government, the Village Representative Body, the Adat Council and other members of the community' (Kesepakatan Lindu 2005: 7). One LKD has been formed for each of the four villages of the Lindu plain, with an attempt at representation from all of the hamlets in each of the villages, encompassing all the settlements of the plain and thus including representatives from among the migrants to Lindu. The Puroo LKD is composed of representatives from among the settlers from Kulawi, while the Tomado LKD draws its representatives not only from the To Lindu residents in the village's centrally located hamlets, but also from several ethnic groups resident in Kanawu – the hamlet located on the eastern side of the lake with a large part of its population composed of Bugis migrants from South Sulawesi, resettlers (local transmigrants) from the mountainous regions in Pipikoro subdistrict, as well as, more recently, Toraja farmers from South Sulawesi. These latter groups are thus also envisaged as stakeholders in park protection. Such inclusion points to the possibility of a different sense of community than the implications of those agreements that have only included the *adat* community of the village. Multi-ethnic participation in the LKD has contributed to fostering an incipient sense of a more encompassing Lindu community (*masyarakat Lindu*), united by a concern for common threats to their livelihood, that extends beyond the indigenous To Lindu.

This Lindu Conservation Agreement thus potentially encompasses all the ethnic groups within the Lindu plain in regard to issues of conservation enforcement, though still relying on the customary mechanisms of the dominant 'indigenous society' (the To Lindu, strictly speaking) in the adjudication of transgressions of park regulations, a role which has also reinforced the authority of the To Lindu *adat* councils to regulate other resources within the enclave itself. The agreement is thus informed by parameters both beyond and within the customary framework. This dual positioning poses contradictions in the operation of the LKD as the main local agency for monitoring compliance with the conservation regulations of the park. The To Lindu elders who serve as members of both the *adat* councils and the LKD also regard these latter organizations as a vehicle for upholding customary regulations. Their dual roles lead them to both promote the LKD as an organization to uphold TNLL regulations for the whole Lindu enclave and also use it as an instrument to promote their precedential rights as indigenous To Lindu to land and resources in the plain, as has been evident in the activities of the LKD to date.

The Lindu village conservation organizations were actually functioning on the basis of the earlier draft conservation agreement before the formal signing of the final agreement on 30 March, 2005. Indeed, their first activity as a group took place in early 2004, when they were taken by TNC staff to visit the areas of the neighbouring Palolo plain devastated by the December 2003 floods, which TNC asserted were a result of the widespread felling of trees by the occupiers in Dongi-Dongi, which had been designated a core zone of TNLL in part due to its watershed functions.[38] Shortly thereafter, armed with this evidence of the environmental consequences of the neglect of conservation regulations, representatives of the LKD of Langko, Tomado and Puroo journeyed in May 2004 with a TNC representative, a forestry policeman (PPA/Polhut), and the village secretary of Tomado to investigate incursions into parkland above Kanawu, especially by the Toraja settlers of Sangali, but also the longer-term Pipikoro residents and others in Katiboli, the two most remote subhamlets of Kanawu.[39]

Although the composition of the team was dominated by indigenous To Lindu elders, the combined LKD survey team included members from the migrant communities. Indeed, one Bugis member from the shore community of Kanawu has been prominent in the LKD's survey activities. The Bugis were particularly concerned about incursions into the forests above their wet-rice lands in Kanawu, which they had been unable to plant with a second rice crop that year due to a shortage of water. Of course, this situation also affected the To Lindu from Tomado who maintained wet-rice fields in Kanawu as well. Having been convinced by the local TNC facilitator that the dearth of water in the dry season was due to the felling of trees by Toraja and Pipikoro migrants to clear new gardens in the watershed area to the east of Kanawu, an area of montane forest that fell outside the Lindu enclave, the Bugis and To Lindu were united in opposing the actions of these more recent migrants; they were determined to use the agency of the LKD to exercise sanctions against this opening of gardens in parkland.

While the opinions expressed by LKD members during the meetings with local farmers conducted by the survey team were motivated by this common consensus – in Agrawal's (2005) terms, the expression of an environmental subjectivity[40] – the framing of the problem by the To Lindu elders participating in the survey as

LKD members also accorded with a different agenda – overtly voicing a converging of interests and a moulding of a unified constituency, but covertly asserting a continuing claim to precedential land and resource rights for the indigenous To Lindu. Echoing the opening address of the TNC facilitator, the Tomado village secretary stressed in his remarks to the initial meeting with Kanawu residents the need for all the people in the Lindu plain to 'do the good thing by continuing to keep watch over the region that we inhabit together, so that it continues to be preserved ... carry[ing] out our activities, both in our gardens and our wet-rice fields in ways that are environmentally friendly [*ramah lingkungan*]'. The policing function of the LKD within the overall management strategy was reinforced by several speakers. The need for all the inhabitants of the plain to preserve the local environment provided the constant refrain of the discussions.

However, while in keeping with this general aim of raising the conservation consciousness of all the inhabitants of the Lindu plain, the interventions of the head of the LKD from Langko, whose role in the *adat* council of that village received prominent mention when he was first introduced, also advanced the interests of the indigenous To Lindu elders as the custodians of the land and hence the leaders whose authority was to be acknowledged in this conservation project. In the conclusion of his opening remarks he carefully labelled all the people present at the meeting and the families they represented as fellow inhabitants of Lindu: 'we [inclusive] all possess (*kita semua punya*) Lindu; it's no longer said that only the [indigenous] Lindu people (To Lindu) possess this Lindu, we all possess it, because – what is the reason? – we have all lived here ...' Despite this inclusive assertion, this To Lindu elder's later contributions revealed the agenda of maintaining the dominant position of the indigenous To Lindu in the conservation project. This position was most evident in his elaboration upon the words of the TNC facilitator, stressing the 'single unity' of the national park and customary territory and the need to align the indigenous Lindu customary 'zoning' (*suaka*) with the national park zonation scheme:

So, my thoughts concerning the customs of my ancestors, this is all *adat* land. If I speak, I have ancestors who lived here in this Olu, for Olu is its name, not Kanawu or anything else, but Olu. So, if I recite the names of all these settlements, I know them all proceeding to Kangkuro, Salumpalili, Tumawu, Tawaiki, Salu Suo, Banbaria, Boya, Lewonu, Sangali, Tae Lampanga, Tae Ropo. I know them all, because of what? Because my ancestors from time immemorial have lived here, my ancestors from time immemorial have sacrificed to extinction their livestock, because of this plain. But now the regulations are different. Gentlemen, my brothers and sisters who have come here, now we no longer think of only ourselves, we think of all of you, Bugis fathers, Toraja fathers, Kantewu fathers, we speak of all of you as Lindu people (*orang Lindu*). And now once we speak of Lindu people in general, then how should we orient our thoughts to preserving this environment, how do we orient our thoughts so that we are all the same, all of us have approximately the same land, so that none of us inhabitants has too much land, that is my proposal ...

While still encompassing all the ethnic groups represented at the meeting as 'Lindu people' (*orang Lindu*), he managed also to advance the claims of the indigenous To Lindu to leading the conservation effort on the basis of their customary knowledge and precedential custodianship of the land and its resources. He asserted the prior rights to the land that he and other indigenous To Lindu possessed, since their ancestors had sacrificed the blood of their livestock upon it – an observance refunctionalized in comments he made later in the meeting to stress the conservation function of buffalo sacrifice in limiting overgrazing and preserving the land, as well as its social function of validating claims to the land by feeding the deceased ancestors and other spirits (e.g. *viata*) who guard it. This custodial relationship was validated by his knowledge of the real, the original names of all the customary territories around the lake. His words thus harnessed the nascent sense of a multi-ethnic community acting in concert to preserve members' common environment, as declared by the TNC facilitator and government officials and even given more explicit articulation by this elder himself, to the project of subjecting all the inhabitants of the plain to the ultimate authority of the To Lindu *adat* council (Acciaioli 2007).

The final words of this elder's intervention during the LKD survey linked the common project of community conservation with the necessity to limit land ownership and use, rendering the distribution of land as more equal. This appeal to the value of egalitarian land tenure in the context of conservation alluded to an earlier declaration by the Adat Council of the Entire Lindu Plain that no inhabitant may cultivate more than 2 hectares of land. This declaration had been aimed squarely at the Bugis and Kulawi settlers, some of whom had opened up to 12 hectares if all their plots devoted to coffee, cacao and other cash crops were counted in addition to their wet-rice fields (Acciaioli 2001: 98–9; Acciaioli and Warren 2005). To Lindu *adat* councils had begun in the new millennium to demand observance of the regulation that all land transactions must be ratified by the head of the plain-wide council or the head of one of the three village-level Lindu *adat* councils. Some members of the *adat* councils of the Lindu plain[41] supported the position that all land within the Lindu *adat* territory could only be owned by To Lindu. No outsiders could purchase land with full private (*hak milik*) rights; non-To Lindu could only acquire use rights (*hak pakai*), as ratified by a local *adat* council. At least one other elder held the yet stronger (and perhaps more consistent) position that indigenous To Lindu themselves could only gain use rights by payment to a relevant *adat* council. In this view all land in the Lindu enclave was ultimately inalienable and controlled by the *adat* councils in perpetuity.

These claims prompted a recent round of community participatory mapping, conducted in cooperation with YTM. Unlike the earlier round of such mapping,[42] this project sought to determine the precise amount of landholdings of both indigenous To Lindu people and migrants. Its result, summarized in a booklet (Sugiharto *et al.*, n.d.) available at the YTM office, revealed a startling asymmetry in land holdings, with migrants, especially the Bugis from South Sulawesi, holding far more land per individual – up to six times more land than individual To Lindu. This report has contributed to the resolve of the Adat Council of the Entire Lindu Plain to limit individual land ownership to 2 hectares, with any land worked above this limit being considered customary land whose use required approval of this council. More recently (June 2007), the newly elected To Lindu headman of Tomado has actually seized some of the land of the Bugis migrants to his village, citing the validation of the Adat Council of the Lindu Plain for his action, although this action continues to be contested.

Of course, this stand has contributed to a disparity of interpretations of the *ganti-rugi* transactions by which the Bugis had been obtaining land from To Lindu individuals since first arriving over a half-century earlier. The Bugis considered that their payments had constituted an outright purchase of these lands; many had already obtained government-issued land certificates (*surat tanah*) for the land that they had so acquired from indigenous To Lindu. *Adat* elders were now arguing that the migrants had only purchased usufruct rights on what remained inalienable customary land, whose range of uses the *adat* councils still had the right to regulate. In contrast, the migrants espoused the state's definition of land not under permanent cultivation as state land (*tanah negara*) rather than customary land. As one migrant had reacted to an earlier attempt by the re-empowered *adat* council in the first years of *Reformasi* to limit the land that could be worked, 'This [land] is the right of us all; the state is the one that owns this property; this is the inheritance of the state, of the Indonesian people, all of it' (Acciaioli 2001: 101). However, the To Lindu elder's oration during the LKD survey team's 2004 meeting with Kanawu residents once more reasserted a conceptualization of the territory of the Lindu plain as a commons under the custodial regulation of the To Lindu *adat* councils, customary bodies whose membership was not open to non-indigenous residents. And this custodianship was predicated on the continuing precedence of indigenous To Lindu rights in land, including the right to set a ceiling on individual land use. In the views of the majority of the elders, the indigenous To Lindu had exclusive right to convert such land to private ownership, although a minority of elders disputed even the possibility of private ownership by indigenous To Lindu. Such an interpretation of the terms of land ownership thus reconfigured the sense of this To Lindu elder's earlier attribution of a common identity as *orang Lindu* to all inhabitants of the plain, regardless of origin.

However egalitarian the imposition of a common limit on access to land as part of a conservation strategy, the exclusive assertion of authority by the Adat Council of the Lindu Plain to impose it, and the differential possibilities of ownership status for indigenous Lindu and migrant farmers effectively recast the terms of community membership by differentiating the rights of various categories of *orang Lindu* to the advantage of indigenous *To Lindu*.

Conclusion: dilemmas of community and conservation

In articulating their 'new paradigm of environmental justice' for Indonesia, Lynch and Harwell (2002: 148) have urged that government recognition of community-based property rights should not be limited only to *adat*-based claims:

> While there are clear legal precedents for recognizing community-based adat property rights, this report does not conclude that only adat institutions should be the beneficiaries of reforms in natural resource management. Many communities have long innovated on tradition (or wish to now do so), which in some cases might have been highly stratified and undemocratic, to devise hybrid forms of governance. In such situations, each community should be given the opportunity to form their own institutions for managing natural resources. Further, many communities are mixed or immigrant or may have been highly mobile and therefore have no clear-cut *adat* rights to territory … From a social justice perspective, it is clear that the law and policies for 'legalizing' these diverse kinds of arrangements need to be developed to accommodate the full spectrum of natural resource-dependent communities.

The World Resources Institute report on democratic decentralization and natural resource management issuing from the Bellagio Conference on Decentralization and the Environment also warns against exclusively targeting 'customary authorities' as the recipients of decentralized powers:

> Customary authorities, however, are rarely democratic. They often inherit their positions, and their degree of local accountability depends on their personalities and local and social and political histories. They may or may not be accountable to local populations … Customary authorities are notorious for entrenched gender inequalities and for favoring divisive ethnic-based membership over the residency-based forms of citizenship so fundamental to most democratic systems. Today there is a troubling convergence of state and donor efforts to find the 'real,' 'traditional' natural resource managers and to empower them to manage the resources. But, giving powers to customary authorities does not strengthen democratic decentralization (Ribot 2002: 12).

Certainly, the applicability to Indonesia of these generalizations, largely derived from complex situations involving divergent histories of colonial recognition and creation of 'tribal' authorities (Mamdani 1996) in Africa, remains debatable. But policymakers need to take seriously the issues raised by those cases and give attention to Lynch and Harwell's (2002: 148) call to complement Indonesian government recognition of *adat* community rights with 'creatively forging new arrangements for non-*adat* and mixed communities, who in many locales are the most disadvantaged'.

The experience of forging and implementing community participation in TNLL illustrates the dilemmas involved in balancing recognition of *adat* institutions that certainly have a demonstrated history of promoting sustainable use of the environment with the contemporary social reality of 'mixed communities'. Initially, the conservation agreements signed by park officials and representatives of villagers living along the park's boundaries exclusively targeted indigenous peoples living there (Khaeruddin 2002). Non-indigenous local peoples were not accorded the same treatment. The village of Dodolo was not granted a conservation agreement; its inhabitants, To Rampi migrants from South Sulawesi, were simply resettled outside the park. In contrast, the To Katu, with a much stronger claim to indigeneity as an offshoot of the long-settled To Besoa, resisted resettlement efforts and were eventually granted enclave status along with a community conservation agreement (Sangaji *et al.* 2004).

While certainly privileging indigenous interests by allocating judicial functions for sanctioning transgressions of park regulations to To Lindu *adat* councils, the Lindu Conservation Agreement has at least attempted to accommodate as well the interests of non-indigenous inhabitants, both spontaneous Bugis migrants and local resettlers from neighbouring Kulawi and Pipikoro. Such inclusion has been most evident in the composition of LKD (village conservation organizations) as executive bodies for implementation of the agreement. As these LKD include members recruited from both indigenous and migrant peoples settled in the Lindu enclave, their joint participation in LKD efforts may be seen as an experiment in forging a nascent community of 'mixed' constitution, contrasting with the earlier 'plural society' (Furnivall 1944) which prevailed on the Lindu plain throughout the era of the New Order. As both McDermott (2001) and Li (2001) have emphasized in their contributions to *Communities and the Environment* (Agrawal and Gibson 2001), communities may coalesce in reaction to the sanctions and incentives imposed by the state and other wider institutions, particularly in regard to such issues as resource management. The appeals made to a wider constituency of *orang Lindu* – a category defined by members' shared habitation (i.e. Ribot's 'residency-based forms of citizenship') of a place defined not simply on internal grounds by those living within it but also on external grounds (i.e. as an enclave) by the state's declaration of a national park encompassing them – in the operation of the LKD signal such an incipient sense of community. Shared subjection to water shortage also operates to forge a common identity between the previously antagonistic To Lindu and Bugis. It is this contribution of identity building to the daily practices of cooperation that allows one to posit an incipient shared orientation toward a new form of local commonweal (Warren and McCarthy 2002).

However, it is an orientation that remains fragile, as the privileging of To Lindu *adat* as the vehicle of community conservation means that membership in this nascent community is negotiated and conducted on different terms for the indigenous To Lindu than for the migrants. This differentiation constitutes the dilemma of 'community' in the practice of community-based conservation in this context. On the one hand, the appeal to *adat* as a basis for articulating community participation with the authorities who manage the park, both the state bureaucracy and the partner transnational conservation organization TNC, has certainly proven its usefulness in a situation where the indigenous peoples long resident in and around the park have forged their identities by reference to this concept. Even the

Bugis at Lake Lindu have learned the lesson of subjecting themselves to indigenous To Lindu *adat* stipulations in order to avert such resource depletion as the fishing out of Lake Lindu – a threat to the commonweal affecting all the peoples of the Lindu enclave and those elsewhere in Central Sulawesi who have depended upon the supply of fish from the lake.

In contrast, the extension of To Lindu *adat* authority to regulation of land use and ownership remains an issue of contention, one that continues to prompt Bugis resistance to the To Lindu *adat* councils and indeed ambivalence about identity as '*orang Lindu*', if submission to these councils without representation in all property and resource matters is an entailment. Correspondingly, To Lindu elders serving on the LKD still feel themselves in an uneasy situation. They must accept the wider ambit of participation in the conservation project and the acknowledgement of settler rights it implies, but also manoeuvre to advance recognition of the authority of *adat* institutions both as arbiter of resource use within the Lindu territory they define by custom and as the basis of adjudication of transgressions against externally imposed national park regulations in force beyond the Lindu enclave. Whether the project of 'community-based conservation' in TNLL will be sufficient to forge and sustain a wider identity for the inhabitants of the Lindu plain as a 'community' oriented by a common commitment to building their local commonweal will depend, in part, upon the strategic ways in which such representative institutions of modernity as the LKD can effect a transformation of *adat* institutions that accommodates a balancing of the interests of all resident groups.

Acknowledgements

Grants from the Australian Research Council and the University of Western Australia enabled me to undertake the research on which this chapter is based. The Asia Research Institute, National University of Singapore, provided the facilities for writing the initial draft of this paper. PMB-LIPI provided research sponsorship; Dr Riwanto Tirtosoedarmo and Dr Johannes Haba of that body were particularly supportive. Members of various organizations with offices in and around Palu – TNC, YTM, WALHI, CARE, AMASUTA and others – responded generously to my requests for assistance, fielding queries and providing documents. I especially appreciate TNC's generosity in allowing the adaptation of one of its maps. All the peoples of the Lindu plain, both indigenous and migrant, have been patient and generous with me, as I pestered them with questions and recorded numerous gatherings. Above all, the participation of my wife, Niniek Toley, a To Lindu from Langko, has been crucial in this research. Comments from the other members of the 'Locating the Commonweal' research team at our 2005 workshop and subsequently, especially from the editors of this volume, have been essential for my recasting of the argument of this chapter. Of course, my construction of all this input remains my own.

Notes

1   Other writers considering the management of common property resources (CPR) have also emphasized the need to look at the diversity of the subjects involved in use of such resources (e.g. Ostrom *et al.* 1999).

2   Contemporary Lindu narratives of origin (Davis n.d. [c. 1973]), however, highlight their original links to the Sigi rajahdom of the Palu valley.

3   See the discussion below concerning the consequences of divergent To Lindu and Bugis understandings of the rights conveyed by *ganti-rugi* transactions.

4   While other species were also introduced in the 1950s to lakes in Sulawesi (Whitten *et al.* 2002: 298), none has thrived like common tilapia (*Tilapia mossambica* or *mujair*, in Indonesian). Tilapia is cultivated as a pond fish in many parts of the world, first having been brought to Java by the Dutch from Africa during the late colonial period. Rather than fostering pond cultivation of tilapia, the Fisheries Department of independent Indonesia chose merely to dump this predatory fish in these lakes with no thought of the consequences for endemic species.

5   Of the fish species listed as found in Lake Lindu in the original draft plan of the Lore Lindu National Park (Watling and Mulyana 1981: 65a), only one, *betok* (*Anabas testudimus*), is listed as 'native'; all the others were introduced from 1951 to 1955.

6   Of course, there were always people who crossed over these categories, especially young To Lindu men who served as subordinate fishermen for Bugis 'bosses', and Bugis who engaged primarily in other pursuits such as wet-rice cultivation and palm sugar processing.

7   Other migrants were mainly resettlers from other parts of Kulawi subdistrict, who had come to Lindu from their home villages in a programme devised by the subdistrict officer (*camat*) to foster opportunities to open wet-rice fields. Such cultivation was not possible in their homeland montane environments where they subsisted mainly from slash-and-burn horticulture, a subsistence activity upon which the Indonesian government frowned (cf. Dove 1985). These resettlers were predominantly, like the indigenous To Lindu, Salvation Army adherents, although some belonged to other Protestant denominations, especially in the settler village Puroo, and a very small minority were Muslim.

8   Undang-undang No. 5 Tahun 1979 tentang Pemerintahan Desa.

9   At that time the To Lindu still were regarded as a *masyarakat terasing* ('isolated societies') as they fit such criteria as distance of their home region from the provincial capital and lack of basic infrastructure in their villages, as set out by the Department of Social Affairs (Departemen Sosial or Depsos). On the effect of Depsos policies regarding *masyarakat terasing*, see Acciaioli (1985).

10 In 1982 I witnessed this sanction exercised against one young man caught stealing fish from another Bugis fisherman's net.

11 PPA is the abbreviation for Perlindungan dan Pengawetan Alam, housed under the Directorate General of Forestry.

12 Surat Keputusan Menteri Pertanian [Ministerial Decree] No. 736/Mentan/X/1982.

13 Cribb (n.d.: 14) notes that the Indonesian government declared the establishment of its first five national parks in 1980, although it had earlier established numerous reserves, some of them carried forward from those established by the Dutch during the colonial period (Cribb n.d.: 3).

14 These included the Jakarta-based Yayasan Sejati, the Palu-based Foundation for a Green Palu (Yayasan Palu Hijau) and the Palu branches of the Indonesian Environmental Forum (Wahana Lingkungan Hidup Indonesia or WALHI) and the Indonesian Institute of Legal Aid (Lembaga Bantuan Hukum, LBH). This period coincided with the incipient blossoming of the NGO movement in Indonesia (Eldridge 1995), as the New Order's surveillance ethos yielded to a more permissive attitude, in tandem with growing global influence of environmental and human rights organizations.

15 The central government eventually ruled that the project could not proceed, as it had not been included within the previous five-year plan.

16 My own PhD thesis (Acciaioli 1989) was used as the single source for this fact.

17 The development of solidarity with other oppressed groups, both within Indonesia and beyond, has been one of the primary aims of the YTM strategy to raise consciousness among the To Lindu as one more example of an 'indigenous people' whose rights were disregarded by the nation-state. YTM had organized a tour to the Kedung Ombo site in Central Java, where To Lindu elders witnessed the effects of a dam project resettlement (Aditjondro 1998), and fostered identification with the struggles of 'indigenous peoples' throughout the world affected by development projects such as large-scale dams and national parks requiring the removal of peoples from their homeland (Clay 1985; Eilers 1985; Aditjondro 1994). One To Lindu informant I interviewed in January 2000 mentioned the analogous situation of Amazonian Indians as we were discussing rumours of the new governor's attempt to resurrect the hydroelectric dam project on a smaller scale (*Formasi*, December 1999). In fact, proposals to build hydroelectric projects, though of smaller scale than the ADB-funded project announced in 1988, continue to the present day (*Antara News*, 9/6/07).

18 Since their successful anti-PLTA campaign, the To Lindu have moved on to become active participants in the national indigenous peoples' movement in Indonesia (Acciaioli 2001: 92, 2002: 217–20).

19 See my discussion below for the implications of the *adat* elders' assertion of non-alienability of cultivated land for conflicting understandings of the *ganti-rugi* transactions, by which Bugis migrants understood that they had purchased land.

20 This land type thus corresponds well with the 'sacred groves', which some environmentalists have maintained are the indigenous equivalents of 'core zones' closed to human access in national parks and other protected zones, thus arguing for the environmental ethic or conservation awareness of 'indigenous peoples' before the imposition of conservation institutions, such as national parks, imported from the West.

21 Laudjeng (1994: 156) uses the term *suaka nu viata*, but I never encountered the term *suaka* applied to this land type, only the term *tana* (land).

22 The function of preserving watershed areas is also mentioned in the NGO literature (e.g. Laudjeng 1994: 156–8).

23 This claim has received some validation from the research of a limnologist from the Indonesian Institute of Sciences (LIPI) who evaluated sedimentation and other ecological aspects of the lake under the auspices of the Central Sulawesi Integrated Area Development and Conservation Project (CSIADCP) (Lukman 2001).

24 In the 1990s YTM activists were among those most involved in the activities of the Jakarta-based adat peoples' movment, Aliansi Masyarakat Adat Nusantara (AMAN).

25 Their emergent role in co-management has been a result of wider trends in the administration of national parks. Since the 1970s, as a result of a confluence of interests between the global environment and indigenous people's movements (Clad 1988: 322), there has been greater recognition that national parks are unviable as isolated preserves if surrounded by degraded lands or by hostile peoples who have long resided there.

26 In fact, the draft management plan of 1981 formulated by WWF had already been based on the division of the park into four internal zones – a development zone, a wilderness zone, traditional use zone, and a sanctuary (or core) zone – as well as a buffer zone (*hutan lindung*) outside the park boundaries proper (Watling and Mulyana 1981: 21). The subsequent draft management plan formulated by TNC (2001) has reconfigured these zones to fit its eco-region concept and current Indonesian legislation mandating zonation (TNC et al. 2001: 24–5).

27 TNLL is only one of four national parks in which TNC is involved as a co-manager in Indonesia. However, it is the only terrestrial park, as the other three – Komodo (East Nusa Tenggara), Wakatobi (Southeast Sulawesi), and Derawan-Berau (East Kalimantan) – are all marine parks, although including smaller islands and coastal areas

of larger islands. In addition, it is involved in drawing up the plans for the proposed Raja Ampat Marine Park in West Papua and is initiating work in Morowali Nature Reserve in eastern Central Sulawesi in preparation for possible conversion to national park status (Fitria Rinawati [field coordinator, TNC head office, Jakarta], personal communication, 15 June 2007).

28 Surat Keputusan Menteri Kehutanan No. 593/Kpts-II/93 of 5 October 1993. A subsequent government declaration in 1999 redefined the borders of the park. In its current form, TNLL stretches across 217,991.18 ha. (TNC *et al.* 2001:2); *Kespakatan Lindu* 2005:2; Sangadji *et al.* (2004:16).

29 Li (2007: 139ff) provides a detailed treatment of the presuppositions and implications of this 25-year plan.

30 Acciaioli (2006: 11–15) reviews the differing emphases of many of these agreements at TNLL, including those brokered by YTM, CARE, CSIADCP, and Yayasan Yambata, See also Li (2007: 132ff.), who analyses many of the contradictions involved in the aims of these community conservation agreements.

31 See Li (2007: 126ff) on the presuppositions underlying the ADB-funded CSIADCP.

32 The agreement itself was accompanied by 'decision documents' from each of the four village headmen of the Lindu plain, implying a clear parallel to the need for 'regional implementing regulations' (*peraturan daerah*) for any laws (*undang-undang*) to take effect locally.

33 The granting of authority in this conservation agreement to the Adat Council of the Entire Lindu Plain and those of the four villages to adjudicate cases of transgression of park regulations follows the conventions of earlier agreements brokered by CSIADCP and CARE.

34 For a discussion of the executive functions of these LKD see the following section.

35 The sum of Rp. 800,000 is far below the actual value of the traditional in kind payments: three water buffaloes, three traditional *mbesa, ikat* cloths from Galumpang or Rongkong in South Sulawesi, and thirty brass plates of Indian origin (*dula*). Actual prices for these items at the time would have been Rp 6 million per water buffalo, Rp 500,000 per *mbesa*, and Rp 100,000 per *dula*, thus totaling Rp. 22.5 million. However much diverging from the actual fine imposed, the statement in terms of these traditional items functions to maintain the customary idiom of adjudication. Elders may also wield the weapon of demanding payment of the actual goods for culprits who are particularly recalcitrant and obstreperous during the *adat* council hearing. Even relatively minor additions to the conventional cash equivalent for such fines may signify much stronger condemnation of the culprit by the *adat* elders.

36 Elsewhere (Acciaioli 2002: 221–6) I have analysed a case of assault handled by one of the village adat councils in Lindu, relating this re-empowerment to the opportunities for adat institutions to be recognized as authoritative institutions under the village-level reforms stipulated in the Regional Autonomy Law 22/1999, as well as the rising power of the national indigenous peoples' movement (*gerakan masyarakat adat*) within this framework.

37 Many suspected that a cash payment may also have been involved – with the *dinas* village headman of Tomado, who was conspicuously absent at the hearing.

38 Li (2007: 142–92) provides an extended critical analysis of the occupation of Dongi-Dongi.

39 Due to space limitations I only mention aspects of the cross-ethnic encounters during the conduct of this survey, discussed in greater detail elsewhere (Acciaioli 2007).

40 See Acciaioli (2007) for a critique of Agrawal's argument that an environmental subjectivity proceeds primarily from participation in state-mandated conservation organizations.

41 Perhaps even more importantly, this position was supported by one of the Lindu elders serving on the executive committee of the Alliance of Central Sulawesi Indigenous Peoples (Aliansi Masyarakat Adat Sulawesi Tengah), the Palu-based 'people's

organization' (*organisasi rakyat*) that worked with local *adat* councils and with the national indigenous peoples' alliance (AMAN) to advance the cause of indigenous rights.

42 The previous round of participatory community mapping in the late nineties, also carried out under YTM auspices, had documented the boundaries of various To Lindu *suaka*, demonstrating the complexity of the To Lindu land use system and their past stewardship of the environment (Acciaioli and Warren 2005). The maps produced, along with ethno-ecological documentation of the complexity of the To Lindu land use system, have, doubtless, contributed to the re-orientation of the park management's approach to community participation through conservation agreements. The reference to 'participatory' fixing of the zones of the park in the conservation agreement with the Lindu people, as well as the declaration of the TNC facilitator during the LKD survey analysed here that park zones would be more closely aligned with To Lindu *suaka* in further revisions of the park management, also has depended upon the contribution of this earlier community participatory mapping effort (Acciaioli and Warren 2005: 590).

Bibliography

Acciaioli, G. (1985) 'Culture as art: from practice to spectacle in Indonesia, *Canberra Anthropology* 8(1&2): 148–74.

Acciaioli, G. (1989) 'Searching for Good Fortune: The Making of a Bugis Shore Community at Lake Lindu, Central Sulawesi'. The Australian National University, unpublished PhD thesis.

Acciaioli, G. (1999) 'Principles and strategies of Bugis migration: some contextual factors relating to ethnic conflict', *Masyarakat Indonesia: Majalah Ilmu-Ilmu Sosial Indonesia*, 25(2): 239–68.

Acciaioli, G. (2000) 'Kinship and debt: the social organization of Bugis migration and fish marketing at Lake Lindu, Central Sulawesi', *Bijdragen tot de Taal-, Land en Volkenkunde* 156(3): 588–617.

Acciaioli, G. (2001) 'Grounds of conflict, idioms of harmony: custom, religion, and nationalism in violence avoidance at the Lindu plain, Central Sulawesi', *Indonesia* 72: 81–112.

Acciaioli, G. (2002) 'Re-empowering the "art of the elders": the revitalization of adat among the To Lindu people of Central Sulawesi', in M. Sakai (ed.) *Beyond Jakarta: Regional Autonomy and Local Societies in Indonesia*. Adelaide: Crawford House Publishing, pp. 217–44.

Acciaioli, G. (2006) 'Indigenous To Lindu conservation strategies and the reclaiming of customary land and resources in Central Sulawesi', *Masyarakat Indonesia* 32 (2): 1–29.

Acciaioli, G. (2007) 'Strategy and subjectivity in co-management of Lore Lindu National Park (Central Sulawesi, Indonesia)', in N. Sodhi, G. Acciaioli, M. Erb, A. K-J. Tan (eds), *Biodiversity and Human Livelihoods in Protected Areas: Case Studies from the Malay Archipelago*. Cambridge: Cambridge University Press, pp. 266–88.

Acciaioli, G. and Warren, C. (2005) 'Constructing the commonweal through participatory community mapping: decentralisation and local resource conflicts in Bali and Central Sulawesi', in *Tanah Masih di Langit: Penyelesaian Masalah Penguasaan Tanah Dan Kekayaan Alam di Indonesia Yang Tak Kunjung Tuntas di Era Reformasi*. Jakarta: Yayasan Kemala and The Ford Foundation, pp. 581–98.

Agrawal, A. (2005) *Environmentality: Technologies of Government and the Making of Subjects*. Durham, NC: Duke University Press.

Agrawal, A. and Gibson, C. C. (1999) 'Enchantment and disenchantment: the role of community in natural resource conservation', http://www.mekonginfo.org/mrc_en/doclib.nsf/0/F52F8186F85933F9472568E80044F906/$FILE/FULLTEXT.html (accessed 12 June 2007).

Agrawal, A. and Gibson, C. C. (eds) (2001) *Communities and the Environment: Ethnicity, Gender, and the State in Community-based Conservation*. New Brunswick, NJ: Rutgers University Press.

Antara News (2007) 'Walhi against hydropower plant project in C. Sulawesi forest area', *Antara News*, 9 June 2007, http://www.antara.co.id/en/arc/2007/6/9/walhi-against-tro-hydropower-plant-project-in-C-Sulawesi-forest-area/ (accessed 21 June 2007).

Baudrillard, J. (1981) *For a Critique of the Political Economy of the Sign*, C. Levin tr. St Louis: Telos Press.

Brandon, K., Kent, H., Redford, H. and Sanderson, S.E. (eds) 1998 *Parks in Peril: People, Politics, and Protected Areas*. Washington DC: The Nature Conservancy and Island Press.

Burger, J., Field, C., Norgaard, B., Ostrom, C. and Poicansky, D. (2001) 'Introduction,' in J. Burger, E. Ostrom, R.B. Norgaard, D. Policansky, B.D. Goldstein (eds), *Protecting the Commons: A Framework for Resource Management in the Americas*. Washington DC: Island Press, pp. 1–15.

Cribb, R. (n.d.) 'The Politics of Environmental Protection in Indonesia', Working Paper No. 48, Clayton, Vic: Center of Southeast Asian Studies, Monash University.

Davidson, J. S. and Henley, D. (eds) (2007) *The Revival of Tradition in Indonesian Politics: The Deployment of Adat from Colonialism to Indigenism*. Routledge Contemporary Southeast Asia Series. London and New York: Routledge.

Davis, G. (n.d. [c. 1973]) 'The People and Legends of Lake Lindu, Central Sulawesi, Indonesia', unpublished manuscript prepared for the Departemen Pendidikan dan Kebudayaan, Tingkat Propinsi, Palu.

Dove, M. R. (1985) 'The agroecological mythology of the Javanese and the political economy of Indonesia', *Indonesia* 39: 1–30.

Eldridge, P. (1995) *Non-Government Organizations and Democratic Participation in Indonesia*, South-East Asian Social Science Monographs. Kuala Lumpur: Oxford University Press.

Formasi (1999) 'Menggagas Kembali PLTA Danau Lindu', *Formasi*, December 1999.

Furnivall, J. S. (1944) *Netherlands India: A Study of Plural Economy*. Cambridge: Cambridge University Press.

Graaf, S. de and Stibbe, D. G. (eds) (1918) *Encyclopædie van Nederlandsch-Indië*, 2nd edn, vol. 2: H-M. 's. Gravenhage and Leiden: Martinus Nijhoff & N.V.V./H.E.J. Brill.

Hardin, G. (1968) 'The tragedy of the commons', *Science* 162: 1243–8.

Harvey, B. S. (1974) 'Tradition, Islam and Rebellion: South Sulawesi 1950–1965', Cornell University, unpublished thesis.

Kaudern, W. (1925) *Ethnographical Studies in Celebes: Results of the Author's Expedition to Celebes, 1917–1920*, vol. 1: *Migrations of the Toradja in Central Celebes*. Göteborg: Elanders Boktryckeri Aktiebolag.

Kaudern, W. (1940) 'The noble families or maradika of Koelawi, Central Celebes', *Etnologiska Studier* 11: 30–124.

Kesepakatan Lindu (n.d. [2005]) Kesepakatan Konservasi Masyarakat Dataran Lindu, Kecamatan Kulawi, Kapubaten Donggala. Unpublished conservation agreement held in The Nature Conservancy, Palu.

Khaeruddin, I. (2002) 'Kesepakatan Konservasi Masyarakat di Lima Desa sekitar Taman Nasional Lore Lindu Sulawesi Tengah: laporan hasil kegiatan', unpublished report, The Nature Conservancy, Palu.

KMAN [Kongres Masyarakat Adat Nusantara] 1999 FACT SHEET (II).

Kruyt, A.C. (n.d. [1897]) Unpublished diary (*dagboek*), Archives of the Hendrik Kraemer Instituut, Oegstgeest, The Netherlands. Kaast IIIA, Dos I, Bundel 2 [containing the account of the trip to Sigi, Kulawi, and Lindu, 1897].

Kruyt, A. C. (1938) *De West-Toradjas op Midden-Celebes*, vol. 1: Verhandelingen der Koniklijke Nederlandsche Akademie van Wetenschappen te Amsterdam, Afdeeling Letterkunde, Nieuwe Reeks, Deel XI, Amsterdam: Uitgave van de N.V. Noord-Hollandsche Uitgevers-Maatschappij.

Laudjeng, H. (1994) 'Kearifan tradisional masyarakat adat', in A. Sangadji (ed.) *Bendungan Rakyat dan Lingkungan: Catatan Kritis Rencana Pembangunan PLTA Lore Lindu* Jakarta: Wahana Lingkungan Hidup Indonesia (WALHI), pp. 150–63.

Lembaga Hadat Sedataran Lindu (n.d. [2003]) Kronologis Kejadian Penangkapan Pencurian Hasil Hutan (Rotan) di Dataran Lindu, unpublished document held at the TNC office, Palu.

Li, T. M. (2000) 'Articulating indigenous identity in Indonesia: resource politics and the tribal slot', *Comparative Studies in Society and History* 42(1): 149–79.

Li, T. M. (2001) 'Boundary work: community, market, and state reconsidered', in A. Agrawal and C. C. Gibson (eds), *Communities and the Environment: Ethnicity, Gender, and the State in Community-based Conservation*. New Brunswick, NJ, and London: Rutgers University Press, pp. 157–79.

Li, T.M. (2007) *The Will to Improve: Governmentality, Development, and the Practice of Politics*. Durham, NC and London: Duke University Press.

Lukman (2001) 'Laporan Akhir: Studi Enclave Lindu', report prepared by CV. Profil Teknik, Palu, for Central Sulawesi Integrated Area Development and Conservation Project (CSIADCP).

Lynch, O. J. and Harwell, E. (2002) *Whose Natural Resources? Whose Common Good?: Towards a New Paradigm of Environmental Justice and the National Interest in Indonesia*. Jakarta: Lembaga Studi dan Advokasi Masyarakat (ELSAM).

McDermott, M. H. (2001) 'Invoking community: indigenous people and ancestral domain in Palawan, the Philippines', in A. Agrawal and C. C. Gibson (eds), *Communities and the Environment: Ethnicity, Gender, and the State in Community-based Conservation*. New Brunswick, NJ, and London: Rutgers University Press, pp. 32–62.

Mamdani, M. (1996) *Citizen and Subject: Contemporary Africa and the Legacy of Late Colonialism*. Princeton NJ: Princeton University Press.

Mappatoba, M. and Birner, R. (2004) *Co-Management of Protected Areas: The Case of Community Agreements on Conservation in the Lore Lindu National Park, Central Sulawesi, Indonesia*. Eschborn: Deutsche Gesellschaft für Technische Zusamenarbeit (GTZ) GmbH (Tropical Ecology Support Programme (TOEB), F-VI/7e).

Maybury-Lewis, D. (1992) *Millennium: Tribal Wisdom and the Modern World*. New York: Viking.

Merton, R. (1949) *Social Theory and Social Structure*. New York: The Free Press.

Ostrom, E., Burger, J., Field, C. B., Norgaard, R. B., and Poicansky, D. (1999) 'Revisiting the commons: local lessons, global challenges', *Science* 3(1): 278–82.

Ribot, J. C. (2002) *Democratic Decentralization of Natural Resources: Institutionalizing Popular Participation*. Washington DC: World Resources Institute.

Sangadji, A. (ed.) (1994) *Bendungan Rakyat dan Lingkungan: catatan kritis rencana pembangunan PLTA Lore Lindu*. Jakarta: Wahana Lingkungan Hidup Indonesia (WALHI).

Sangadji, A. (1996) *Menyorot PLTA Lore Lindu*. Palu: Yayasan Tanah Merdeka.

Sangaji, A. (2000) *PLTA Lore Lindu: Orang Lindu menolak pindah*. Yogyakarta: Pustaka Pelajar kerjasama dengan Yayasan Tanah Merdeka, ED Walhi Sulawesi Tengah.

Sangaji, A. (2002) *Politik Konservasi: Orang Katu di Behoa Kakau*. Bogor: Penerbit KpSHK.

Sangaji, A., Hamdin, M., Sugiharto, Lumeno, F., Lahigi, S. (2004) *Masyarakat dan Taman Nasional Lore Lindu*. Jakarta: Yayasan Kemala and Yayasan Tanah Merdeka.

Schweihelm, J., Wirawan, N., Elliott, J. and Khan, A. (1992) *Sulawesi Parks Program Land Use and Socio-Economic Survey: Lore Lindu National Park and Morowali Nature Reserve*, Report prepared for the Directorate General of Forest Protection and Nature Conservation (PHPA), Ministry of Forestry, Republic of Indonesia, and The Nature Conservancy.

Scott, J. C. (1998) *Seeing Like a State: How Certain Schemes to Improve the Human Condition Have Failed*. Yale Agrarian Studies, New Haven: Yale University Press.

Sugiharto, Hasan H., and Yabo, N. (n.d.) Final Studi: Masalah Agraria di Dataran Lindu,

118   G. Acciaioli

Kecamatan Kulawi, Kabupaten Donggala, Sulawesi Tengah, unpublished report held at Yayasan Tanah Merdeka Office, Palu.

The Nature Conservancy (TNC) and Direktorat Jenderal Perlindungan Hutan dan Konservasi Alam, and Taman Nasional Lore Lindu (2001) Draft Management Plan: 2002–2007. 4 volumes. Unpublished document held at the TNC office, Palu.

Wahyono, A., Patji, A. R., Laksono, D. S., Indrawasih, R., Sudiyono, Ali, S. (2000) *Hak Ulayat Laut di Kawasan Timur Indonesia*, Yogyakarta: Penerbit Media Pressindo.

Warren, C. and McCarthy, J. (2002) 'Customary regimes and collective goods in Indonesia's changing political constellation', in S. Sargeson (ed.) *Collective Goods, Collective Futures in Asia*, Asian Capitalisms. London and New York: Routledge, pp. 75–101.

Watling, Dick and Yaya Mulyana (1981) *Lore Lindu National Park Management Plan 1981–1986*, report prepared by the World Wildlife Fund – Indonesia Programme (Bogor) for the Directorate of Nature Conservation, Directorate General of Forestry, Republic of Indonesia.

Whitten, T., Mustafa, M., Henderson, G. S. (2002) *The Ecology of Sulawesi*, 2nd edn, The Ecology of Indonesia Series, vol. 4. Singapore: Periplus.

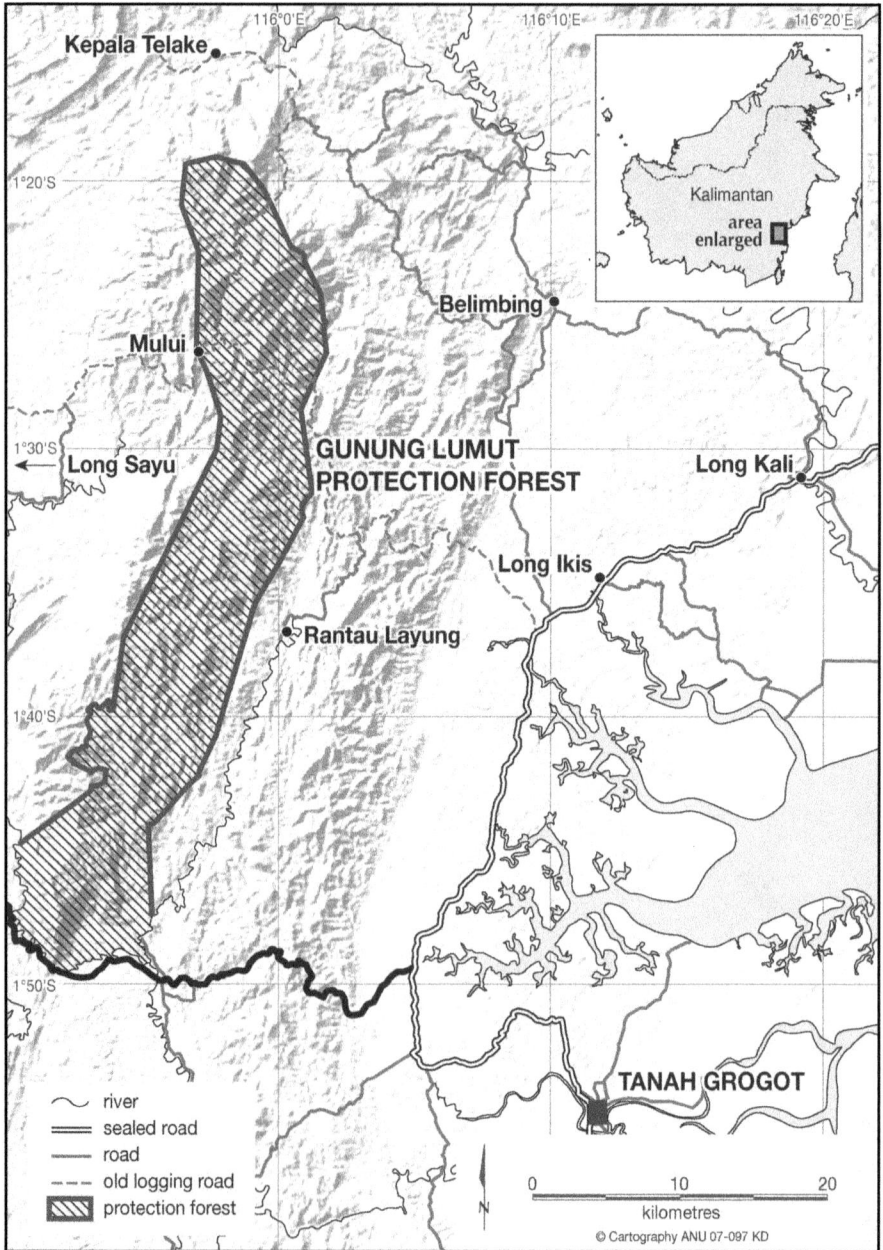

Map 5.1 Gunung Lumut Protection Forest, East Kalimantan Province.

5 Community, *adat* authority and forest management in the hinterland of East Kalimantan

Laurens Bakker

Community and identity at the margins of the state

Regional autonomy brought politics literally to the front doors of many Indonesians, for whom decision-making in the official sphere of governance had been dominated by central government authority. These people, albeit mostly through NGOs and other organised representatives, rose on the tide of decentralization to claim their influence in regional politics. They are not major political players as they lack the capital, connections and other resources available to other participants in the political process (see Rosser *et al.* 2005), but Indonesia's masses have gained increasing influence in the local politics of the regions (Pratikno 2005: 32–3; Morrell 2001).

In this context of dynamic social and power relations my concern is with the communities in the densely forested mountains of the Pasir district of East Kalimantan and the ways in which they argue and maintain collective claims to land and forests vis-à-vis other parties. On the edge of Pasir's mainly coastal society, the Gunung Lumut mountains, the tip of the Meratus mountain range which stretches northwards from the province of South Kalimantan, form the hinterland of the district. (Map 5.1)[1] Their remoteness and poor infrastructure severely limit contact between the mountain communities and the government, and among the mountain communities themselves. Nonetheless, each group holds ideas and assumptions regarding the other, and snippets of news and rumours travel to and fro. These communities broadly share common experiences vis-à-vis the state and the normative reference points provided by *adat*, a term commonly associated with 'tradition' or 'custom', although its relevance lies primarily in its meaning and usage in the present.

Throughout most of Pasir's history, these mountain communities kept a low profile. They are mostly inhabited by subsistence farmers who, with little government control of their actions, use the land and forest according to customary (*adat*) practice. As state influence and government control are still comparatively weak in the mountains, Pasir villagers state that local *adat*, to a greater or lesser extent, regulates and mediates their lives. Like state law, *adat* is primarily about maintaining social order, yet it allows considerable leeway for local detail and negotiations. No two mountain communities thus share precisely the same *adat*, although underpinning principles regulate inter-village relationships. Strictly speaking, *adat* is subordinate to national law, but the mountain communities have often profited from their remoteness and lack of governmental engagement to apply *adat* norms alongside, or instead of, state law. However, whenever state authority made its presence felt directly, there was little option but to accede to its power.

When the national Department of Forestry issued logging concessions in the *adat* territories of various mountain communities in the 1980s, these met with little resistance. News had reached Gunung Lumut from surrounding areas that the loggers could count on police or army protection and demonstrators might get arrested or even jailed. As the mountainous terrain prevented logging on its steep slopes, the communities subsequently made their fields and gardens there or in areas that had been logged over. This did not mean that the *adat* claims were given up, though: many individuals felt that they had been treated improperly by their own government. However, at the time claims and complaints based on *adat* carried very little, if any, weight.

In 1998 news of the fall of Suharto and the onset of regional autonomy inspired new hopes and fears in the mountains. Local *adat* activists, some linked to the indigenous peoples' alliance, AMAN (*Alliansi Masyarakat Adat Nusantara*), entered the mountains with tales of a *masyarakat adat* (*adat* communities) movement focused on recognising indigenous rights under national law. Although *adat*-derived land rights and claims are intrinsically interwoven with Indonesian history and state ideology (see Supomo 1953; Burns 2004: 242–51), de facto recognition or even protection has been uncertain at best (see Bedner and Van Huis forthcoming). The guardianship of the nation by the state, and thus of all *adat*, equated the government to the highest *adat* authority and the interest of *adat* communities with the common good of the nation (Daryono 2004; Warren and McCarthy 2002: 77–9). In this view, recognition of special rights for a small community threatened the rights and interests of the rest of the population.

Until recently, the official stance with regard to today's *masyarakat adat* communities saw them as a burden for the state. 'Isolated' and 'backward' groups in need of government assistance to keep up with the nation's development, they lagged behind in becoming proper Indonesian citizens (cf. Persoon 1998). Li (1999: 2), writing on upland communities living on the edge of 'normal' Indonesian society, points out that popular perceptions are less one-sided. *Adat* communities are associated with poverty, ignorance, disorderliness and a stubborn refusal to adapt, but also seen as free and living in integrity with their natural surroundings. *Adat*-oriented NGOs emphasise these latter positive, green connotations (cf. Tsing 1999; Katoppo 2000; Li 2001: 656–8), which are shared by numerous (inter) national indigenous peoples and environmental organisations, and hence offer valuable opportunities for alliance.

The choice for communities to join the *adat* movement is, however, not a self-evident one. When the 1979 Village Government law instructed the implementation of a uniform national (Javanese) structure of village government, this law had little formal space for *adat*, and introduced new authorities in its place.[2] Although the new state-based authority structure often held a difficult position in strongly *adat*-oriented communities, it effectively replaced local *adat* governance in other areas. When the decentralization laws once again authorised diversity of village government, a return to *adat* was thus no self-evident choice. In many mountain and forest communities, *adat* and its associations swiftly returned to prominence (cf. Li 2000: 163–9; Acciaioli 2003: 224–5; Roth 2006; Henley and Davidson 2008), but some of the results directly contradicted the green image NGOs associated with *adat* regimes. The unclear extent of the regional administration's new authorities over forest land combined with regional power struggles to turn Kalimantan's timber sector into a boom industry that defied the Department of Forestry's authority and national forest law (cf. McCarthy 2000, Casson and Obidzinski 2002; Resosudarmo 2007). Forest communities, persuaded by profit and the promise of work, have become engaged in logging in their *adat* areas. Provided with a logging licence from the district government (Resosudarmo 2003: 236–40; McCarthy 2004), or on their own account (Rhee 2000: 35–7) communities enter into agreement with logging companies willing to pay for the right to operate in the area. The new administrative freedom of the village (see Antlöv 2003), as well as rumours of new 'community forestry' rights (see Safitri 2005) inspired a sense of autonomy and a right of initiative that was concerned with forest as a disposable asset rather than as a communal responsibility.

The important point in either case was to assess rights to the claimed *adat* area vis-à-vis encroaching outside powers. Whereas the New Order government apparatus turned a deaf ear towards such claims, the empowerment of the regional governments that followed *reformasi* brought the ears of the administration much closer to the demands of the population. Tsing (1999: 160–3) points out that representing Indonesian tribal communities offers new opportunities for local leaders to distinguish themselves. By balancing the interests and support of both outside organisations and the community these leaders maintain a powerful position that is vulnerable to contestation, but allows for creative intervention in the life of the community.

Campbell (2002: 114–18) asks whether the emphasis on traditional *adat* institutions should be seen as the vanguard of a new phase of empowerment of local *adat* processes, or whether the focus should be on how communities can most effectively access and use the new governance structures to manage local resources and gradually replace the customary process. I focus here on the dynamic interaction between continuing local *adat* processes and new structures in defending community interests towards assessing their effectiveness in building sustainable futures. The inroads made during the New Order by earlier government influence over decision-taking authority in the election of *adat* leaders in some areas (Sardjono and Samsoedin 2001: 120–2; Eghenter 2006: 165–6) or through the status government support provides individual *adat* leaders (Sillander 2004: 309–10) shows that *adat* and government authorities can both complement and oppose each other. Likewise, *adat* authority is not necessarily uncontested even within the local group: traditional *adat* leaders tend to be senior men, whereas women (cf. Blackwood 2001; Ihromi 1994), the younger generations (Jacobsen 2002) and immigrants (Schulte Nordholt 2007) are groups that have considerably less influence in the adat sphere.

This chapter offers a review of local dynamics in two neighbouring Pasirese mountain communities struggling to maintain or reassert *adat* land and forest resource claims. Communal interests are at the core of both communities' strategies in their deployment of *adat* identity and pursuit of good local governance. Yet although ostensibly quite similar conditions apply, the interplay with outside powers and different presentations of the self lead to notably different interpretations of 'community' and *adat* authority in modern Indonesia, and different approaches to forest protection and use.

The mountains of Pasir

Pasir is the southernmost district of East Kalimantan. Its territory covers some 11,600 square kilometres bordering the districts of Pasir Utara Penajam and Kutai Barat to the north, the provinces of South and Central Kalimantan to the south and west, and the Strait of Makassar to the east. The district's main geographical features are a flat stretch of fertile land along the coast, which gives way to the steep northern stretch of the Meratus mountain range, known as Gunung Lumut. Pasir has a population of 176,000, most of whom are farmers living on the flat, coastal land. The district's capital is the small city of Tanah Grogot. A number of large villages have developed along the provincial road that connects the provinces of South and East Kalimantan. Small villages are scattered throughout the district, but mainly on the plain and along the coast. Many of these villages grew in the 1970s and 1980s when large oil palm plantations were set up in the area, which needed workers.

Pasir has been a migration destination for centuries. The plantation industry added a considerable number of Javanese migrants, as well as groups from Sumatra, Central Sulawesi and Nusa Tenggara. In the early twenty-first century, the coastal plain around Tanah Grogot is multi-ethnic, with sizeable minorities of Paserese, Buginese and Javanese. The indigenous population of Pasir, the Paserese (*Orang Paser*) do not stand out among the Javanese and Buginese in the coastal area, where Muslim Malay influence has been strong for centuries. Away from the coast, however, the plain rises into the steep Gunung Lumut Mountains. Although the area covers some 25 per cent of Pasir's territory, only around 10 per cent of its population lives there.[3] Very few migrants settled in the mountains. The terrain was not suitable for extensive plantation cultivation, and the logging companies that worked in the area needed mobile, not settled, employees. The steep inclines and lack of good roads, combined with a population known for its fierceness and black magic skills, made the mountains an unpopular destination among migrants. Communities living here are ethnic Paserese with a few individual migrants, often men married to local women, in their midst. Here people believe themselves to be related to Dayak groups living further inland, and although most Paserese in the mountains are Muslims, many still adhere to animist beliefs as well.

The various Paserese communities claim specified territories as their ancestral *adat* lands. These territories border one another, thus leaving no unclaimed land between them. The villages are governed by both official (state-recognised '*dinas*') village heads and customary *adat* leaders, and many individuals feel that they have little need for more government involvement in local affairs. The villagers practise swidden cultivation in which fields are cleared in the forest, used for one or two years, and then left fallow, allowing the forest to regenerate.

Among the mountain communities the district administration is a real, yet remote, authority. Past experiences with the government-approved logging companies have instilled the notion that the central and district governments hold more power than the community.[4] In the New Order period, these communities had no leverage with the district government and hence saw no option but to passively watch the logging of their forests. Most accepted the compensation offered by the companies in exchange. Although regarded as too low, these payments were considered to be better than nothing at all. After the loggers moved on, the communities reclaimed control of the land, but with lingering distrust towards government. As no repercussions or objections followed, the general feeling in Gunung Lumut is that the district government tacitly agrees to these *adat*-based land claims. The political climate since *reformasi* and decentralization and the current attitude of the government towards the mountain communities seem to support this notion. Nonetheless, each community carefully monitors what goes on in its territory: not only might the government take land, encroaching neighbours can be a threat as well.

Protecting Mului's forest

The village of Mului is located some three hours by car from the main provincial road. It stands on a hilltop with two rows of houses alongside a logging road that passes through the village. The village is named after the Mului river, which crosses some 8,600 hectares of *adat* land claimed by the community. The relatively abundant land and forest sustain the practice of swidden agriculture using slash and burn methods. The forest and unused land in the vicinity of Mului mean that fields are almost never further than half an hour's walk from the village. Rice and vegetables are grown mainly for local consumption, but if transport is available surplus is occasionally sold in the market towns.

For much of their history the Orang Mului[5] did not live in a village, but in individual houses spread across their territory.[6] Following the implementation of the 1979 Village Government Law, the people of Mului were required to settle in a new village together with the people of the nearby Swan and Solutong rivers, who were traditionally subordinate to the leaders of Mului.[7] The village was to be named Swan Solutong and its administration consisted entirely of Swan and Solutong people. Many Orang Mului refused and instead built their own village at a distance of some two kilometres from Swan Solutong (see Map 5.2). This settlement had no formal village status and was led by Mului's *adat* head, Pak Lindung. In 1987 the uneasy situation led to an open conflict, when the Orang Mului built a new school in their own village. On the advice of the Swan Solutong village head, the local government refused to provide a teacher and rebuked Pak Lindung for Mului's obstinacy. In reaction a large number of the Orang Mului

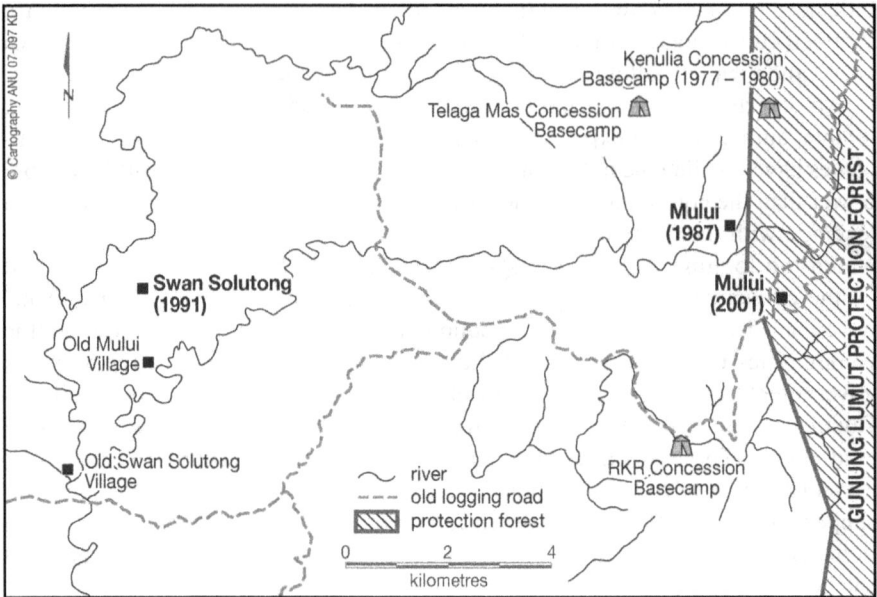

Map 5.2 Mului settlements, Pasir District, East Kalimantan Province.

led by Pak Lindung left their village and built new houses along the banks of the Mului river, some ten kilometres away. In 1991 Swan Solutong was designated as a transmigration site and Javanese and Bugis migrants were brought to the area in order to assist in its development. The government asked the people from Swan Solutong and those from Mului to settle in the new Swan Solutong village. The Mului people living along the river were still unwilling to forget the traditional hierarchy between the villages and declined, but the villagers of Swan Solutong including some 15 Mului families that had remained in the village accepted. They were given the same assistance as the newly arrived migrants and received houses, seeds, utensils, clothing and money.

   The Mului people living upriver decided that they also wanted to partake of this assistance and that they should not be isolated, but would not either become subordinate to the Swan Solutong government. This uncompromising attitude initially earned them a reputation among the local government for foolhardy stubbornness in clinging to a 'primitive' way of life that was doomed to disappear, but gradually more positive perspectives on continuity of traditions became part of the group's public image. In the late 1990s, with the onset of decentralization in full swing, the Mului were assisted by PADI, an indigenous peoples' NGO from Balikpapan, in writing a request to the provincial governor for aid to resettle a former log yard that was located within the boundaries of their traditional territory along a nearby logging road. This approach bypassed the Swan Solutong village and Pasir district government. The governor agreed to the request and instructed the

Tanah Grogot welfare department to set the project up and carry it out. By 2001, a new village of 52 uniform houses, with a school and a mosque, was populated by the Orang Mului who had been living along the river, by then some 130 persons.

With the new village came new problems. In 1983 a sizeable part of Gunung Lumut had been declared a protection forest (*hutan lindung*) by the Minister of Forestry. The district level forestry department had failed to inform other district departments of this change in status (and the welfare department had neglected to inform the forestry department of the intended resettlement). As a result, the new village of Mului turned out to be built within the borders of the protection forest.[8] This status severely restricts human interference in the area, and does not allow for habitation.[9] However, the forestry department reported their objections only three years after the settlement was built and the department lacked the means to take direct action to remove the newly settled Mului people. The village thus remained in an illegal location. The people of Mului, meanwhile, stressed not only that the protected forest overlaps with their *adat* territory, but also the green approach their *adat* has towards the environment.

They have a reputation in that regard. When, in the early years of decentralization, disagreement between local governments and the Ministry of Forestry on the validity of locally issued logging permits brought logging companies to negotiate directly with communities, Mului caused a stir in the region by turning all loggers down. I was given two different reasons for this. The first, by the people of Mului, was that the logging would go against their *adat* and, since they had a say in it now, they would do what they could to prevent it. The second reason, given in a neighbouring village, was that the forest of Mului had already been logged in the 1980s and did not contain good wood. The neighbours believed that the people of Mului simply wanted too much money. Strategically speaking, however, the stance of the Orang Mului was a wise one. Essentially living in an illegal location, logging the protection forest would have added to their reputation as self-interested malcontents, whereas refusing logging operations in the area because of their *adat* associated them further with the growing environmental and indigenous peoples' movements.

Whereas many other villages allow small-scale loggers to work in their *adat* forests, the people of Mului forbid everyone to work in theirs. To cut down a few trees for one's own needs is permissible if the plan has been discussed and does not meet with objections from the community, but no one may cut down trees for private financial gain. In this the status of wood differs from that of other forest products that can be gathered by anyone and sold off if the individual wants to do so. The prohibition on logging was established when the people saw the forests of neighbouring villages that did allow logging disappear at an alarming rate.[10] Financial enticements and a weak position under national law made many communities agree to the felling of large stretches of forest, whereas a relative abundance of forest and vague boundaries meant that no clear *adat* regulations against such decisions could easily be drawn upon.

In Mului the forest is the main provider of all requirements – food, building materials or products to sell when cash is required. Logging the forest would provide a lot of cash in a short time, but destroy future harvests for the current generations and for their children. The forest is, as the assistant *adat* head called it, their 'insurance' for when disaster strikes and a direct supply of resources is needed.[11] Although the decision not to log leaves Mului less well-off financially than some of its neighbours, the steady flow of forest products and the self-supporting abilities of the community means the people have little urgent need for cash.

This stance, and the people's stubborn refusal to move or to cooperate with outside investors attracted both government and NGO attention. In a recent film (*Balai Penelitian dan Pengembangan Kehutanan Kalimantan*, 2005) by the provincial forestry department, the department's commentator describes the people of Mului as *masyarakat adat* who had been living in the area for hundreds of years before it was designated as a protected forest, and goes on to have Jidan, Mului's assistant *adat* head, explain why it was really for the best of all that the people did not join the Swan Solutong resettlement. Jidan states:

[They said] In the future, things might get difficult, many [people] will come. We said ... I said – I alone answered at the time – we ask, I said, [to live] just at the TPK above.[12] Although the place is far from water, the environment there will still be guarded, we will guard it, I said. Ya, [it will] not [be] disturbed by outside activities. I mean company people, people who damage the protected forest. I said we, starting now, automatically guard this forest. If it is not guarded, we ourselves will guard it to help the government that guards it. If, I said, we were to go very far from Gunung Lumut, Gunung Lumut would not be safe.

Another film made by PADI (2005), presents the same message in a much more direct way. The opening scene has Jidan sitting in the forest stating 'The forest is our right. In the Paserese language that means "*alas adalah milik kain*"', which is the title of the film. Further on we see a gathering of the villagers and Pak Lindung states, to general consent 'thus we do not like it when [Gunung Lumut's forest] is logged by companies or for illegal income'. Pak Jompu, one of the village elders, adds that forest usage needs to be agreed to by the people of Mului in a public deliberation session. 'If the forest is logged without proper and honest consultation, the forest will be destroyed' he says 'we wouldn't get anything. We will not get anything when our forest is controlled by other parties.' Jidan is sitting in the background wearing an AMAN t-shirt.[13] Although the two movies largely show the same picture, the message varies. In the first one, the Orang Mului help the government guard the protection forest of Gunung Lumut; in the second one emphasis is on their own rights and interests in that forest. Both get the point across, but the different approaches of the makers make one wonder whether each would agree to the line of argument in the other's film.

Formally Mului is a hamlet subordinate to the official village government of Swan Solutong. Formal government has little influence in Mului, however. First because Pak Lindung, the *adat* head and Jidan, the assistant *adat* head are both influential and outspoken individuals who live in Mului, whereas the official village head lives in Swan Solutong. Second, because the population prefers to decide matters through communal discussion to following an individual leader's opinion, regardless of whether that leader's position is based upon governmental or *adat* authority.[14]

Although they govern their daily village affairs according to their own *adat*, the Orang Mului realise that they cannot ignore the state. The nearest government official is the village head of Swan Solutong, a village with about four times as many inhabitants as Mului, of which the majority are transmigrants with no understanding of local *adat*. Yet the village head of Swan Solutong has for several years now been a local Orang Paser, the current one being a stepson of Pak Lindung. Preferring to have one of their own as the local government representative, the Orang Paser of Mului and Swan Solutong voted together *en bloc* to elect a Paser candidate as village head, and so have managed to retain the strongest possible position for local *adat*.

Application of *adat* means that it must be continuously adapted, updated and redefined in consideration of recent developments. An example is the punishment that had to be established for selling logs from Mului's forest. The usual way of punishing someone who sells communal property for his own gain without the consent of the community is by fining him. A fine may consist of goods, foodstuffs, or – since the perpetrator is likely to have recently come into possession of a substantial amount of it – money. The two *adat* leaders, some elders and other villagers made up an *ad hoc* council and, basing their decision on a historic case, concluded that the fine should be two times four *reals*, a *real* being thought of as an ancient coin, either of silver or gold. Nobody knew the present value of a *real*, but by referring to the scarcity of *reals* in the area it was quickly decided that a *real* would at least equal one million modern Indonesian *rupiah*.[15] In the discussion, Pak Jompu related how a nearby village had asked government assistance in solving a conflict they had with a logging company that worked in their *adat* forest without their consent. A government mediator arrived and when it was established that the village had an *adat* fine of Rp 1,000 for every tree that was felled, the loggers' representative could hardly withhold his laughter and gladly promised that his company would pay this fine each time they transgressed local *adat*. The mediator helped the parties settle on a fee of Rp 15,000 per cubic metre of wood that the loggers extracted from the forest.[16]

The council in Mului therefore concluded that it was of vital importance to adapt historical fines to their present equivalent, as well as to the perpetrator's identity and profits. The council initially intended to establish a standard fine, but dropped that idea when it was argued that there were too many relativities to be taken account of among perpetrators. The fine should be higher than the money made by the perpetrator, but a logging company would laughingly pay a fine that a local villager would not be able to pay in a lifetime. The composition and operation of the *adat* council thus is a relatively informal affair. The *adat* head needs to be there to sum up the final conclusion; nobody else's presence is essential, but all community members are free to join in. That does not mean, however, that all voices are equal.

Charismatic and tenacious speakers as well as older people with a lot of experience are held in high regard. This goes for both men and women, although few women and younger people are inclined to engage in these long discussions. One must have time, as well as be sufficiently involved or interested. Young people and women often have daily chores to attend to and simply do not have the time for day-long deliberations if not absolutely essential, whereas older men often are exempted from mundane tasks or are able to delegate these to their wives or children.

The territory of Mului borders the territories of various other Paser villages (Pinang Jatus, Kepala Telake, Rantau Layung and Long Sayu). The local population has a clear notion of the borders of their *adat* territories, but these differ from the village borders registered by the local government. This rarely leads to problems as the villagers refer to *adat* borders rather than to the official borders, but the different stances with regard to logging are a potential source of conflict. The protected forest area, for instance, is a government construction that overlaps with the *adat* land of various villages. The borders of the protected forest have been marked out with white poles following Mului's relocation, but these poles are not controlled. Not all villages whose *adat* forest is (partly) taken up by the protected forest consent to its protected status and at various locations logs are taken from *adat* territories in the protection forest.

In August 2004, loggers from Rantau Layung were working only a few kilometres away from the border between the two *adat* territories. As the village of Mului is situated almost at the border of its territory, they were in fact working very near to Mului itself. This worried Pak Lindung since loggers have a reputation for looking at trees rather than at borders. He sent word to the loggers that he wanted them to show him exactly where they were working and they sent a truck to Mului to pick him up and show him the site. It was agreed by both parties that the logging was taking place in the *adat* forest of Rantau Layung, but Pak Lindung reminded the loggers to stay away from Mului's forest and retreat to two kilometres behind the border. After some negotiation the loggers agreed not to come closer to the border than one and a half kilometres.

At about the same time, the village council of Swan Solutong decided to give one of their villagers permission to start a small logging operation in the nearby forest. According to the official government map this forest belongs to the territory of Swan Solutong, but it falls within Mului's *adat* territory. In Mului the people were told that this individual wanted to fell two or three trees to be able to pay for repairs to his motorbike; but when two men went to investigate they found that six trees had already been felled, two additional persons had been hired, chainsaws and a truck had been brought in, and a camp had been set up. This caused great concern and the loggers were told to take the felled logs and then stop immediately. The next day Pak Lindung went to a market in a village along the provincial road and there ran into his stepson, the village head of Swan Solutong, who told him that he had not sanctioned the logging and had not been consulted about it. Nor had there been an official meeting of the full village council. The village head told his stepfather that he would sort things out when he got back to the village. This proved, however, to be unnecessary, as the logger had in the meantime visited Jidan on his motorbike and told him that as he was a migrant, he had been unaware that the forest was part of the *adat* territory of Mului and that he would take the trees he had already cut, eight by now, and otherwise leave the forest alone, to which Jidan consented. This decision was agreed to by Pak Lindung. Jidan and Pak Lindung explained to me that it was difficult to apply *adat* to immigrants, as these could successfully claim not to know it and were not part of the *adat* community. Things were easier if immigrants were more involved in local society, for instance through marrying a local woman or through business relations with local Orang Paser. As yet this was not the case in this instance. By letting him off they showed the Orang Mului to be generous and forgiving towards non-locals. However, the immigrant was now considered to be sufficiently knowledgeable regarding Mului adat and a second transgression would not be dealt with so lightly. The case was also thought of as instructive to other immigrants.

A few days later a car stopped in Mului from which alighted a delegation from the neighbouring village of Long Sayu, consisting of several prominent villagers, the *adat* head, and some friends of theirs from the village of Batu Kajang. Between Mului and Long Sayu lies a forest that belongs to Long Sayu; but as the people of Long Sayu originate from Mului, the people from Mului have a right to gather forest products in this forest. The people from Long Sayu had decided to open this forest to logging, but were required by *adat* to inform the people from Mului and, for the sake of good relations, ask their consent. Like Swan Solutong, Long Sayu had been subordinate to Mului in the past, yet Mului and Long Sayu had maintained close relationships throughout the years. The main dilemma the people of Long Sayu faced was not whether they would be allowed to carry out the logging or not (they had already decided themselves that they would); and in any case, national legislation invested the official (*dinas*) village government with the authority to take such decisions.[17] They counted upon the Orang Mului to be unwilling to risk a deterioration of relations between the two villages, but had come in numbers and prepared for a stiff round of discussion to be sure.

The people from Mului listened and politely argued against the entire idea of logging, but as this did not change their visitors' minds Pak Lindung concluded that he did not agree with Long Sayu's plans and would be much happier if they forgot about it, but that if they wanted to go ahead he would not stop them. This caused an uproar among some of the younger men. Jehan, a younger brother of Jidan, angrily raised his voice to argue that Pak Lindung was wrong to consent and that logging in that forest posed a further threat to the forest of Mului and its products, as the gardens and fruit trees in Long Sayu would doubtless be destroyed. Whom, he asked, would the people from Long Sayu turn to for wood, rattan and fruit once their own were gone? They would appeal to Mului he was sure, as the communities are related. A heated discussion raged for about an hour, the younger men unwilling to agree with Pak Lindung and Jidan. Then Pak Jompu, attracted by the noise, entered the house. A very senior, rather deaf (the debate was heated) elder in Mului, he listened to the arguments and spoke his mind: the logging could not be stopped, but the people from Long Sayu had to be very precise with regard to the border; the Orang Mului would keep a close eye on it. Jehan angrily left the room, but returned later to sit sulking against the wall. Neither he nor his friends voiced any other objections and the two groups parted on good terms. For Jehan and his friends, going against the venerable elder Pak Jompu was almost impossible. Going against Pak Lindung was possible through Jidan, the assistant *adat* head and Jehan's own brother. To a brother he could be quite direct and critical but less so to Pak Lindung, who was his uncle and the father of his wife. Jidan thus served as a convenient intermediary. Pak Jompu, however, was another matter. A more remote senior relative to Jidan, Jehan and Pak Lindung, Pak Jompu could not be criticised through an intermediary. Doing so directly would have been very impolite, especially in front of the people from Long Sayu, and Jehan was unwilling to go that far.[18]

The people from Mului have a reputation among their neighbours as good and tenacious debaters, especially when they are undivided. In January 2005 a delegation from the neighbouring village of Belimbing arrived. They wanted to plant oil palms along the border of their respective territories and came by to make sure both communities agreed on where that border was. A hill stands on the border, and the people from Belimbing argued that the border was thus the top of the hill. This was countered by Jidan and others, who argued that the hill indeed marked the border, but that it stood on the territory of Mului. The border, they said, lay at the foot of the hill on Belimbing's side. The matter got interwoven with other issues, and the two groups engaged in a long negotiating process. By the end of the second day all matters had been agreed upon, and the border determined on Belimbing's side, a little upwards from the foot of the hill. The people from Mului did however concede to the people of Belimbing the right to collect forest products and hunt on all of the hill.

Pak Lindung had a stroke in the summer of 2005, and lost much of his joy in life. When I returned in the summer of 2006 he had passed away. Jidan proudly told me that he now was the *adat* leader of Mului. This was disputed, although not to his face, by other, older, members of the community who were supported by, amongst others, Jidan's brother Jehan. However, to the outside world Mului presented a united front and continued the course of identifying themselves as a *masyarakat adat* community guarding the forest. Pragmatically speaking, they carry out the policy of protecting the forest that covers most of their territory. Although they manifest a keen independence, it is fully in line with forestry policy. This position ensures that they do not attract negative government attention and ensures a lack of government interference in local matters. It is difficult to say whether the same line of action would have been followed if the forest did not have the reinforcement of protected status under state law. The protection this status offers makes it worthwhile for the Orang Mului to invest in the maintenance and future of 'their' forest rather than cash in while they have the chance. On the other hand the latter action would pose difficulties for them since logging, especially in a forest that is protected by state regulation would substantially weaken their 'green' conservationist image and their credibility as *masyarakat adat*, preserving the forest because they depend on it. They would also risk the cultural independence and the degree of authority over forest management that this status validates.

Kepala Telake's riches

Some 27 kilometres to the north of Mului lies the village of Kepala Telake. The villages are separated by a number of steep mountains, thick forest and rivers. No road connects the two: travellers wanting to go from Mului to Kepala Telake may choose to walk for two days or go by car, descending first from Mului to the inter-provincial road, from where another logging road runs to Kepala Telake. A number of villages are located along this road, but after the first hour or so signs of habitation first become scarce and then disappear. To reach Kepala Telake takes three hours. The logging road, by 2005 an overgrown dirt track, ends as the village's main street.

Before my first visit I had expected Kepala Telake to be something like Mului, but even more remote and, based on local stories, even more traditional. Thus I was very surprised when we turned the last bend of the road early in the evening and found that the village had electric lighting and many of the houses had two floors and glass windows. Although by far the remotest village in the Gunung Lumut area, Kepala Telake is also one of the most prosperous. A conscious and selective usage of the opportunities that presented themselves upon decentralization allowed the population to make the most of their *adat* territory and forest.

At the end of the twentieth century, Kepala Telake was a tiny village of 15 families. Most young people moved away to the towns and villages at the foot of the mountains since Kepala Telake had no school and was isolated for most of the year. The village could be reached from the plain only in the rainy season, when the water in the river was high, and the boat trip would take three days. In 2000, a logging company received a permit from the district government to work in the area and constructed the road that ends in the village. The road shortened the trip to the plain to a mere three hours, and brought news of the changes in Indonesia's politics and the new administrative reforms. In the spirit of reform the villagers dismissed their unpopular village head, who had been appointed rather than elected and had made himself highly unpopular through corrupt practices and a general untrustworthiness. From their midst they elected a new village head, a trader named Pak Abbas who regularly travelled to the villages of the plain.

From 2001 onwards, Kepala Telake's economy developed spectacularly. The community claims some 56,200 hectares of land as their *adat* territory. Most of it is forest, but the territory also contains some deep limestone caves where birds' (*burung walet*) nests are collected which are in great demand by the Asian market.[19] Pak Abbas used to buy these nests from his fellow villagers and sell them on at the markets on the plain. When he became village head, he brought efficiency in the trade that considerably benefits the community. The trade in swifts' nests is not without danger. The caves are located half a day on foot from the village and gangs of armed robbers have attacked small groups of collectors on several occasions in the past, one time even killing a villager. Pak Abbas organised the construction of a sturdy, seven-metre high guard tower in front of the main cave. It overlooks the approach and the entrances to the other caves. This tower is continuously garrisoned by around 10 villagers armed with spears and machetes, and equipped with jerry cans of gasoline and battery acid to make impromptu bombs. All guards are paid a daily fee equivalent to a day's pay in the plantations of the plain by Pak Abbas as part of his trading operation. Those light and brave enough to scale the cave walls to collect the nests come to the cave once a month. The harvest is divided among climbers and helpers (who often include guards) and sold to Pak Abbas for prices between Rp 1 and 2 million per kilo.[20] The men of the village take turns working at the caves. All have their fields to tend to and most have families, which can make it difficult to be away for weeks at a time.

Pak Abbas clearly has a double role here. As a village leader he has to protect the interests of the community whereas as a private trader, he has his own profit to think of. As yet Pak Abbas appears to balance the two quite well. Satisfying himself with a lower profit than coastal traders, the immense profits made from swifts' nests allowed him to become the richest man of the village and gain the praise and support of the village population. In 2001, Pak Abbas requested the district authorities to formally recognise the exclusive *adat* right of the Kepala Telake community to collect swifts' nests in the caves. The officials asked him to bring along Kepala Telake's *adat* head to explain this *adat* right and its validity. This posed a problem. The *adat* head had died some years before and his traditional successor, his son, had moved to the plain and had no interest in the position. The community elected a new *adat* head from their midst. He was a young man who had migrated to the plain but had recently returned when news of the village's developments reached him. He knew very little of Kepala Telake's *adat*, but a group of elders who had never left the village took it upon themselves to act as his advisors and teachers. The *adat* head managed to convince the bureaucrats in Tanah Grogot of the contemporary strength of *adat* in Kepala Telake, and of the validity of their claim to the caves. The community received a monopoly on the exploitation of the caves on the condition that the government would receive a revenue tax of 25 per cent. This settlement has been very beneficial to Kepala Telake, as some form of recognition is now in place and the government's endorsement deters robbers at least as much as the guards in their fortress do.

By 2004, Kepala Telake's blossoming economy had inspired a considerable number of young people to return and the village had grown to 66 households. This younger generation had lived and married in villages on the Pasir plain, and a number of them brought non-Paser spouses to the village. With them came a number of non-Paserese friends and skilled artisans, who were attracted by Kepala Telake's wealth and requested permission to settle in the village. All three of Kepala Telake's carpenters, for instance, are Javanese and continuously engaged in the construction of new houses. A number of sturdy Bugis are nowadays included among the caves' watchmen.

As in Mului, the community's remote location makes visits by government officials extremely rare, which leaves the community pretty much to arrange matters as they please. Yet in Kepala Telake it is the official village head rather than the *adat* head who has the primarily leadership role in the community. He does so, however, in close deliberation with the community and long consultation sessions are held before decisions are taken. The new *adat* head has the task of mediating in local conflicts and chairing meetings in which requests to open up fields in the territory are discussed by the community. The relative abundance of *adat* land usually makes these requests unproblematic affairs, but especially the requests of immigrants are thoroughly considered as their unfamiliarity with the area's agricultural history may unwittingly have them request usage rights to land that is already part of another person's plot.

Logging in Kepala Telake differs radically from the situation in Mului. Whereas all of Mului's territory was logged during the authoritarian New Order regime, including large stretches of the area that would later become protected forest, logging in Kepala Telake only started in 2000. In 2002, Pak Abbas and the village council issued a village regulation in which it is decreed that every logging operation working in Kepala Telake's *adat* territory has to pay a fee per cubic metre to the community.[21] The height of the fee is to be determined in mutual deliberation. The logging company that constructed the road to the village had its district level permit revoked by the central government in 2004 on the ground that they had entered Gunung Lumut's protected forest, an accusation that is still being disputed by the company with Kepala Telake's support. The central government issued two new permits to the area in 2002, in which decision the community of Kepala Telake was not involved or consulted. One was issued to a company currently operating to the north of Kepala Telake's territory in the neighbouring district of Kutai Kartanegara, and one to the company that worked in and around Gunung Lumut in the 1980s. The company to the north will operate in a remote part of Kepala Telake's territory with fee rates yet to be determined; but having agreed to acknowledge Kepala Telake's *adat*-based authority over the area, villagers do not regard it as a threat to their fields and gardens. The other company has received restricted permission to work around the village. The *adat* head negotiated with them and they agreed that the loggers would only work in the regenerated area where they had already logged two decades earlier. In exchange, their fee was set at a meagre Rp 20,000 per cubic metre.[22]

Two other logging operations are run by people from Kepala Telake, and both pay a fee of Rp 50,000 per cubic metre to the community. These two operations work with village permits, and have access to the large old growth trees and expensive woods from which the outside government-licensed company is excluded. They laughingly refer to themselves as 'illegal loggers' since the Forestry Department would label them just that; but in the opinion of the community these people have more rights to log the local forests than any company with government permits. Both operations are relatively small, each harvesting some ten cubic metres of ironwood per month. Buyers from the plain, who come up by truck to the village, buy the wood for Rp 500,000 per cubic metre. Each operation provides work to a number of young men in the village and suffices to keep the village's three small sawmills in operation.[23]

The issue of sustainability does not play a direct role in Kepala Telake decision-making. Indirectly it comes to the fore in that people wish to have their rattan gardens, fruit trees, and other possessions in the forest preserved. The local loggers are especially hampered by this, as they are expected to respect all these properties in their operations and pay indemnifications if they damage anything.

The money the community receives from the various logging operations is used for community projects. All houses in the village are connected to a communal generator that provides the village with electricity in the evening. The fuel required, a fee for the machine's operator, and the wages of the two teachers working at the village school make up the community's invariable monthly expenses.[24] Surplus money is used for various one-off projects. During my stay in the village a pipeline network delivering water to all the houses of the village was in the final stages of construction. Essentially a simple affair of plastic pipes leading from a man-made pond above the village to all the houses, this project is nonetheless an unusual example of community initiative in Indonesia.

At the same time the village's streets were paved with stones brought up from a nearby river. Whereas such work would frequently fall under community service (*gotong royong*), the size of the project and the intensity of the work made it unlikely for the project to succeed if all labour had to be provided on a voluntary basis. Workers thus were paid a fee from the village treasury, and many continued the work for one or two weeks. Nowadays the overgrown logging track leading to Kepala Telake turns into a stone paved road just before entering the village. It passes a large number of new houses and old houses in various stages of extension, some saw mills, a few shops, and then comes to a brand new school building and a new office for the village head.

Towards the outside world the image of a remote, traditional community is carefully maintained. Not only does this help Pak Abbas in securing district government aid money for the development of the village, it also supports the community's extensive claim over *adat* territory. For instance, when negotiations were initiated with the logging company that will enter the territory from the north, the people from Kepala Telake took the opportunity to clearly establish the boundary with the community living in that area. The two groups erected a wooden effigy of a protective spirit at a central point of the border to commemorate their agreement, and held a ritual feast as prescribed by local *adat*.

Yet although *adat* has an important role in the daily life of Kepala Telake, and the people consider themselves as living according to *adat*, most are reluctant to refer to themselves as *masyarakat adat*, and have their reservations against joining the movement.[25] Together the village secretary and the *adat* head explained that the present usage of the forest and the swift caves would be difficult to sustain if Kepala Telake had to comply with the image of *masyarakat adat* popularised by AMAN and other NGOs. The presence of logging companies could be ascribed to the dominance of the state, but they felt that the local logging operations and the legal efficiency with which Kepala Telake had monopolized the exploitation of the swifts' caves went against the strict ideas on nature conservation promulgated by many of these NGOs. Moreover, they were quite content to cooperate with the district government and felt uneasy about the protesting and positioning they would have to undertake if they were to join the movement. They felt that the community had decided for itself how it used its *adat* and *adat* resources, and that joining the *masyarakat adat* movement would hamper the community's development. Moreover, Kepala Telake's economic situation attracts immigrants who are willing to adhere to local *adat*, but would feel excluded and put off if the community redefined itself as *masyarakat adat*. At present, adherence to *adat* means asking the *adat* head's permission to open up fields in the territory and respect his answer. The *adat* head, being asked such permission by everyone, practically functions as a local land agency and knows where everyone's fields, gardens and former fields are. Land that has been used in the past and has been overgrown again may still be subject to a customary claim. Migrants will not know this and think of it as free land, whereas the *adat* head will know differently. *Adat* permission thus gives migrants local legitimation of their land usage. Not being full members of the *adat* community, migrants' claims would lose their certainty if that community decided to start courting *masyarakat adat* identity under which the migrants risk exclusion and inherent loss of rights. Such full dependency would certainly discourage new migrants from joining the village and strengthening the local economy.

The Kepala Telake community feels certain of their *adat* claims, but especially the younger generation feels the need of keeping up good relations with the district government in order to sustain them. As many of them have lived outside of the mountains, they have realised that their *adat*-based resource access is a privilege that survives due to its remoteness from the plantations of the plain. Although the people of Kepala Telake are happy to be 'among themselves', many people aim to obtain a similar standard of living to that they know from the plain. Led by the village head, the village secretary and the *adat* head, Kepala Telake thus seeks to maintain the privileges of an *adat* community in having the sole right to exploit the natural resources of their *adat* territory and combine this with the benefits of developing large-scale market-oriented natural resource projects within their territory.

During my last visit in 2005, a plan had been launched by villagers to convert part of the forest into a palm oil plantation that would operate similarly to the birds' nest caves and the sawmills. Although received enthusiastically, the plan was not carried out due to a lack of funds, the poor state of the road and the remoteness of a production plant which would make export too expensive. In the summer of 2006, I met the village head again at a district government office. He enthusiastically told me that the plan might be carried out after all. Together with the heads of other villages along the road to Kepala Telake he had been lobbying for road improvements with the district government, which would greatly increase their possibilities for development. In the autumn, the district head visited all these villages to see their conditions with his own eyes. His visit coincided with the announcement by the district government of a plan to revitalise Pasir's agricultural sector with 20,000 hectares of oil palm and 10,000 hectares of rubber. The villagers jumped at the opportunity and requested that the district head award them the rubber concession, which has logistical advantages over oil palm. As of August 2008 the rubber concession has not been allocated, but government assistance would greatly improve the possibilities for Kepala Telake to develop a large plantation. The potentially problematic issue for Kepala Telake's *adat* territory would be avoided by locating the entire plantation on recently logged land that falls within the official village boundaries.

Concluding remarks

Successfully upholding communal claims to land and resources necessitates choices with regard to the authority evoked and the allies chosen. The commonweal in today's Gunung Lumut thus involves the articulation of national and local government polities and of nationwide networks operating in support of *adat* communities, not to mention market forces and the appeal of development options, adding strong non-local influences to what remains an essentially local process. In this chapter I have shown how two neighbouring communities who maintain similar claims, both attempting to build upon their local resources and circumstances, use different strategies in responding to the limitations and possibilities introduced by outside forces. Both invoke customary communal rights to a territory of land and the natural resources found upon it.

The communities maintain their relevance in safeguarding collective economic interests and the local identity, authority and privileges of the group, while accommodating state decisions such as the establishment of the protection forest area in the case of Mului and the issuing of commercial logging licences in that of Kepala Telake. Thus the local commonweal at Gunung Lumut is about collectively defined sustainability in the sense that each community attempts to maintain those natural resources under its control in such a way as to ensure future benefits and protect against the insecurity of a 'free for all' exploitation of resources by more or less effectively regulating forest access and use. The state is able to impose some limiting conditions in even the remotest locations, but these communities have much more room to impose their own conditions in the use of the forests for which they claim *adat* authority than they did under the New Order.

Moving between the inside and outside perspectives in the process of representation demands political insight and diplomacy of communities' representatives, but is decisive for their image and identity as well. Pak Lindung and Jidan of Mului, both rugged-looking men and not averse to strong statements, fit the *masyarakat adat* image much more than Pak Abbas' diplomatic trading negotiations and those of Kepala Telake's inexperienced *adat* leader. Both groups of leaders need to have and maintain the support of their respective communities. In Mului this occasionally leads to heated debates, challenging the correctness of decisions. Kepala Telake's leaders have more or less divided their fields of authority into *adat* and non-*adat* related issues; thus removing *adat* from sight where it is potentially inconvenient but keeping its potentially legitimating arguments at the ready to fall back upon.

The histories of the communities and the discussion of the recent cases taking place there are illustrative of the locally driven character of this process and of how remoteness and high levels of local solidarity vis-à-vis the outside world enable it. The broader political ecology of their situation is adapted in relations with non-local influences, notably the government, commercial interests and *masyarakat adat* NGOs, but rights and privileges are maintained and given meaning in contacts with neighbouring communities. *Adat*, which is given as the basis for communal claims in all of Gunung Lumut's communities, thus comes to the fore as an argument to substantiate or refute claims and to resist undesirable influences from outside. In Mului *adat* has been put forward since the first proposed resettlement of the community, and its *masyarakat adat* image proved a useful tool when the forestry department protested its new location. In Kepala Telake no such objections to government policies has ever taken place, but here as well *adat* is the community's source of legitimation for claims to natural resources and ensuing rights such as logging fees.

In Gunung Lumut the commonweal thus consists of an intricate mixture of local practices that are honed and reshaped in a dialogue with outside influences. The essence is to maintain the communal claim and govern through the customary norms that underlie it. In principle it is the welfare of the community rather than that of individuals which is the main concern of *adat*, but as the national state poses limits to the communal claim, so does it offer opportunities and trade-offs. The question for the future in both Mului and Kepala Telake is to what extent this forward looking priority of common interest will be maintained through the inclusive and negotiative *adat* grounded processes that underpin collective decision-making in the two communities to date. Critical to this will be the degree to which local interests and identities are able to establish a mutual accommodation within wider spheres of governance.

Acknowledgements

This paper is based on research carried out in the districts of Pasir and Nunukan by the author from 2004 to 2006. The research is part of the Indonesian-Dutch INDIRA project and of the Indonesian-Dutch Tropenbos Gunung Lumut Biodiversity research project. The author would like to thank Fakultas Kehutanan of Universitas Mulawarman, Lembaga Ilmu Pengetahuan Indonesia, Tropenbos International Indonesia and the Van Vollenhoven Institute of Leiden University for their kind support of the research. Financial support was provided by the Royal Netherlands Academy of Arts and Sciences (KNAW), the Treub Foundation, the Netherlands Foundation for the Advancement of Tropical Research (WOTRO) and the Adatrechtstichting.

Notes

1 Tsing's (1993: 8, 14–17) usage of 'marginality' captures this well: the mountain communities exist in the margins of the Paserese society of the coastal plain where the government is located; the coastal plain (and the government) exist on the margins of life in the mountains.
2 UU 5/1979, which was replaced by UU 22/1999 on administrative decentralization.
3 Personal estimate based on data of the Pasir district statistics agency.
4 The logging licences to the area were issued by the central government, yet in the mountains the district government was perceived as equally responsible in the matter.
5 Literally: 'Mului people', the community sees itself as a sub-group of a sub-group of Paserese. The actual number of Paserese sub-groups, which differ in dialect and *adat*, is unclear but Yusuf's (2004: 3–4) estimate seems to be the most accurate to date. 'Paser' is the local equivalent of 'Pasir' in Indonesian. Hence Orang Paser who live in the district of Pasir.

6 This rendition of Mului's history is based on accounts of past events given by Orang Mului.

7 Local tradition has it that the people of Mului arrived first in the area, and gave permission to the ancestors of the people of the Swan and Solutong rivers to settle in the area as well. This precedence of the area's first settlers gives the people of Mului a senior status, according to *adat*.

8 The protection forest was established by Ministry of Forestry Decree No. 24/Kpts/Um/1983.

9 Protected forests are supposed to function as 'green lungs' that convert carbon into oxygen. Only activities that are considered as not interfering with forest conditions are allowed, such as collection of non-timber forest products (but not hunting) or environmental tourism (see also Riyanto 2006: 203–7).

10 A vivid warning is a stretch of land belonging to Swan Solutong, lying between that village and Mului. It has been logged thoroughly in the past and became covered in *alang-alang* grass rather than recuperating forest. The land is quite infertile, and has been given out to migrants who planned to use it for oil palm after it was no longer possible to grow food crops on it. But as they lacked the financial means to invest in oil palms, some of the migrants have already left for other areas.

11 An assistant *adat* head (*wakil kepala adat*) is not common in Gunung Lumut, and rare throughout Indonesia. A few reasons were given for there being one in Mului. Initially I was told that the *adat* head was an old man, who might not live much longer. Although true, Pak Lindung was a strong, muscled and healthy man. Pak Lindung himself pointed out that as Mului's *adat* had become essential in defending the community's land against intrusions from outside, he simply needed the support of another expert as he could not be everywhere at the same time. A third reason, given by Pak Lindung's brother, was that Pak Lindung had received death threats from angered neighbors and illegal loggers on a few occasions. Being a fiery personality, Pak Lindung instantly countered these threats by stating that he would kill them first, but having an assistant *adat* head spread the risk. In general, Pak Lindung has the final say in interpreting adat, but the assistant head has all the authorities of a second-in-command.

12 *Tempat penimbunan kayu* (TPK), a log depot, referring to the log yard mentioned earlier, which is located along the top of a hill.

13 Yet the Orang Mului do not cooperate automatically with *adat* NGOs. Elsewhere (Bakker 2005) I discuss how the people of Mului openly opposed an *adat* Paser NGO from Tanah Grogot that had been donated machinery left behind by the logging company that had worked in and around Mului. The machinery represented several tons of scrap metal, and thus a hefty sum. The NGO intended to sell it off and use the profits for its projects, but the people of Mului considered the metal as their *adat* property since it lay on their *adat* land.

14 The authority of an *adat* leader has always been limited. He has a great knowledge of local *adat*, its history and its various rules, but his authority is like that of the chairman of an assembly. Executive domination of state instituted government structures on the other hand gave state supported village heads much more individual authority until the Reform Era. See Warren (1993) for similar examples of local understandings of *adat* and state-based leadership in Bali.

15 The East Kalimantan historian Assegaff (1982: 149) writes that in the first half of the nineteenth century the sultan decided to coin his own money since the diverse kinds of foreign money in circulation were confusing and inefficient. The sultan's coins were called *real* and made of silver. The value is unclear: a *real* equalled 2 *batu*, or 4 *suku*, or 10 *tetep*, all silver coins; or 25 *uang*, 40 *gobang* or 100 *picis*, which were copper coins.

16 Such mediation would never have taken place under the New Order. As loggers nowadays need to pay considerably to get police or army support, it is often more efficient for them to reach a deal with the local population. Government mediators are

sometimes brought in to speed up the process and give the company official support. The expenses to obtain the assistance of a government official for one or two days are considerably less then bringing in a group of soldiers for a period of time.

17 Their main dilemma was to decide whether it would be more profitable to allow a large logging company to extract the wood in exchange for fees or to use small-scale loggers who might pay a better price but are not as reliable.

18 Similar disagreements occur often, and it is important to recognise that leaders in most contexts are not vested with unqualified authority. In the case of the remnant logging equipment left on Mului *adat* territory mentioned earlier (see footnote 14), Pak Lindung and Jidan had agreed after contentious negotiations to allow the scrap metal to be sold through the *adat* NGO. But upon return to the village from that meeting the *adat* leaders were met by outraged villagers who forced the NGO representative to tear the agreement up (see Bakker 2005).

19 These birds are the White-nest Swiftlet (*Aerodramus fuciphagus*) and the Black-nest Swiftlet (*Aerodramus maximus*).

20 One kilo may contain as much as 100 nests. A Chinese restaurant in Balikpapan sells birds' nest soup containing one nest per portion for Rp 250,000 per bowl. By the time the nests reach Balikpapan they have passed hands at least three times. Most nests are exported.

21 Kepala Telake Village Regulation No. 1 of 2002. As far as I am aware this is the only regulation ever formally drawn up by the village council. The bureaucratic procedure and the amount of work involved in passing one make *adat*-based deliberation far more efficient. As this regulation concerned outsiders it was based upon the village administration's formal authority rather than *adat* claims, which are more open to contestation.

22 The fee for the company in the north had not yet been determined, but was expected to be considerably higher.

23 The money made inspired yet another villager to buy a small truck and start a daily shuttle service to the lowland markets. A few fishermen, an intrepid greengrocer and a small number of mobile snack sellers drive up to the village on motorbike from the plain a few times a week to sell their goods along the road. Often they start in Kepala Telake, since people there pay the highest prices.

24 These amount to Rp 1.4 million.

25 I need to point out, however, that the term was not known to about 20 per cent of my respondents. Many of the village leaders, among them the *adat* head, and younger people did not see themselves as *masyarakat adat*. Some of the older villagers, notably three out of the seven that assist the *adat* head, on the other hand felt sympathy for the concept and would support a *masyarakat adat* movement in Kepala Telake.

Bibliography

Acciaioli, G. (2003) 'Re-empowering the "art of the elders": the revitalisation of adat among the To Lindu people of Central Sulawesi and throughout contemporary Indonesia', in M. Sakai (ed.), *Beyond Jakarta: Regional Autonomy and Local Societies in Indonesia*. Adelaide: Crawford House Publishing. pp. 217–44.

Antlöv, H. (2003) 'Village government and rural development in Indonesia: the new democratic framework', *Bulletin of Indonesian Economic Studies* 39 (2): 193–214.

Assegaff, A. S. (1982) *Sejarah Kerajaan Sadurangas Atau Kesultanan Pasir*, Tanah Grogot: Pemerintah Daerah Kabupaten Tingkat II Pasir, Tanah Grogot.

Bakker, L. (2005) 'Resource claims between tradition and modernity: *Masyarakat adat* strategies in Mului (Kalimantan Timur)', *Borneo Research Bulletin* 36: 29–50.

Bedner, A. and Van Huis, S. (forthcoming) 'The return of the native in Indonesian law: indigenous communities in Indonesian legislation', *Bijdragen tot de Taal-, Land-, en Volkenkunde* 164 (2/3): 165–193

Blackwood, E. (2001) 'Representing women: the politics of Minangkabau adat Writings', *The Journal of Asian Studies* 60 (1): 125–49.

Burns, P. (2004) *The Leiden Legacy: Concepts of Law in Indonesia*. Leiden: KITLV Press.

Campbell, J. (2002) 'Differing perspectives on community forestry in Indonesia', in C. Colfer and, I. Resosudarmo (eds), *Which Way Forward? People, Forests, and Policymaking in Indonesia*. Washington: Resources for the Future.

Casson, A. and Obidzinsky, K. (2002) 'From New Order to regional autonomy: shifting dynamics of "illegal" logging in Kalimantan, Indonesia', *World Development* 30 (12): 2133–51.

Daryono (2004) 'The co-existence of state land law and local legal practices (adat): implementation of Basic Agrarian Law', *Jurnal Studi Indonesia* 14 (2): 118–36.

Eghenter, C. (2006) 'Social, environmental and legal dimensions of *adat* as an instrument of conservation in East Kalimantan', in F. M. Cooke (ed.), *State, Communities and Forests in Contemporary Borneo*. Canberra, ANU E Press, pp. 163–80.

Henley, D. and Davidson, J. (2008) 'In the name of *adat*: regional perspectives on reform, tradition and democracy in Indonesia', *Modern Asian Studies* 42 (4): 815–52.

Ihromi, T. O. (1994) 'Inheritance and equal rights for Toba Batak daughters', *Law and Society Review* 28 (3): 525–37.

Jacobsen, M. (2002) 'On the question of contemporary identity in Minahasa, North Sulawesi Province, Indonesia', *Asian Anthropology* 1: 31–58.

Katoppo, A. (2000) 'The role of community groups in the environment movement', in C. Manning and P. van Diermen (eds) *Indonesia in Transition: Social Aspects of Reformasi and Crisis*. Singapore, Insititute of Southeast Asian Studies.

Li, T. M. (1999) 'Marginality, power and production: analysing upland transformations', in T. M. Li (ed.), *Transforming the Indonesian Uplands: Marginality, Power and Production*. Amsterdam: Harwood Academic Publishers, pp. 1–44.

Li, T. M. (2000) 'Articulating indigenous identity in Indonesia: resource politics and the tribal slot', *Comparative Studies in Society and History* 1: 149–79.

Li, T. M. (2001) 'Masyarakat adat, difference and the limits of recognition in Indonesia's forest zone', *Modern Asian Studies* 35 (3): 645–76.

McCarthy, J. (2000) 'The changing regime: forest property and *Reformasi* in Indonesia', *Development and Change* 31: 91–129.

McCarthy, J. (2004) 'Changing to gray: decentralization and the emergence of volatile socio-legal configurations in Central Kalimantan, Indonesia', *World Development* 32 (7): 1199–223.

Morrell, E. (2001) 'Strengthening the local in national reform: a cultural approach to political change', *Journal of Southeast Asian Studies* 32 (3): 437–49.

Persoon, G. (1998) 'Isolated groups or indigenous peoples', *Bijdragen tot de Taal-, Land-, en Volkenkunde* 154 (2): 281–304.

Pratikno (2005) 'Exercising freedom: local autonomy and democracy in Indonesia, 1999–2001', in M. Erb, P. Sulistiyanto and C. Faucher (eds), *Regionalism in Post-Suharto Indonesia*. London: Routledge Curzon, pp. 21–35.

Resosudarmo, I. (2003) 'Shifting power to the periphery: the impact of decentralisation on forests and forest people', in E. Aspinall and G. Fealy (eds), *Local Power and Politics in Indonesia: Decentralisation and Democratisation*. Singapore: Institute of Southeast Asian Studies, pp. 230–44.

Resosudarmo, I. (2007) 'Closer to people and trees: will decentralisation work for the people and the forests of Indonesia?', *European Journal of Development Research* 16 (1): 110–32.

Rhee, S. (2000) 'De facto decentralization during a period of transition in East Kalimantan', *Asia-Pacific Community Forestry Newsletter* 13 (2): 34–40.

Riyanto, B. (2006) *Hukum Kehutanan dan Sumber Daya Alam*. Bogor: Lembaga Pengkajian Hukum Kehutanan dan Lingkungan.

Rosser, A., Roesad, K. and Edwin, D. (2005) 'Indonesia: the politics of inclusion', *Journal of Contemporary Asia* 35 (1): 53–77.

Roth, D. (2006) 'Losing in court, winning in the field: land reform between customary tenure, project law and state regulation in Indonesia', paper presented at the conference, Les Frontières de la Question Foncière [At the frontier of land issues]. Montpellier, May 6.

Safitri, M. (2005) '*Tenure Security*, Sebuah arena pluralisme hukum di kawasan hutan negara', in Yayasan Kemala *et al.* (eds), *Tanah Masih di Langit: Penyelesaian Masalah Penguasaan Tanah dan Kekayaan Alam di Indonesia yang Tak Kunjung Tuntas di Era Reformasi*. Jakarta: Yayasan Kemala.

Sardjono, M. A. and Samsoedin, I. (2001) 'Traditional knowledge and practice of biodiversity conservation: The Benuaq Dayak Community of East Kalimantan, Indonesia', in C. Colfer and Y. Byron (eds) *People Managing Forests: The Links between Human Well-Being and Sustainability*. Washington: Resources for the Future, pp. 116–34.

Schulte Nordholt, H. (2007) 'Bali: an open fortress', in H. Schulte Nordholt and G. van Klinken (eds), *Renegotiating Boundaries: Local Politics in Post-Suharto Indonesia*, Leiden: KITLV Press, pp. 387–416.

Sillander, K. (2004) *Acting Authoritatively: How Authority is Expressed through Social Action among the Bentian of Indonesian Borneo*. Helsinki: Swedish School of Social Science, University Of Helsinki.

Supomo, R. (1953) 'The future of adat law in the reconstruction of Indonesia', in P. W. Thayer (ed.), *Southeast Asia in the Coming World*. Baltimore: The John Hopkins University Press, pp. 217–35.

Tsing, A. (1993) *In the Realm of the Diamond Queen*. Princeton: Princeton University Press.

Tsing, A. (1999) 'Becoming a tribal elder, and other green development fantasies', in T. M. Li (ed.) *Transforming the Indonesian Uplands: Marginality, Power and Production*. Amsterdam: Harwood Academic Publishers, pp. 159–202.

Warren, C. (1993) Adat *and* Dinas*: Balinese Communities in the Indonesian State*. Kuala Lumpur: Oxford University Press.

Warren, C. and McCarthy, J. (2002) 'Customary regimes and collective goods in Indonesia's changing political constellation', in S. Sargeson, (ed.), *Collective Goods, Collective Futures in Asia*. London: Routledge, pp. 75–101.

Yusuf, H. M. (2004) *Adat dan Budaya Paser*. Samarinda: Biro Humas Pemerintah Propinsi Kalimantan Timur.

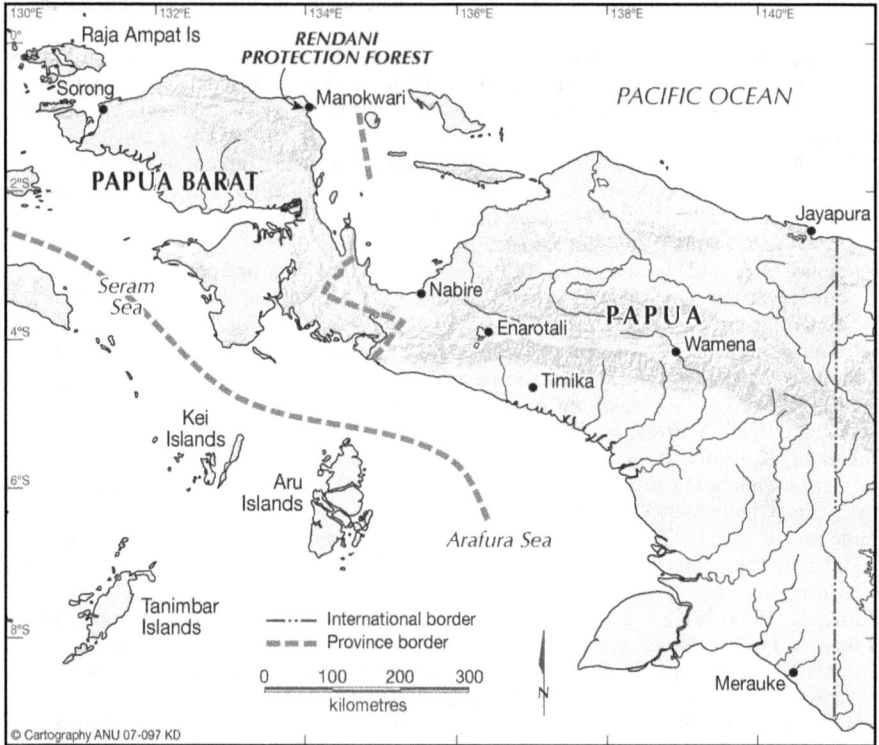

Map 6 Papua and West Papua (Papua Barat) Provinces.

6 Forests for the people?

Special autonomy, community forestry cooperatives and the apparent return of customary rights in Papua

Hidayat Alhamid, Chris Ballard
and Peter Kanowski

'Forests for People' was the cheerily optimistic slogan employed at the Eighth World Forestry Congress held in Jakarta in 1978, which also issued the Jakarta Declaration, subsequently proclaimed as a 'turning point in the history of forestry and in the evolution of forestry's contribution to social and economic development in general and to the well being of rural people in particular' (FAO n.d., Eighth World Forestry Congress 1978). Yet during the two decades that followed, until the end of the Suharto era, benefits from the exploitation and management of Indonesia's forests were decidedly not for the 'people' (Barber 1989). The period of political reformation or *reformasi* since 1998 has witnessed moments of similar optimism, with the promise of regional autonomy and a redistribution of the benefits from natural resource exploitation from the centre to the provinces and districts. But are these expectations any more realistic than those of 1978? The case of Papua's forests provides some indication of the possible trajectories for change in the post-Suharto period; and yet Papua is also a special case, attracting specific attention and provisions from Jakarta in the form of a Special Autonomy law. How has *reformasi* played out in the forestry sector in Papua in terms of the flow of sustainable benefits to the 'people' and what, if anything, can be identified as 'special' about Papua's Special Autonomy?

On 21 November 2001, Law No. 21/2001 on Special Autonomy (*Otonomi Khusus* or Otsus) for the Province of Papua was enacted by the president after being passed by the Indonesian Parliament. Many observers, both in Indonesia and overseas, hailed this move as a significant step forward in finding a peaceful solution to the long period of antagonistic relations between the people of Papua and the central government of Indonesia in Jakarta. One of the defining characteristics of Law No. 21 is the protection of the rights of indigenous people over natural resources, encompassing the forestry sector. This chapter describes and seeks to understand changes to forest management and forestry practices in Papua since the enactment of Special Autonomy (Otsus).

The actual implementation of Otsus for Papua has been greatly delayed, and we can do no more here than indicate some of the probable future directions and challenges for this unfolding process. However, a brief review of the conception and negotiation of Otsus and of the details of its final form sets the stage for consideration of the re-emergence of *adat* or traditional custom as the figure of the queen in the rhetorical chess of Papuan politics. A closer analysis of the way in which appeals to *adat* are being deployed in the forestry sector identifies one of the principal outcomes of Otsus thus far: the emergence of indigenous customary community cooperatives or Kopermas (*Koperasi Masyarakat Adat*) as a new vehicle for both industry and indigenous aspirations. For its case material, this analysis draws primarily on doctoral fieldwork by the first author during 2001 in and around the Rendani Protection Forest, adjacent to the city of Manokwari in the Bird's Head Peninsula region of Papua (Alhamid 2005; see also Alhamid, Kanowski and Ballard, in press).

Special Autonomy (Otsus) in Papua

Otsus has its genesis in the social and political forces driving 'decentralization' and 'autonomy' in Indonesia generally, and in Papua in particular, since Suharto stepped down in May 1998. In this new era of decentralization, natural resource governance issues are fundamental, especially in Papua, where many people still depend on the forest and the land for economic activity and livelihoods. Thus, in addition to its commitments to human and cultural rights, Law No. 21/2001 also altered the arrangements for the governance of natural resources and the structure of distribution of natural resource-derived revenues between Papua Province and the central government.

Otsus for Papua Province has the stated intention of improving the prosperity of Papuan *adat* communities, through granting of wide authority to regulate the benefits of natural wealth to provincial, district and municipal governments as well as to the *adat* communities of Papua. The introduction of Otsus acknowledged that 'the management and benefits of natural wealth in Papua Province have not been used optimally for raising the living standard of the indigenous community' (Law 21/2001). Otsus also aims to augment the economic, social and cultural potential of indigenous Papuan communities by returning to them a significant role, through their representatives, in the formulation of regional policy and in the choice of appropriate development strategies. According to one of its architects, academic Agus Sumule (2003), the Special Autonomy Law for Papua (UU 21/2001) is exceptional as the only Indonesian legislation that explicitly acknowledges the restoration of basic rights to indigenous communities, and is in stark contrast to the long history of undermining the basic rights of Papuans, and *adat* rights in particular, under Suharto.

From this perspective, Otsus might be regarded as a 'reconciliation tool' (Sumule 2003), designed to address a history of injustices in social services, politics and the economy between Jakarta and Papua. Sumule also notes that Otsus was designed to address the impasse between the politically determined alternatives of autonomy (*otonomi*) or independence (*merdeka*) for Papua. In this context, Otsus became a means to appease the independence movement, and thus became a tool for 'pacification' rather than for 'reconciliation' (Sumule 2003).

Otsus in action: *politik setengah hati*, or half-hearted policy

While decentralization in Indonesia has been widely welcomed, its implementation has been fraught, as actors in both the regions and in central agencies have struggled to interpret the new laws and regulations, and to use them to their advantage to the extent possible. The intensity of this struggle has recently been documented for the case of forest resources in two districts in East Kalimantan by Resosudarmo (2007). Similar tensions have been evident in Papua, exacerbated by the context of Papua's history and politics within the Indonesian nation. Nor does decentralization offer any guarantee of natural resource sustainability (Hidayat 2000, Simorangkir 2000, Casson 2001, Barr and Resosudarmo 2002, Casson and Obidzinski 2002). Resosudarmo's (2007) research in East Kalimantan, for example, suggests that – in the particular conditions that have prevailed in Indonesia since the promulgation of decentralization laws in 1999 – it may actually have exacerbated forest exploitation.

Amongst the reasons for this are that local entities and *adat* communities may themselves be undemocratic and unaccountable, controlled by a small and powerful local elite or by outside interests seeking opportunities for expanding illegal logging operations. In any case they could not be expected to have developed the necessary capacity for commercial forest management in this very short timeframe (see, by way of example, Casson and Obidzinski 2002 on illegal logging in Kotawaringan Timur and Berau; as well as Resosudarmo 2007).

Similarly, a preliminary assessment of Otsus, by faculty members of the state university in Jayapura, Universitas Cenderawasih (Uncen 2003), concluded that the implementation process in Papua was far from effective, at a time when swift results were politically essential. The report identified critical failings of implementation including, most importantly, delays in the establishment of representative institutions, as a consequence of which no operational regulations such as Provincial Regional Regulations (Perdasi) and Special Regional Regulations (Perdasus) for regional governance had been issued at the time of the review. Under Special Autonomy provisions, responsibility for enacting regional government regulations lies with the Papuan Regional Legislative Body (*Dewan Perwakilan Rakyat Papua* [DPRP]) together with the governor, after receiving the consideration and agreement of the Papuan Peoples Council (*Majelis Rakyat Papua* [MRP]). The MRP, which is composed of *adat*, church and women's leaders, is designed to protect the *adat* rights of indigenous Papuans, and serves to guarantee to Papuans some degree of customary guidance of the political process (Sumule 2002a). However, at the time of the review, the MRP had still to be established.

The 2003 Uncen report broadly reiterated the concerns of an earlier German Technical Assistance mission on Support for Decentralisation Measures (GTZ-SfDM 2001). Written nearly three years after the fall of Suharto and the initiation of the nationwide political debate over decentralization, the GTZ report identified a number of issues likely to obstruct or prevent effective decentralization, including:

- the lack of clarity in the distribution of functions between different levels of government;
- an ineffective system of supervision of regional governments by the central government;
- the lack of clear responsibilities of the provinces;
- the failure of the current intergovernmental fiscal system to ensure some degree of equalization between resource-rich and resource-poor regions, and a mismatch between the assignment of expenditures and the assignment of revenues;
- the lack of policy coordination with sectoral laws and regulations, leading to contradictory regulations, for instance in the forestry and in the mining sectors;
- the strong role of 'money politics' in the election of heads of regions both in districts and provinces;
- an unsatisfactory accountability mechanism that focuses on the annual report of the heads of regions to their regional councils, and;
- the lack of capacity at the regional level to fully implement the new decentralization framework, and the lack of programmes at the level of central government that support capacity-building in the regions.

The implementation of Law No. 21/2001 has also altered the structure of the distribution of revenues between Papua Province and the central government, to that shown in Figure 6.1.

Papua's Special Autonomy law includes provisions for 70 per cent of oil and gas royalties to be channelled to the territory (for the first 25 years, and 50 per cent thereafter), as well as 80 per cent of mining, forestry and fisheries royalties. As a result of the implementation of Otsus, the provincial revenue increased from Rp 777 trillion in 2001 to Rp 1,382 trillion in 2002 (Uncen 2003). However, according to the Uncen Report (Uncen 2003), the provincial government remained unable to implement decentralization planning effectively because of the lack of

Items	Papua Province (%)	Central Government (%)
Land and Building Tax	90	10
Individual Income Tax	80	20
Forestry	80	20
Fisheries	80	20
Mining	80	20
Oil	70	30
Gas	70	30

Figure 6.1 The proportional distribution of revenues received by Papua Province and the Central Government, 2002.

input from the community, reflecting the failure to establish the enabling representative institutional structures (the DPRP and MRP) under Law 21/2001. As Resosudarmo's (2007) research in East Kalimantan demonstrates, whilst these new bodies can be more responsive to local pressures than their predecessors, they are also susceptible to capture by vested interests.

The involvement of local people in Otsus

Barr and Resosudarmo (2002) argue that decentralization in Indonesia, including Otsus, has largely been a political manoeuvre in response to separatist movements and the dissatisfaction of resource-rich regions with the centralization of resource revenues under the Suharto regime. As a result, decentralization policies were rushed through before the strong institutions necessary for a stable, functioning democracy could be established. The whole process has been *ad hoc* in nature, with little coordination among national, provincial and district governments. This early assessment has been largely confirmed by Resosudarmo's (2007) more recent work in East Kalimantan.

It is hard to discern any significant difference in the levels of public participation or public accountability in Papua between the periods before and after Otsus. The aim of improving the prosperity of *adat* communities remains merely a promise, and life remains substantially unchanged for the vast majority of Papua's people. School tuition and health services, for example, remain as they were before Otsus (Somba 2003). Somba (2003) also reports the widely voiced claim that only those with links to the local elite have been able to enjoy the disbursement of the autonomy funds for Papua.

Commenting on this situation, Sumule (2003) argues that Otsus has dissolved into a fight for money, in which no consideration has yet been paid to how to use the enlarged influx of funds to improve the prosperity of Papuan communities. His discussion of regional budgetary planning shows how the projects introduced under the banner of Otsus, which purport to empower indigenous Papuans, have been misused by local elites. Sumule observes that it is hard to conclude on this evidence that projects that have been designed using Otsus funds have been directed to empower indigenous Papuans. He argues further that this situation has been compounded by the lack of capacity in the Papuan government system, noting that professionalization of the government system in Papua will take a long time to develop if no external technical assistance is provided, and if there is no system to hold public servants accountable for their actions.

The undermining of Otsus, which began with delays in drafting and enacting the governmental regulations required for the establishment of the Papuan People's Assembly (MRP), was completed by the issuing of the controversial Presidential Instruction No. 1/2003 to split Papua into three provinces, without the approval of the MRP as mandated by the Papua autonomy law (ICG 2006). The process of diluting or sabotaging Otsus culminated in the inauguration of Abraham Octavianus Atarury as the new acting governor in West Irian Jaya Province (since renamed Papua Barat) in November 2003.

Although the MRP has since been established (in November 2005), the harsh conclusion drawn by many Papuans observing the central government's reluctance to acknowledge this customary authority is that Papua's Special Autonomy law is no longer a reconciliation tool but simply another toothless document on the Indonesian legal bookshelf, intended largely to extend the longstanding practices of control and exploitation of Papua by Jakarta (see SKP 2006).

Otsus and forest management in Papua

As noted previously, Otsus, which notionally involves the transfer of a range of powers from Jakarta to Jayapura (and now also to Manokwari, capital of the new Papua Barat Province), was originally intended to support the legal recognition of Papuan *adat* rights (Sumule 2003). Papuans have often heard the rhetoric that the Special Autonomy law would introduce potentially important opportunities for them to participate in forest management, as well as to enjoy more equitable access to economic benefits. However, such goals are fundamentally undermined if the Papuan *adat* communities are not recognized as legitimate management bodies. The problem revolves around the question of whether or not forest management under Otsus can increase the wellbeing of Papuans in terms such as those measured by the Human Development Index (Bappenas, BPS and UNDP 2001), and whether it can do so on a sustainable basis. How much power would the central government actually return to Papuan *adat* communities, especially in relation to such a valuable resource as timber?

Our reading of the draft Special Regional Regulations in the Forestry Sector (*Perdasus Kehutanan 2002*) suggests that the position of indigenous Papuans and of *adat* communities remains largely the same as that existing under the old forestry law in Indonesia, with only minor changes. Some degree of legal recognition of *adat* rights is countenanced in Section 11.2 of the draft, where it is stated that forestry planning should 'take local specifics and aspirations into account', and in point B of the considerations, which states that:

> ... sustainable forest administration should be able to accommodate the dynamics of customary communities' aspirations and participation.

But when this ideal is transformed into practice, such recognition disappears, submerged once again under state control. In Section 4.1 of the draft, for instance, it is stated:

> All forests within the territory of the Province of Papua including all the richness contained therein, are under the state's control for the maximum welfare of the people, the management of which shall be regulated by the Governor as the representative of central government in the region.

The role of the governor in this formulation simply mimics that of the state in Article 33 of the 1945 Indonesian constitution.

On the positive side of the ledger, the scope of environmental units of management may be broadened to include more than just one administrative district: the new environmental management units can potentially be defined on such bases as watersheds, cutting across administrative boundaries and thus acting to limit the effects of local economic pressure on natural resources. Controversially, however, this clause could also permit the governor to adopt the laissez-faire practices of the central government during the New Order period (see Alhamid 2005, chapter 3, on forest management in Indonesia).

Law 21/2003 on Special Autonomy was seen by many Papuans as providing a space for *adat* and for the acknowledgement of *adat* rights to land as well as to other natural resources, through provisions such as those in Article 43 (Protection of the Adat Rights). Many local communities in Papua understand regional autonomy primarily as the return of their *adat* rights. The attitude of most people is that 'This is our turn to enjoy the benefits from our own resources.' Thus, a team member from WWF Sahul Irian Jaya reported that when a WWF team visited a village near Sorong (Bird's Head region) in 2000 to discuss the long-term negative impacts of logging, they were confronted with complaints that they had come only at the point when natural resources had been restored to local control (Interview 2001).[1] The resolution adopted by the *adat* community representatives at the 2001 'Workshop on Revisiting the Forestry Management Policy in Irian Jaya' held in Jayapura neatly summarizes this view:

> Forest resources ... can only be managed in a sustainable manner if [they are] given back to the customary community. We would act as the main player ... while the government, tertiary education and non-government organizations facilitate us to develop our capacities to manage the resource properly ... We are [open] for collaboration with honest and responsible [members of the] business community. (Quoted in Sumule 2002b: 8)

This attitude on the part of Papuan *adat* communities has been reinforced by the approach taken by district governments (*Pemda kabupaten*), which have sought to raise revenue from natural resources in order to increase district or regional income (*Pendapatan Asli Daerah*). With decentralization, the role of investors has become more prominent, as funds are in short supply and district governments are expected to depend on locally generated revenues. For investors, dealing with district governments is much more convenient than with higher levels of government. District governments demand a set of payments and fees but do not enforce laws of the higher levels of government, or insist on the payment of taxes and the reforestation fund required by the central government (see Figure 6.2).

Adat communities in Papua have also been targeted by brokers or illegal loggers as 'investors in forestry businesses', in order to exploit their own tribal forests. In their efforts to attract investment, district governments have drastically liberalized permit procedures. Large timber companies, including Malaysian loggers operating illegally in a variety of ways (EIA/Telepak 2005), regard these policy incentives as an exceptional opportunity to further increase their wealth. *Adat* communities, on the other hand, see this option as an opportunity to become involved in decision-making processes and to obtain quick cash. As Otsus has facilitated access to investors, so Papuans have now become even more aware of the monetary value of their resources. Many have raced to claim as much forest as possible in order to convert it into cash in the short term; for example, in the Rendani forest case study reported by Alhamid (2005), some individuals or lineages within clan groups who had previously shared title to particular resources have claimed rights to those resources so that they can negotiate their sale to logging businesses.

The struggle for a more democratic sharing of benefits and a more diffuse local control over resources has thus contributed to increasing deforestation, conflict and violence. In 2002, the Governor of Papua stated that around 3.6 million hectares of forest in Irian Jaya had been logged in the previous five years (West Papua Net 2002). He identified the principal cause of this dramatic increase in the rate of deforestation as related to the collaboration of *adat* communities and forest concessionaries under the umbrella of the newly instituted Kopermas cooperatives (*Koperasi Masyarakat Adat*, Customary Community Co-operative).

Determining whether or not there has in fact been an increase in deforestation under Otsus is no simple task. Figure 6.2 shows the production of sawn timber and logs in the Province of Irian Jaya from 1996 to 2001 according to official Forest Department statistics.

As Figure 6.2 shows, production of 'legal' logs and timber has dropped dramatically since 1997, as in Indonesia more generally. However, almost half of the total timber production is considered illegal, and the level of harvest of 'illegal' logs has increased significantly. *Bisnis Indonesia* (5 March 2003) reported that the level of illegal timber exported from Papua was around 600,000 cubic metres per month

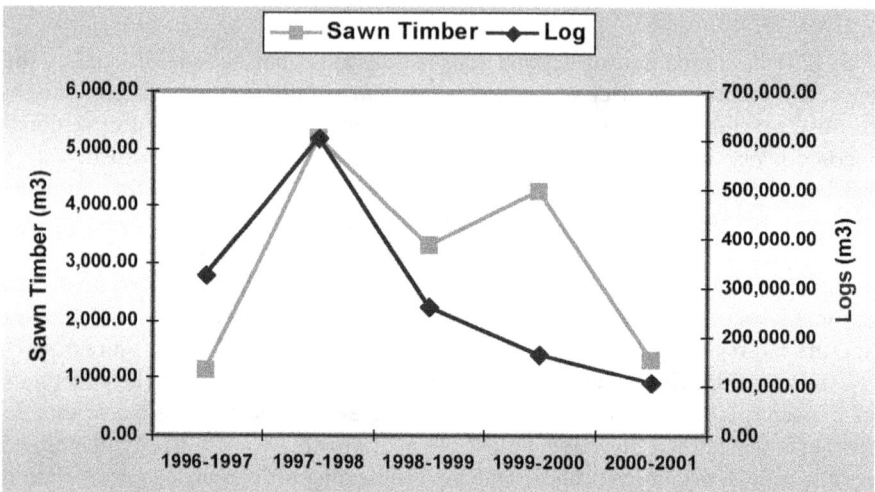

Figure 6.2 Production of 'legal' sawn timber and logs in Irian Jaya, 1996–2001.

in 2002. If we assume a conservative timber price of about Rp 1 million per cubic metre,[2] the income received from illegal logging can be estimated at around Rp 600,000 million per month or Rp 7.2 trillion per year. This amount is around three times the budget of the Indonesian military/TNI (Rp 2.8 trillion), or three times greater than the Papua provincial budget in 2002 (Rp 2.4 trillion). A more recent and more closely researched report estimates that the monthly total of illegally exported *merbau*[3] logs from Papua is closer to 300,000 cubic metres (EIA/Telapak, 2005), but this is still almost half of the *annual* total legal log production from Papua of 730,306 cubic metres (as of 2004, consisting of 373,869 cubic metres from Papua Province, and 356,437 cubic metres from the newly constituted West Irian Jaya Province) (Down to Earth 2006).

Otsus and the resurgence of *adat*

Papuans in many parts of the province have low levels of awareness and understanding of national political structures, and consequently often neglect to consider national policy in making their own decisions. For example, local claims of customary or *adat* rights to land within the area of the Rendani Protected Forest near Manokwari clearly illustrate the reluctance of these customary landowners to accept the claims of the state as expressed in the 1945 Constitution. Instead, their loyalties revolve principally around *adat* systems rooted in family, clan and tribe.

Within Papua, the power of *adat* varies from place to place. It ranges from being completely dominant in the hinterland or isolated areas, such as the Central Highlands, to being quite weak in some coastal and urban areas with a long history of contact and multicultural populations, such as the Rendani area where there has been substantial immigration from other parts of Indonesia since the 1970s. Across the island, Papuans adapt *adat* rights to meet their present and specific economic and environmental needs. However, *adat* is not imposed on a community by its leaders; rather, it is kept alive through people's actions and reproduced through their daily activities. Tribal leaders (*kepala suku*) are thus widely regarded as icons, and acknowledged as custodians, of *adat*. *Adat* has become an institution of mediation between the spiritual and the living world, between individuals in conflict, and often between people and the state where land is concerned (Li 2001). Province-wide *adat* movements in Papua have recently made their influence felt at the local level as well. Here, *adat* revitalization movements have led to a sense of empowerment and some *adat* leaders have sought to regain their old positions by sponsoring *adat* rituals and rebuilding *adat* institutions. The Papuan customary council (*Dewan Adat Papua*), for example, is a new institution that has been created to protect *adat* community rights in Papua (Howard *et al.* 2002).

On the negative side, *adat* has the potential to promote tribal fragmentation. Otsus, which strengthens district government and provides a space for communities to return to *adat* as a basis for claims to control over land and resources, has been understood by some communities as an opportunity to return to the tribalism of the past. *Adat has* gained in importance as people seek alliances and political position through ethnic mobilization and competition for power. For example, the Ayamaru tribe, of which former Governor Jaap Solossa was a member, has held a disproportionate number of senior bureaucratic positions in Irian Jaya. *Adat* has also been appropriated as a slogan to be wielded in the pursuit of political power. A common rhetorical statement, often heard at the beginning of speeches by district heads, the governor, or heads of government departments, is that they wish to ground their government on the principles of *adat*. But is this affirmation merely symbolic and invoked solely for self-interested purposes, or will it also generate concrete or practical consequences related to the rights to control resources?

Following the enactment of the Otsus legislation, 'Papua' has emerged as an identity and a banner with which to secure positions in the bureaucracy. 'Papuanization' has become a common phenomenon in the competition for jobs in Papua, and Papuans now hold most of the positions in the provincial and district bureaucracies. But does this also mean that Papuan communities without direct representation in the bureaucracy have acquired more power or access in decision-making processes and in controlling activities on their own tribal land?

Despite the presence of the National Land Bureau (*Badan Pertanahan Nasional, BPN*), dualism in control over natural resources continues wherever *adat* communities still maintain customary control over the land and its resources. The cash economy has transformed the social constitution of communities, with the result that the unity of these communities is often weakened, and conflicts among clans have become a regular occurrence in most of the villages around Papua. In Rendani, conflicts have even spread to lower levels, for example between lineages that had formerly held shared rights over particular resources – an axis for conflict that was previously unthinkable. A shift towards permanently individualized ownership has also occurred over time because of factors such as the adoption of permanent agriculture and the cultivation of perennial tree crops, population pressure, and changing views on the nature of property itself. This shift is encouraged by the greater security of tenure granted private property rights under Indonesia's national land law (UUPA 1960).

The solution to land conflicts is widely seen to lie in the control of land by communities constituted through *adat*. But will this entail a return to common property, in the traditional sense, or does it mean that *adat* communities will be recognized as playing a more limited role as land-controlling institutions? It does not seem likely that people will want to give up their current individual rights to land; yet, on the other hand, they also wish to hold on to *adat* as a broader symbol of identity and tradition.

Kopermas: symbol of a resurgent *adat*?

In discussion with a senior provincial forest official in Jayapura on the topic of whether forest management can contribute to reducing poverty among indigenous Papuans, the question arose of ways to sustain the production of non-timber forest products (NTFP). The attitude that local welfare could only be increased using

NTFP is an old paradigm that has been advanced, since the Dutch colonial period, largely to exclude local communities from timber management (Bari 1974).

Three of the principal NTFP exploited in the past in Papua are massoy (*Cryptocarya massoy*), copal (a resinous extract from trees of the Araucariaceae family), and cinnamon (*Cinnamommum culilawane*). Currently, the most significant NTFP in Papua is gaharu (also known as agarwood, aloewood, or eaglewood), a trunk, branch or root of a particular species (often *Wikstroemia* spp.) that has been modified chemically by a fungus (Maai and Suripatty 1996), and which commands very high prices because of its use in perfume and incense (Gunn *et al*. 2003). Gaharu collection in Papua is conducted by local people in groups, each group usually consisting of three to five people. The group spends three to seven days looking for gaharu in the forest. When gaharu is found in a particular tree, the tree will be felled, using axes or knives. If the fungus is present, one tree usually produces around 0.5 to 4 kg of gaharu (Maai and Suripatty 1996). The price of gaharu in 2001 was about Rp 2.3 million/kg for the super class (*Kompas*, 2001), although only a small proportion of this is usually captured by the collector (Gunn *et al*. 2003). Current harvesting practices and levels of harvesting are highly unsustainable, but these could be redressed by progress that has been made in understanding the ecology of the species producing the gaharu, the recent development of artificial methods to enhance production of the commercially valuable heartwood, and wider communication of the value that harvesters might expect to capture (Gunn *et al*. 2003).

However, besides bringing in money, the gaharu industry has also contributed – in conjunction with the timber trade – to a dramatic increase in the incidence of HIV/AIDS and other sexually transmitted diseases, especially in Merauke and other southern Papuan districts. Gaharu traders and military personnel began bringing in sex workers from Java and Sulawesi in 1996, and the commercial boom in the wood trade was followed by an explosion in the number of HIV/AIDS positive cases (Aditjondro 2004; Ama 2002; Nainggolan 2002). Papua now has the highest rate of reported AIDS cases in Indonesia at 19.19 per 100,000 (36.2 times the national average), followed by the City of Jakarta with a rate of 3.9 (7.36 times the national average), and Bali with 1.3 (2.45 times the national average). The Indonesian national average is 0.53 per 100,000 (*Kompas*, 2001).

Realizing, as have others more generally (e.g. Belcher and Schreckenberg 2007), that NTFP-based programmes had achieved only limited success in alleviating rural poverty among indigenous Papuans, the government, supported by many NGOs, proposed involving local communities in the timber business. National NGOs such as AMAN (The Indonesian Alliance of Adat Communities) and KPA (Consortium for Agrarian Reform) are strong supporters of a return to *adat* rights over people's own tribal forests. Employing the same NGO rhetoric, the Young Entrepreneurs Association of Indonesia *Himpunan Pengusaha Muda Indonesia* (HIPMI) also pushed for the return of timber rights to local people. A number of NGOs, however, voiced their suspicions of a hidden agenda on the part of HIPMI which, as an association of profit-seeking companies, appeared an unlikely champion of indigenous rights.

The permit given to the local communities to become involved in a logging business is called the Licence for Forest Product Collection or IHPHHMA (*Ijin Hak Pemungutan Hasil Hutan Masyarakat Adat*). The policy enabling award of such licences may prove to be the most important decision made in the forestry sector during the Otsus period. The licence is given to a community institution called Kopermas. Elsewhere in Indonesia, Kopermas is an abbreviation of *Koperasi Peranserta Masyarakat* (Community Collaborative Co-operative). However, in Papua the abbreviation stands for *Koperasi Masyarakat Adat* (Customary Community Co-operative), emphasizing the *adat* status of the business owners, even though – as we suggest here – *adat* is actually being deployed as a front for commercial interests. Since 1999, a limited right to manage forests in Papua has been transferred to the Kopermas system under the Ministry of Forestry Decree No. 538/KPTS-II/1999 on Timber Utilization Permits (*Ijin Pemanfaatan Kayu* [or IPK]).

Under the Kopermas scheme, indigenous Papuans are recognized as the rightful owners of their lands and are allowed to log their forests if they apply for logging licences from the Forestry Department. Under the same scheme, however, the tribal forest is also being exploited by logging companies and local military commanders, who buy the traditional landowners' forest rights at extremely low prices and then sell the timber directly to international buyers. The procedure for proposing that a forest area be managed under Kopermas is illustrated in Figure 6.3.

If the forest area proposed by a Kopermas is located inside an existing forest concession (*Hak Pengusahaan Hutan* [HPH]), then that Kopermas has to submit an application to the Provincial and District Forestry Services (*Dinas Kehutanan Propinsi Papua* and *Dinas Kehutanan Kabupaten*) through the forest concession holder. If the proposed forest area is located outside the HPH concession boundaries, then the Kopermas has to submit its application to the district head via the Provincial and District Forestry Services.

A Kopermas application for a Licence for Forest Product Collection (IHPHHMA) must include the following elements:

- a plan of work and management;
- a map of the location at a scale of 1:100,000;
- a recommendation from the relevant District Forestry Services;
- identity data for the proponents (the cooperative's licence);
- a letter of recommendation from the local community consenting to the exploitation of the forest;
- a statement letter from the district head confirming the existence of the local community;
- a letter guaranteeing that the Kopermas will pay the required fees for the Reforestation Fund or (*Dana Reboisasi* (DR) and the Forest Resource Rent Provision (*Provisi Sumber Daya Hutan* [or PSDH]).

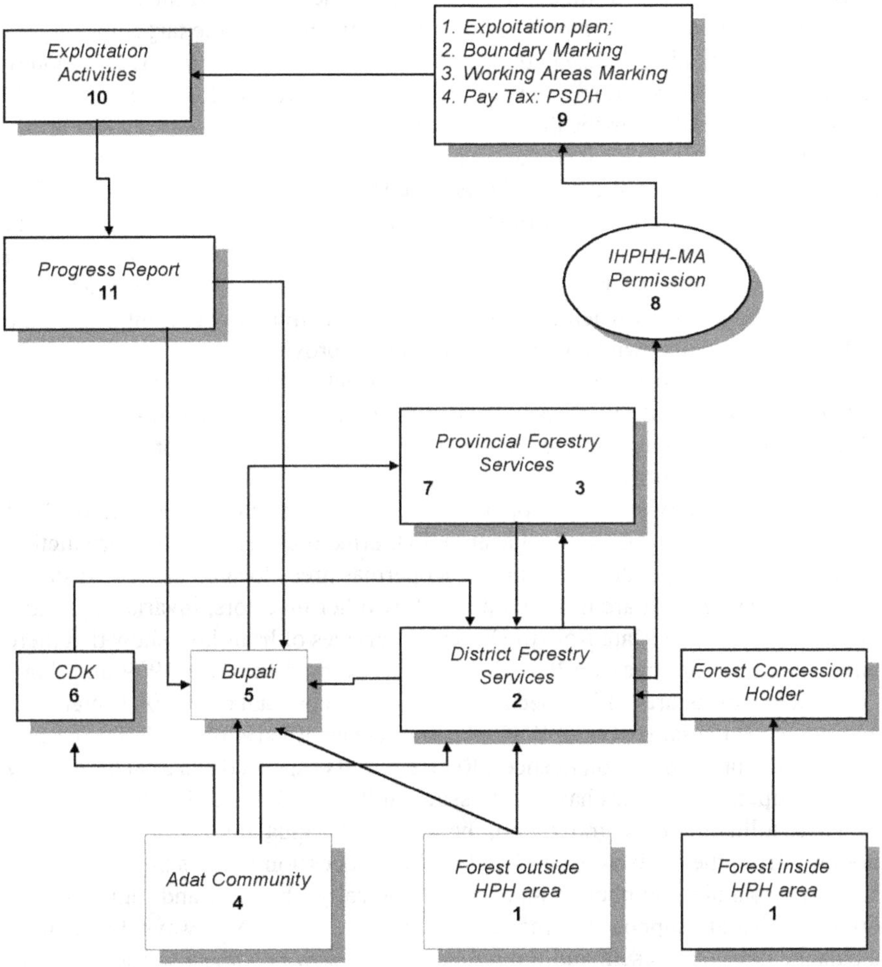

Figure 6.3 The procedure for proposing a KOPERMAS in Papua.

Abbreviations:

CDK	Cabang Dinas Kehutanan, Branch Office of the Provincial Forestry Service
HPH	Hak Pengusahaan Hutan – Concession for commercial selective harvest of natural forest
IHPHH-MA	Ijin Hak Pengusahaan Hasil Hutan – Masyarakat Adat (Adat Community Concession for Forest Product Collection)
PSDH	Provisi Sumber Daya Hutan, Forest Resource Royalty

Very few rural Papuan communities, the vast majority of whose members live in poverty, would be able to meet the above requirements without the assistance of well-resourced outsiders. In practice, Papuans have become the target of logging companies, particularly those under Malaysian control, which offer people money to set up a Kopermas and in return acquire the timber at very low rates.

According to data from the Provincial Forestry Services, there were 32 Kopermas active in the forestry sector in Papua in 2000 (Dinas Kehutanan Propinsi Papua 2001). The area of forest that can be assigned to a Kopermas ranges from 100 to 250 hectares. According to the Ministry of Forestry Decree on Kopermas, this area should then be converted to plantation. In practice, however, most Kopermas aim simply to harvest the timber from the forest. The inequity of the present arrangement is starkly evident when the rights of an *adat* community seeking to operate independently on its own land are contrasted with provisions under the Kopermas system. If an *adat* community seeks to independently exploit an area greater than 250 hectares, approval can be issued only by the central authorities in Jakarta. Each HPH concession holder, on the other hand, can request a forest concession of up to 50,000 hectares in area.

The logging activities of a Kopermas located outside the boundaries of HPH areas are often managed by an adjacent HPH, effectively as a satellite production area for the HPH operator. In contrast, Kopermas areas located at some distance from an existing HPH are usually operated by other investors, invariably using a clear-cutting system. Data from the Forestry Services of Irian Jaya show that there have been 66 Kopermas IHPHHMA licences granted between 1999 and April 2001, each with an area of 250 hectares, for a provincial total of 16,500 hectares. In addition, over the same period, 19 Kopermas were awarded a further 5,500 hectares under IPK timber utilization licences. Recent reports suggest that a total of as many as 300 Kopermas licences have been issued in Papua (Down to Earth 2006).

Thus, following decentralization, the forests of Papua are now under pressure not only from the 61 established HPH timber concession holders but also from a steadily expanding number of Kopermas. Ironically, the local and regional governments that are supposed to manage the forest in sustainable ways, by creating a balance between present and future needs, have now become a serious threat to the existence and sustainability of the forest. With the aim of boosting regional and district government incomes, local governments have increasingly come to regard the forest as a 'timber storeroom', to be utilized solely to produce quick cash. In the short term, the regional government benefits by receiving taxes and fees from forest exploitation, not to mention the incentive of bribes and election funds offered to officials. As a consequence, there has been a significant increase in deforestation throughout Papua since the implementation of this component of Special Autonomy.

From the point of view of those interested in exploiting Papua's forest for profit, Kopermas has been a perfect vehicle since it has all the trappings of a community-based economic development model established through partnership between corporate interests and *adat* communities. This model has the power to silence critics because of its apparent reliance on democratic principles, and its seeming protection of the rights of *adat* communities to participate in forest exploitation, while simultaneously reversing the poor image of the HPH system, which has notoriously excluded local communities as partners in logging.

From the perspective of investors, the Kopermas system has some additional advantages over the HPH approach, including:

1   Overcoming land rights disputes. The risk that local communities or individuals will claim back the land is reduced because they are involved as board directors of the Kopermas company.
2   Investors do not have to pay for lobbying fees and do not have any difficulties in establishing and maintaining relationships with the *Dinas Kehutanan* (Forestry Local/Provincial Service), which is a notoriously corrupt institution (Nafi 2004), because this task is now performed by the *adat* community.
3   Although, in principle, prices are agreed between loggers and the *adat* community, the determination of the wood price is often established unilaterally by the investors. Communities in a relatively remote province such as Papua have very weak access to market information and often cannot easily check the accuracy of the market price offered.
4   By involving the community in the logging process, investors can avoid the risk and uncertainty of price changes by transferring these risks to the *adat* community.
5   As community businesses, Kopermas are not required to comply with normal Indonesian labour standards, and production costs can thus be lowered by reducing the salaries and wages paid to employees.
6   In practice, Kopermas are also not obliged to undertake reforestation because planning assumes that such logged-over areas will be converted to cash crops, such as cocoa and oil palm. By contrast, under HPH, the Indonesian Selective Cutting and Planting System (*Tebang Pilih Tanam Indonesia* [*TPTI*]) theoretically requires replanting of a logged forest. The orientation of Kopermas to land conversion will automatically increase the immediate level of timber production.

Kopermas at Rendani

The case of the Kopermas established at Rendani, in the Manokwari area, illustrates some of the complexities of the Kopermas system, and identifies some of its potential dangers for the *adat* community, as well as for the forests. In 2001, a member of the local *adat* community at Rendani, DM (full names are abbreviated here), who is also employed at Manokwari's Local Forestry Service (*Cabang Dinas Kehutanan*), took the initiative to establish Kopermas 'Niauw Syoribo'. He did this with the support of his father, TM, an *adat* leader of the local community. This Kopermas was established in collaboration with CV Fajar Papua Indah, a local logging company owned by Javanese investors, to produce sawn timber. The mechanism that was agreed between DM and Fajar Papua Indah was that the company bought logs from the Kopermas at a price of Rp 800,000 per cubic metre. Figure 6.4 provides an approximate breakdown of the total cost of production of a cubic metre of timber at Rendani.

Since most local people at Rendani know how to operate chainsaws, they conducted the entire process of logging (except for transportation) under the supervision of DM. However, not all of the local people wanted to work for the Kopermas as they did not seem to enjoy working with DM, for reasons articulated by BK: 'DM is cheating us. The Boss [the director of Fajar Papua Indah] paid chainsaw operators about Rp 250,000, but DM paid us only Rp 100,000. Instead of fighting, I would prefer not to work with him' (Interview 2001). Because many of the Rendani did not wish to work with DM, he then invited people from Ifanbeba, Gusminan and Ransiki to work as chainsaw operators and to transport the timber from the forest to the main road.

According to DM, the profit gained by the Kopermas in 2001 was only around Rp 5 million. This low level of profit was attributed by DM to the small volume of timber produced, of only about 100 cubic metres. However, DM admitted in interview that he had personally received an income of around Rp 10 million, twice the profit received by the Kopermas. In a separate statement, DM suggested that the real operational cost was closer to around Rp 630,000 per cubic metre, significantly lower than the Rp 800,000 figure quoted publicly. If this is true then he may have been receiving a profit of about Rp 170,000 per cubic metre of timber. The sawn timber was then sold by Fajar Papua Indah to a buyer in Surabaya for a price of about Rp 1.9 million per cubic metre. If the transportation expense from Manokwari to Surabaya was around Rp 650,000 per cubic metre (based on data collected from discussions with staff from PT Pelabuhan Indonesia by Hidayat Alhamid in 2001), then the net profit gained by Fajar Papua Indah would be around Rp 450,000 per cubic metre. It is apparent from these figures that the profit gained from their Kopermas by the community is only a fraction of the total profit being enjoyed by other parties. Moreover, in order to realize their profit of Rp 5 million, equivalent to approximately US$600, the Rendani people have to tolerate the destruction of some 1,500 hectares of their tribal forest.

Expense components	Cost (Rupiah)
Chainsaw operator	250,000
Transport from forest to road (approx. 1 km)	150,000
Truck hire	100,000
Price of timber per m3 paid to landowner	150,000
Profit to Kopermas	50,000
Profit to community co-ordinator (DM)	100,000
Total expense of timber production per m3	800,000

Figure 6.4 Breakdown of total cost per cubic metre of timber produced by Kopermas 'Niauw Syoribo', 2001.

Apart from the economic motivation, the establishment of Kopermas in Papua is also linked to a defence of the claim of *adat* rights over tribal land. It might appear as though Otsus has encouraged people to reject state ownership and forced the state to accept their rights of ownership. But the facts in the field indicate that the euphoria of democracy, culminating in the demand for the return of resource rights to *adat* communities, has merely resulted in the loss of state control over some forest areas to the advantage of particular individuals and corporations who have used this situation to increase their incomes through logging, either legally through Kopermas or illegally.

Even though, by definition, Kopermas is a community institution, in reality the initiative for Kopermas licence applications typically emanates from pressure and persuasion by various means from external brokers and investors in alliance with local elites. The basic problem lies not in the question of whether the forest can or cannot be logged to produce 'quick cash', but rather of whether logging on some more sustainable basis can increase the welfare of the community or not. It is clear that there will be little or no positive long-term benefit to *adat* communities if the Kopermas system persists in its current form; the forest will be exploited, but poverty will still characterize *adat* communities. If this happens, *adat* communities will not only be stripped of their forest but also of their culture and heritage. Nevertheless, under the prevailing political situation, to halt the operations of a Kopermas would be tantamount to declaring war on local communities.

Otsus, *adat* and the Trojan Horse of Kopermas

The principal development objective of Otsus, which is to improve local people's welfare through combating the three main challenges of 'poverty, ignorance, and backwardness' (*kemiskinan, kebodohan, keterbelakangan*), reflects a continuation of earlier ideas: Papuan communities are familiar with earlier incarnations of this rhetoric, in the form of the Presidential 'Backward Village' project (*Inpres* 2/1993 *Desa Tertinggal*, or IDT) or the Social Safety Net (*Jaring Pengaman Sosial*, or JPS), and are justifiably sceptical about claims for their results. What marks Otsus out as potentially 'special' or different from previous central government programmes is the range of political concessions to Papuan political autonomy, and particularly the recognition of *adat* rights, as incarnated in the MRP. The politics of the MRP and its final form and functions aside, the most challenging question now confronting rural Papuan communities is the definition or redefinition of *adat*, in a new moral economy of community empowerment.

Even though Otsus has introduced some improvement in local access to forest resources, this has not translated into an improvement in the quality of life for indigenous Papuans. Kopermas, both allegedly and superficially a tool developed to reduce rural poverty, may prove instead to be a device that will bring absolute poverty to *adat* communities, while at the same time creating greater benefits and profits, but only for outsiders. In the past, many rural Papuan communities realized that they had been classified as living under the poverty line only when census officials came to their villages. Obviously, they still had sufficient food from their gardens, forests and rivers. But now, as the forest is being converted to generate a small amount of cash that cannot support them indefinitely, these communities really will be trapped in absolute poverty.

Indigenous forest management, which is supposed to guide the relationship between communities and forests, is now operating in an uncertain environment in which traditional norms, values and practices are struggling for recognition. People are fighting for their rights to control and use the forest, but there appears to be little recognition on their part of their own reciprocal duty or responsibility to maintain the sustainability of the forest. One case study at Rendani (Alhamid 2005) revealed that many of the traditional structures for sustainable resource management had broken down as a consequence of pressure on land from immigrants, the need but limited options for engaging with the market economy, and the loss of traditional authority regimes, in part due to their non-recognition by the state. In large part, the degradation and loss of indigenous forest and indigenous forest management practices may be due to there being little or no opportunity to exercise traditional *adat* practices since Indonesia took over Papua in 1963. Under such intrusive state control, indigenous Papuan people have lost their incentive to maintain their forests.

Adat institutions and *adat* forms of leadership were severely eroded under the New Order. The modern political system of the *desa* (village) established by the 1979 Village Government Law marginalized *adat* leaders (*kepala suku*) by transferring many of their functions to village heads (*kepala desa*), who are government functionaries under the control of a subdistrict head (*camat*). The resulting dualism in village leadership created the potential for conflict, confusion and tension. A simple step employed by many communities to address this problem was to combine the two institutions by selecting the *adat* leader as *kepala desa*. An important consequence of this move, however, has been a decrease of *adat* authority. When *adat* leaders accept the role of *kepala desa*, *adat* almost invariably loses its role as an independent and strong institution. By becoming *kepala desa*, *adat* leaders repositioned themselves as subordinate to state political institutions, and became particularly vulnerable to co-option and collusion. While certain offices of *adat* institutions (such as clan or lineage leaders) still enjoy some respect in their own spheres, it is the *kepala desa* who is empowered by the state.

Rights to land and to forest resources are obviously implicated in this contested transformation in customary authority. A steady process of increasing individualization of claims to land has been experienced amongst many rural communities in Papua. This is evident at Rendani, where private or individual gardens have now been established without reference to clan authority, and even in defiance of claims from other lineage members. The introduction of cash crops such as cocoa and some perennial plants has further strengthened the tendency to privatization of land. While the land, according to the Rendani community, is still under the control of lineage leaders, and no one will openly admit that it has become privatized, in practice family and individual forms of ownership have clearly been asserted.

The rhetoric of Kopermas holds particular appeal to landowners with newly recognized *adat* rights and newly claimed individual rights. Those best positioned through their rights to *adat* authority and their claims to individual holdings have led the race to form Kopermas cooperatives. Thus, individuals from the local Rendani community such as DM assume rights over communal forest resources when they act as intermediaries in their own interest between outside entrepreneurs and the local community. These new local entrepreneurs are able to exploit their experience as local government employees (often as staff of the Local Forestry Service) and their established personal business networks, which give them unparalleled influence to negotiate on behalf of the villagers, as well as to promote profit-seeking from resource exploitation.

While the mechanisms of *adat* have been significantly weakened, the principles of *adat* as a basis for advancing the rights of a community to control its own resources remain strong. However, without appropriate mechanisms in place, communities remain vulnerable to the new alliance of outsiders and local elites represented by Kopermas. Kopermas, though, or something very like it, is in Papua to stay. Repealing the decree establishing Kopermas would be widely regarded as an attempt to roll back the recognition of *adat*, and would do little to solve the central problem of resource sustainability and long-term community welfare. Local communities, strengthened by the apparently untouchable figures behind both Kopermas and illegal logging in Papua, and with the support of the commercially implicated security forces and local government, would fight back, as in the case of the 'MV Africa' in Sorong (EIA/Telapak, 2005: 9).[4]

One possible solution might be to create a 'Kopermas plus', an overhauled and revised form of Kopermas involving the genuine collaboration of local communities and governments, with counterparts from universities and conservation NGOs such as WWF or Conservation International, to log the forest in acceptable and low impact ways. The involvement of non-profit advisors in supplying advice on sustainable community forest management would be of significantly greater benefit to the community in the long term than continuing to allow loggers to exploit the forests in unsustainable ways. This is one approach that could offer some hope of bringing immediate social and economic needs and longer-term environmental protection into some correspondence. Practically speaking, efforts at preventing local people from depleting their own forests, whatever the ecological rationale, have to make economic sense to the local community if they are to work.

Notes

1 Interviews referred to in this chapter were conducted by Hidayat Alhamid.
2 According to a staff member of the Forestry Local Service (Cabang Dinas Kehutanan) of Manokwari, the price of timber in Surabaya market in June 2001 was around Rp 1.9 million per cubic metre (Interview 2001).
3 Merbau is the common and trade name for three *Intsia species* – *Intsia bijuga, Intsia retusa*, and *Intsia palembanica*. The first two of these are found in Papua Province (EIA/Telepak 2005)
4 In this notorious case, the MV Africa, loaded illegally with 12,000 cubic metres of merbau logs, was impounded by maritime police at Sorong in 2001, but subsequently released by the police department. The police chief of the time was later arrested for his role in this evasion.

Bibliography

Aditjondro, G. (2004) 'Dari gaharu ke bom waktu HIV/AIDS yang Siap Meledak: ekonomi politik bisnis tentara di tanah Papua', *Wacana, Jurnal Ilmu Sosial Transformatif*, No. 17/Tahun III: 83–111.

Alhamid, H. (2005) 'Forests for the People? Indigenous forest management under decentralisation: a case study of the Rendani Protection Forest, Papua, Indonesia', unpublished PhD thesis, Canberra: The Australian National University.

Alhamid, H., Kanowski, P. and Ballard, C. (in press) 'Forest management and conflict at the Rendani Protection Forest, Papua', in B. Resosudarmo and F. Jotzo (eds), *Development and Environment in Eastern Indonesia: Papua, Maluku and East Nusa Tenggara*, Singapore: ISEAS.

Ama, K. K. (2002) 'Menyebar HIV di pedalaman Papua: barter perempuan dengan gaharu,' *Kompas*, 18 November 2002.

Bappenas, BPS and UNDP (2001) *Indonesia Human Development Report 2001. Towards a New Consensus: Democracy and Human Development in Indonesia*. Jakarta: Bappenas.

Barber, C. V. (1989) 'The state, the environment, and development: the genesis and transformation of social forestry policy in New Order Indonesia', unpublished PhD thesis, Berkeley: University of California.

Bari, A. (1974) 'Potensi hutan Irian Jaya dan prospeknya', *Irian, Bulletin of Irian Jaya Development* 3 (3): 1–50.

Barr, C. and Resosudarmo, I. A. P. (2002) *Decentralisation of Forest Administration in Indonesia: Implications for Forest Sustainability, Community Livelihoods, and Development*. Bogor: Center for International Forestry Research.

Belcher, B. and K. Schreckenberg (2007) 'Commercialisation of non-timber forest products: a reality check', *Development Policy Review* 3: 355–77.

Casson, A. (2001) *Decentralisation of Policymaking and Administration of Policies Affecting Forests and Estate Crops in Kotawaringin Timur District, Central Kalimantan*. Bogor: Center for International Forestry Research.

Casson, A. and K. Obidzinski (2002) 'From New Order to regional autonomy: shifting dynamics of "illegal" logging in Kalimantan, Indonesia', *World Development* 30: 2133–51.

Dinas Kehutanan Propinsi Papua (2001) *Laporan Tahunan, Tahun Anggaran 2000–2001*. Jayapura: Dinas Kehutanan Propinsi Papua.

Down to Earth (2006) 'The future for Papuan forests', *Down to Earth* 69: 11–13.

EIA/Telapak (2005) *The Last Frontier: Illegal Logging in Papua and China's Massive Timber Threat*. London and Bogor: Environmental Investigation Agency and Telapak.

FAO (n.d.) 'Forestry: the Jakarta declaration (Eighth World Forestry Congress, October 1978)', http://www.fao.org/docrep/x5565E/x5565e06.htm (accessed 30 May 2007).

GTZ-SfDM (2001) 'Decentralization in Indonesia since 1999: an overview'. Deutsche Gesellschaft fuer Technische Zusammenarbeit (GTZ) – Support for Decentralization Measures (SfDM), http://www.gtzsfdm.or.id/dec_in_ind.htm (accessed 30 May 2007).

Gunn, B., Stevens, P.,Singadan, M., Sunari, L. and Chatterton, P. (2003) 'Eaglewood in Papua New Guinea', RMAP Working Paper 51. Canberra: Australian National University, rspas.anu.edu.au/papers/rmap/Wpapers/rmap_wp51.pdf (accessed 5 October 2007).

Hidayat, S. (2000) *Refleksi Realitas Otonomi Daerah dan Tantangan ke Depan*. Jakarta: Pustaka Quantum.

Howard, R., McGibbon, R. and Simon, J. (2002) *Resistance, Recovery, Re-empowerment of Adat Institutions in Contemporary Papua*. Jakarta: Civil Society Strengthening Program, USAID.

ICG (2006) *Papua: The Dangers of Shutting Down Dialogue*. Asia Briefing No.47. Jakarta and Brussels: International Crisis Group.

Kanwil Kehutanan Propinsi Irian Jaya (2000) *Statistik Kehutanan Propinsi Irian Jaya*. Jayapura: Kanwil Kehutanan Propinsi Irian Jaya.

Kompas (2001) 'Kepala Polda agar Tangkap "Otak" Perambah TNGL', *Kompas*, 24 April 2001.

Li, T. M. (2001) '*Masyarakat adat*, difference, and the limits of recognition in Indonesia's forest zone', *Modern Asian Studies* 35: 645–76.

Maai, R. R. and Suripatty, B. A. (1996) 'Pengaruh wadah penyimpanan dan kelas diameter terhadap pertumbuhan stump wikstroemia polyantha', *Buletin Penelitian Kehutanan* 1. Manokwari: Balai Penelitian Kehutanan.

Nafi, M. (2004) 'Puluhan kasus korupsi pengolahan Hutan', *Tempo*, u.d.

Nainggolan, N. (2002) 'HIV/AIDS di Papua (1). 'Belrusak bikin rusak masyarakat', *Suara Pembaruan*, 25 November.

Nakashima, E. (2006) 'Papuans idle after buzz of prosperity falls silent', *Washington Post*, 21 May 2006.

Ngeethe, N. (1998) 'The politics of decentralization through decentralization in Kenya: policy and practice with emphasis on district focus for rural development', Occasional Paper No. 45. Des Moines: University of Iowa- International Programmes.

Perdasus Kehutanan (2002) Preliminary Draft, Special Provincial Regulations of Papua [un-numbered] of 2002 on Forestry in the Province of Papua. Jayapura: Provincial Government, Papua Province. English translation available at http://www.papuaweb. org/dlib/lap/sullivan/perdasi-perdasus/forestry.rtf

Resosudarmo, I. A. P. (2007) 'Has decentralization improved local forest governance? Case studies from East Kalimantan, Indonesia', PhD thesis, Canberra: Australian National University E Press.

Ribot, J. C. (2002) *African Decentralisation: Local Actors, Powers, Accountability*. Geneva: United Nations Research Institute on Social Development (UNSRISD) Programme on Democracy, Governance, and Human Rights.

Simorangkir, B. (2000) *Otonomi atau Federalisme: Dampaknya Terhadap Perekonomian*. Jakarta: Pustaka Sinar Harapan; Harian Suara Pembaruan.

SKP (2006) *Sekilas Informasi (Januari-Maret 2006)*. Seri Papua Aktual No.5. Jayapura: Sekretariat Keadilan dan Perdamaian Keuskupan Jayapura.

Somba, N. D. (2003) 'Life as usual after Papua autonomy', *Jakarta Post*, 2 August.

Sumule, A. (2002a) 'Majelis Rakyat Papua: the Papuan People's Assembly and its significance in protecting the rights of the indigenous people of Papua', *Development Bulletin* 59: 73–6.

Sumule, A. (2002b)' Protection and empowerment of the rights of indigenous people of Papua (Irian Jaya) over natural resources under special autonomy: from legal opportunities to the challenge of implementation', RMAP Working Paper No.36. Canberra: Resource Management in Asia Pacific Project, Australian National University, http://rspas.anu. edu.au/papers/rmap/Wpapers/rmap_wp36.rtf (accessed 30 May 2007).

Sumule, A. (2003) *Satu Setengah Tahun Otsus Papua: Refleksi Dan Prospek*. Manokwari: Yayasan Topang.

Uncen (2003) *Pokok-pokok Pikiran Sekitar Kebijakan Pengembangan Propinsi Papua*. Abepura: Democratic Center, Universitas Cenderawasih.

UUPA (1960) Undang-undang no. 5 tahun 1960 tentang peraturan dasar pokok-pokok agraria. Republic of Indonesia, 1960.

West Papua Net (2002) 'Hutan Papua Rusak 3,6 Juta Hektar', http://www.westpapua.net/ news/02/06/210602-hph.htm.

World Forestry Congress, Eighth (1978) *Proceedings of the Eighth World Forestry Congress, Jakarta, Indonesia, 1978*. 7 vols. Jakarta: PT Gramedia. Vol. 1: 3–5.

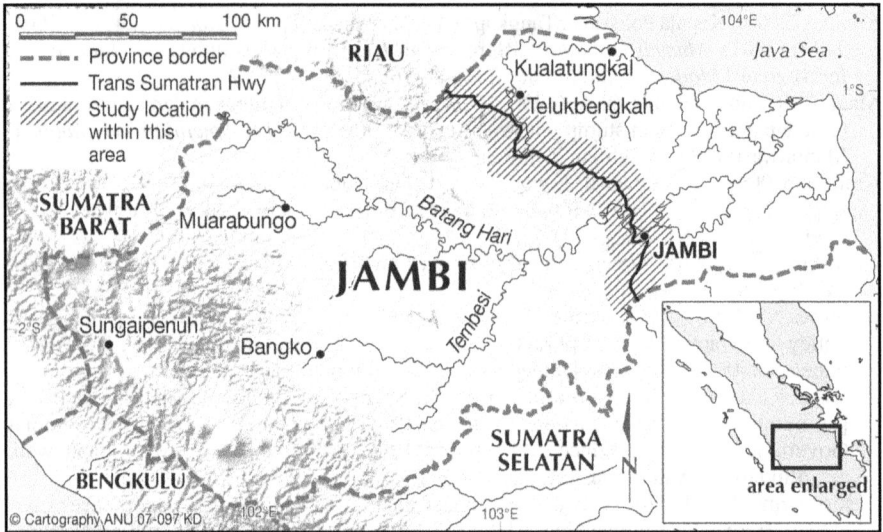

Map 7 Jambi Province, Sumatra.

7 'Where is justice?'

Resource entitlements, agrarian transformation and regional autonomy in Jambi, Sumatra

John F. McCarthy

On 16 August 2005, the day before Indonesia's Independence Day, I sat in a coffee shop in a village in the hilly subdistrict of Peluang in Eastern Sumatra. Here a Melayu farmer described how, along with a large group of local farmers, he had opened an area of land three years earlier to plant oil palm. Although he lacked land tenure certificates under state law, alongside fellow villagers he felt assured that the district government would allow for this traditional claim over abandoned timber concession land. Later he learnt that the Ministry of Forestry had allocated the same area of land to the conglomerate PT Pulp and Paper to plant an acacia monoculture for an industrial tree plantation. Now, given his want of legal certificates, he faced eviction by a militarized police brigade (*Brimob*) flown in especially for this purpose from Jakarta. Gazing at the Indonesian flags that hung around the village, he told me this was no time to celebrate independence. 'We are not independent,' he fumed. 'Maybe the company can celebrate, but we still feel colonized.'

This farmer's response points to the difficulty that Indonesia, despite recent reforms, still faces in reconciling the economic interests of powerful corporate actors and their political partners, the revenue needs of the state, environmental sustainability and the cry for agrarian justice among rural populations in the face of a rapid agrarian transition. It also illustrates the reality that rural Sumatra, like many other areas of Indonesia, is a place torn by conflicts. These include land conflicts that pit farmers, attempting to regain control of their customary village lands or extend their smallholdings, against timber conglomerates and palm oil plantations.

In historical terms these disputes arise from a structural transformation changing long-standing customary systems of local resource control. Between 1968 and 2005, the total area under oil palm cultivation in 'outer island' Indonesia increased from 120,000 hectares to about 4.1 million hectares, much of this in Sumatra, the historical centre of the plantation sector in Indonesia (*Jakarta Post*, 21 January 2005). From the late 1980s the Indonesian government also launched an ambitious programme to build timber plantations (HTI). By the end of 2005, there were 209 operational HTI which, according to departmental statistics, on paper had extended their activities over a total area of eight million hectares.[1] During the last decade transmigrant farmers, rural elites and a limited number of customary landowners with access to knowledge and capital have also been converting local land – including agricultural fallows, areas that were, until recently, forested, and farming land purchased from impoverished local farmers – into oil palm. Taken together with the insecurity faced by farmers locked into older, more extensive agricultural practices, these changes have increased land scarcity, undermined existing economic structures, and intensified the uncertainty of Melayu farmers' incomes.

Following the transition away from authoritarian rule, reforms under regional autonomy legislation aimed to make government decision-making more responsive to local concerns, improve the position of regional actors with respect to resource access and distribution, and more effectively deal with encompassing conflicts over agrarian and natural resources. To be sure, the political opening associated with these changes led to a struggle over resource entitlements – between the state-guaranteed order established during the New Order, and an alternative local moral economy associated with the remembered (but now overshadowed) principles of a former customary *adat* order.

In analysing these changes, a useful distinction can be made between endowments, the property or resources that actors actually have, and entitlements, the claims over resources or benefits that actors can legitimate under existing legal or institutional arrangements.[2] While endowments refer to property which a person already possesses, entitlements emerge from the processes where actors negotiate and claim access to and use of resources. Such negotiations necessarily involve power relationships, identity claims, and debates over meaning; they also are mediated by the matrix of existing institutional arrangements (Leach *et al.* 1999).

When laws, legitimizing discourses and institutional arrangements change, endowments inherited from the past do not immediately shift – unless the new system of entitlements is worked out on the ground. Although Melayu resource endowments were inherited from the past, under the New Order a new system of legal entitlements came into play that effectively disregarded the parallel *adat* (customary) regime. With the granting of plantations and timber concessions, corporate actors came to command control over extensive agrarian resources in the local domain. With the Reform Era and regional autonomy, the question emerged as to what extent an emergent order of entitlements would allow villagers to reclaim some of their old *adat* endowments or otherwise assist them to make the transition into more productive smallholdings. Alternatively, would the New Order pattern of entitlements finally settle back (more or less) into place without providing local farmers with improved and sustainable livelihood options?

The rapid and locally intensive transition into oil palm raises serious sustainability questions. With oil palm developments transforming whole landscapes, villages and regional economies, the transition is associated with a process of agrarian differentiation that has significant impacts on socially shaped relationships. Are the impacts on village living conditions and land tenure systems sustainable over the long term? If village and regional livelihoods become overly dependent on a single commodity subject to global market fluctuations, are the risks for village and regional livelihoods too high? Further, intensive oil palm production is associated with significant cumulative ecological impacts – including biodiversity loss and declining ecological functions from pollution, deforestation, carbon emissions from fire, increased vulnerability to pest infestation, and possible impacts on human health. On a larger scale, can the planned expansion of intensive oil palm cultivation be continued at the expense of the forests which only decades ago covered 70 per cent of Indonesia's landmass without a significant negative impact on the environmental services forests offer at both local and global scale?

This chapter reflects on these issues in Jambi Sejati, a district in Jambi province, on the east coast of Sumatra, predominately populated by people of Malay (Melayu) ethnicity.[3] The following section describes the concepts of common property and interest embedded in shifting village *adat* regimes and the effect of state policies.

Background

In a similar fashion to other parts of the world, farmers practising shifting cultivating in Jambi traditionally embraced diversity, developing 'a multi-stranded and spatially fragmented approach to building livelihoods' (Wilson and Rigg 2003: 695). This involved a diverse portfolio of activities, including fruit and vegetable production from gardens, the production of agricultural commodities for export markets – particularly rubber – for cash income, and rice production in shifting plots, supplemented with the collection of non-timber forest products in surrounding forests (Mubyarto 1992). This was a means of spreading the risks associated with ecological variations and fluctuations in the market economy through time and space (see Dove 1993).

The system of resource control worked in the following way. Behind the extensive coastal mangroves and peat swamp forests lay a hinterland of agricultural and forested land cultivated by people of Melayu ethnicity. Here villagers engaged in swidden agriculture and rubber cultivation. Until the logging concessions began to construct roads during the 1980s, farmers from villages located along the rivers would travel up tributaries to reach their farming lands and carry agricultural goods down to market by boat. Villages tended to develop at strategic nodes along the rivers, where farmers could readily access lands. Over time, particular villages had developed territorial rights to lands found along particular stretches of river and along the tributaries accessed conveniently from their village. The borders between neighbouring villages were formed by watersheds and territorial (*ulayat*) *adat* rights of a particular village were identified according to the aphorism 'as long as the water drains to the river' (*asal air jatuh di sungai*). Given that the area between these rivers and tributaries was extensive, the *adat* lands of each village in Peluang were also far-reaching. The colonial administration systematized and mapped this customary order of entitlements under territorial units known as *marga* which managed 'adat community' territories under the colonial system of indirect rule.

In the Suharto period, three major changes affected this local land use pattern. As in other areas of 'outer island' Indonesia, first, state policy created a land tenure and forestry regime that systematically overlooked indigenous tenurial assumptions (McCarthy 2000). Second, changes in village government under 1979 legislation led to a systematic eclipse in *adat* forms of authority, taking away the formal role of *adat* in the management of local resources. Yet, while this led to the decline in *adat* forms of authority, the idea of a local moral economy embedded in *adat* has persisted. Third, the area was transformed into a resource frontier where the commercial value of resources became paramount (McCarthy 2006). Subsequently, as New Order resource policies were applied, other social, cultural and ecological values embedded in the old *adat* moral economy were disregarded.

With the extension of these state policies into the frontier areas of Jambi, the system of resource entitlements shifted along with changing laws, power relations and strategies of capital accumulation and revenue generation. Local actors could no longer simply negotiate access to land and forest resources within a frontier space previously allowed for by the colonial socio-legal dispensation. A new order of resource entitlements had come into being that worked in ways that did not accord with long established, local resource rights. As the New Order state began to allocate rights, it displaced established *adat* entitlements and facilitated the transfer of *adat* territory to corporate actors under legal arrangements that were supported by actual or threatened action by security agencies. Yet local conceptions of *adat* rights tied to local identity remained salient in village communities as alternative concepts of resource entitlement, even while the state oversaw a process of enclosure and effective privatization of extensive areas of *adat* lands that had previously been at the disposal of villagers.

In a number of respects this had a remarkable effect on Melayu communities. First, with the hiving off of village common pool resources to timber concessions, forest depletion ensued and forest products became scarce. Second, with the expropriation of dry agricultural swidden (*ladang*) lands for plantation developments and transmigration settlements, villages had lost extensive surrounding areas required for swidden and rubber cultivation. Land shortages emerged, a problem exacerbated by population growth and in-migration. Rubber cultivation had been integrated into a specific land use framework where villagers did not intensively manage 'jungle rubber' gardens, but rather used old, low productivity clones over extensive areas. 'Jungle rubber' gardens – dissociated from the extensive swidden shifting cultivation system, squeezed into smaller areas, and in many cases needing replanting – were unable to support community livelihoods on their own. Taken together, these changes had dismantled the multi-stranded, extensive forms of livelihood strategies that villagers had pursued over many decades. Meanwhile, villagers lived alongside oil palm plantations located within their former customary territories that had been taken without appropriate compensation and that failed to employ village people in significant numbers. At the time when the government had supported the development of transmigrant 'satellite' oil palm smallholdings, Melayu villagers had been offered smallholdings in the Nuclear Estate and Transmigrant (PIR-Trans) development. The Melayu villagers mostly refused to join projects they saw as designed for 'poor' Javanese. A decade later they watched jealously as these transmigrants farming in former *adat* territories became better off than the original Melayu inhabitants.

From 1998 these accumulated grievances and unresolved claims surfaced in conflicts with plantations and spontaneous occupations of some areas. I will now discuss the development of village conflicts with two oil palm estates, disputes with their roots in the New Order period, as well as the possibilities opened by 'satellite' smallholder oil palm developments, before finally considering disputes that emerged from developments in the regional autonomy period.

Conflicts over New Order era oil palm estates

During the Suharto period a large oil palm group pocketed a plantation licence (HGU) over 9,000 hectares of land in Jambi. Although this land fell within the boundaries of eight villages, it had been classified as part of the nation's 'forest estate' and accordingly was allocated to the company, PT Minyak Manis. Local officials describe how they had been involved in a rather careful process that sought to compensate villagers. Given that this land was considered part of the state administered 'forestry estate' under national law, villagers only received compensation for rubber trees. The state would not compensate villagers for their land rights or for the loss of secondary forest and fallow swidden areas even though these fell within *adat* territory.

During the political upheavals of 1998, villagers joined together into a farmers group (*Kelompok Tani A*) to take action against PT Minyak Manis. Farmers involved in this group maintained that the *adat* lands allocated to the company had been taken over without sufficient compensation. 'Our land was taken over arbitrarily during the era of ex-president Soeharto,' Yusril, the leader of a 2001 demonstration, was quoted as saying. 'In those days, the central government was trying to boost the output of the palm oil industry and strongly supported plantation owners, without caring about the problems suffered by local farmers' (*Indonesian Observer*, 27 January 2001). In 1998–1999 a series of demonstrations had culminated in the burning down of the plantation office and equipment, reportedly costing Rp 16 billion.[4]

According to provincial officials, in the early days of *reformasi* (1998–1999) government officials felt overwhelmed by conflicts that found their source in New Order resource allocation policies. By 2004 the state was beginning to regain its confidence, and various levels of government were becoming more 'tough' (*tegas*) in handling these conflicts. They now demanded evidence of pre-existing *adat* rights in plantation lands from protesters. Given that these areas had long since been converted to plantations, that the state had already compensated villagers for their tree crops, and that *adat* rights lacked legal certification, protesters faced difficulties substantiating their grievances. A district government official noted that, although informally he could understand *adat* claims, as villagers lacked a state recognized basis for proving their claims, the government would not compensate villagers. Compensation would amount to recognizing prior claims, overturning the assumptions underlying state regulation of land affairs, and would open a floodgate of claims. Consequently, negotiations dealt with the *de facto* demonstration of claims and avoided the question of legally legitimated rights.

By 2004, according to a member of the district legislative assembly (DPRD), these land conflicts had become the single biggest issue occupying the time of the assembly. Virtually every plantation company in the district had a dispute with surrounding communities. The Jambi Sejati district assembly had set up a commission to handle these conflicts. However, as the DPRD member noted, the conflicts were intractable. With conflicting maps developed by different levels of government, and with a legal system that failed to provide a clear and con- sistent framework for recognizing these *adat* rights, efforts to resolve conflicts could only proceed via pragmatic efforts to accommodate villagers through a programme that would benefit them, such as the Farmer Group A (*Kelompok Tani A*) initiative discussed below. In effect this approach worked to weaken the position of villagers: in the absence of hard evidence and access to the formal legal system, conflicts could be solved only informally, and there was no provision for long-term security.

The district administration now plays a key role in efforts to mollify villages in these various disputes. But these solutions often raise their own problems. When the government set about resolving the PT Minyak Manis dispute, as a part of the settlement, Farmer Group A was granted an area of 1,500 hectares from lands outside the forestry estate and 2,500 hectares of forestry land in the village of Sungai Badak to develop a 'satellite' (plasma) area for smallholders. The leaders of Farmer Group A listed villagers who would receive allotments of 'satellite' land, issuing cards to many members of Farmer Group A that would entitle them to receive a land allocation. Farmer Group A also obtained logging permits from the district government to log the areas through the cooperative established in their name; the leadership of Farmer Group A then oversaw the logging of this area. Although PT Minyak Manis promised to develop oil palm smallholdings for cooperative members, the development stalled during the economic crisis period because the mother conglomerate claimed that it was unable to provide credit. Instead, PT Minyak Manis supplied 100,000 oil palm seedlings, a bulldozer, a car, and Rp 100 million in capital. Those leaders controlling Farmer Group A obtained these assets. While they either sold or distributed the 100,000 seedlings, it remained unclear where the other assets ended up. Meanwhile, some people from Sungai Badak village, including the village head, said they would not accept Farmer Group A obtaining satellite lands in their village territory because it represented a loss of its own customary *adat* lands. In any case, the conflict between Farmer Group A and Sungai Badak village was avoided by default. In the absence of credit and an effective partnership with PT Minyak Manis, the Farmer Group A 'satellite' development failed to proceed.

Many of the farmers from the five villages that have sought compensation from PT Minyak Manis were disheartened by this experience. With the support of the environmental advocacy organization, WALHI Jambi,[5] and a consortium of Jambi NGOs, a group of farmers from one village formed a farmer's group (*Gelompok Tani B or GTB*) that continued to struggle for compensation. GTB and their NGO supporters organized a series of demonstrations in the district capital and Jambi City, making representations to the legislative assemblies at both levels and the district head and governor. These villagers also occupied areas of company land from 2000 to 2003. In response, *Brimob* officers, reportedly receiving payments from the plantation, repeatedly arrested villagers and attempted to intimidate them, at times firing shots into the air. At the time of writing GTB had yet to succeed in their efforts to obtain compensation.

A second case with roots in grievances associated with New Order land acquisitions emerged on the other side of the same villages. Here the state had also 'freed up' land that was 'surrendered' by the village heads for PT Banyak Sawit's Nuclear Estate and Transmigrant (PIR-Trans) development.[6] This area largely encompassed forest lands that included less extensive areas under cultivation by surrounding villages. Since much less of the land in this area had been utilized for farming, PT Banyak Sawit never faced the same degree of contention as PT Minyak Manis. Nonetheless, a dispute emerged with surrounding villagers when a local PT Banyak Sawit employee disaffected with the company leaked word that PT Banyak Sawit had taken possession of 2,006 hectares above what had been allocated to it under its plantation concession permit. Subsequently, the six surrounding villagers began to make land claims on the basis that the company had taken village *adat* land above the limit of its legal allowance and at the expense of villages.

PT Banyak Sawit had clearly breached its legal rights and faced pressures from all sides. The district government sought 'a compensation contribution' and taxation back payments from PT Banyak Sawit for the profits derived from these excess lands (*Jambi Ekspress*,18 July 2003). Meanwhile, surrounding villagers sought to obtain their share of this land. Eventually, the company was forced to divide up 974.5 hectares among 700 villagers, with most claimants receiving 0.83 hectares of productive oil palm (*Jambi Express*, 7 October 2003). Even though the parcels of productive oil palm land received by these villagers were small and the processes of allocation contentious, the villagers who received them attested that it made a significant addition to their meagre livelihoods. This increased the desire of Melayu villagers for further redistributions and for developing their own oil palm smallholdings. Later it was revealed that PT Banyak Sawit was still 1,032 hectares above its legal HGU concession area, and this time the district government demanded that PT Banyak Sawit return this land to the district government. The district government aimed to use the area as a source of income and to avoid the highly contentious process of deciding which villagers should obtain redistributed land. Villagers demonstrated outside the district assembly and spontaneously occupied an area of PT Banyak Sawit land (*Jambi Express*, 2 October 2003). In the absence of a sympathetic response from district government, NGOs threatened to take the case to court (*Jambi Express*, 10 February 2004). Nonetheless, at the time of writing the district had been able to resist these claims to return this excess land to villagers.

A comparison of these two cases points to elements that need to come together to support villagers' claims against powerful outside actors. In the first case, legally speaking, PT Minyak Manis was on firmer ground: the state had granted it legal concession rights over what was formally declared to be state land (*tanah negara*). Challenging the definition of state land clearly was beyond the capacity of villagers or local NGOs. Consequently, they lacked the means of asserting their claims for tenurial rights over the PT Minyak Manis area in a fashion that the state was prepared to recognize. As in other cases, when faced with a powerful opponent and unable to threaten legal sanctions, villagers had a weakened ability to negotiate a solution (World Bank 2004). The company elicited local government support in its attempt to defuse the conflict via an accommodation. Apparently 'captured' by the cooperative managers, the government's concessions in any case failed to address the substantial grievances. In the second case villagers initially had the support of the district government. PT Banyak Sawit clearly had exceeded its legal rights, and faced the real threat of legal sanction. So there were considerable incentives for the company to settle the dispute via negotiations outside of the courts. A section of PT Banyak Sawit's lands were redistributed to farmers. Later a dispute emerged between villagers and the local government over a second parcel of land. However, this time villagers faced a powerful opponent without (at the time of writing) strong external support.

These disputes point to the reality that local outcomes emerge from, and as a result of local processes involving direct action, clientelist relations and *ad hoc* processes that emerge in response to disputes. It cannot be doubted that while state structures and laws that provide for legal entitlements shape these processes, they often do so without addressing underlying, embedded notions of *adat* rights tied to local identities and livelihood strategies. Previously these notions of *adat* rights were embedded in local moral economies that underpinned the institutional arrangements governing local 'commons'. After the enclosure of these 'commons', the various localized processes emerged in the absence of effective state legal structures. Such processes can certainly force a slow down and perhaps even a compromise. However, in the absence of powerful local NGOs, they typically fail to bring about fundamental change in the developmental trajectory. In other words, they do not articulate with formal decision-making processes in a fashion which leads to the resource rights of marginalized local groups finding institutionalized expression in an enduring form. This is because the lack of clear legal legitimacy of local claims curtails the negotiating power of village actors. Furthermore, these localized processes are readily dominated by informal power relations that lack accountability and transparency. Accordingly, the enhanced capacity of marginalized local actors to take action during the period following the demise of the authoritarian regime has not seen substantively better outcomes for marginalized customary landholders. As an alternative to pursuing a dispute with an oil palm estate, as I will discuss now, farmers could join a 'satellite' oil palm development.

Customary landholder grievances with cooperatives

The idea of a cooperative as an association of persons who join together in an egalitarian fashion to carry on an economic activity of mutual benefit has long held a symbolic position in Indonesian nationalism.[7] In the early twentieth century Indonesia's cooperative movement had emerged in resistance to colonial policy as part of the efforts by Indonesian nationalists to develop the economic independence of ordinary people. After 1949 the newly independent Indonesian government sought to encourage and promote cooperatives as legal entities owned and controlled by their members. In subsequent legal formulations cooperatives had the twin objectives of improving the welfare of members, and that of the society in general, to participate in developing the national economic system.[8] Subsequently cooperatives became the principal policy tool for soliciting the participation of ordinary people in national economic development. Indeed, in 1998 the national People's Consultative Assembly's (MPR) decree on regional autonomy had stressed the role of cooperatives, affirming that the management of natural resources should be carried out in 'an efficient and open fashion that provides extensive opportunities to cooperatives, small and medium size enterprises'.[9]

Earlier, when the government had attempted to increase smallholder participation in oil palm developments in the 1990s, it had initiated the Prime Cooperative Credit for Members (KKPA) scheme. In 1995, under this scheme credit was made available at concessionary rates for developers who needed to establish a separate company to partner a cooperative of customary landholders. While cooperative members were required to contribute the land, the developer had responsibility for supplying the capital. In a similar fashion to the Nuclear Estate-Transmigration scheme, 20–30 per cent of the land made available for oil palm development would be provided for the company's nucleus estate. The remaining 70–80 per cent was allocated to customary landholders working 'satellite' areas (Casson 2001). By emphasizing that private estates should be based on a 'partnership' relationship with local people, the scheme aimed 'to maintain smallholder involvement and protect the rights of local landowners' (Potter and Lee 1998).

On the one hand, these policies could provide the capital, knowledge and technology that local landholders lacked, which had effectively locked them out of the transition to oil palm production. On the other hand, it is possible to argue that cooperative credit (KKPA) schemes extended earlier contract agriculture policies that, by allowing plantations both to gain access to community land and to control smallholder production, created a 'captive orientation' of smallholders towards nuclear estates (Dove and Kammen 2001). This dependency emerges because customary landholders are locked into an arrangement with the cooperative that developed their 'satellite' smallholding in partnership with the oil palm nucleus estate. During the four years before the oil palms can be harvested, smallholder participants receive subsistence level support from the cooperative. Later they must repay this, 'plus the costs of land clearance, planting, fertilisers and other inputs – including cooperative administration – from the harvest'. Further, 'farmers have little bargaining power over the price for their crop where palm mills are scarce and owned by the plantation company', or where they are obliged by cooperative arrangements to sell their crop to the company. International fluctuations in crude palm oil prices and instability of the Indonesian currency further affect smallholders. Consequently 'smallholders can remain in debt for decades without ever knowing when they will be free of this burden' (Down to Earth No. 63, 2004).

In Jambi Sejati a number of village and subdistrict actors began to form cooperatives and work on initiatives to take advantage of this credit scheme.

However, the sudden collapse of the banking system in 1997 prevented several oil palm developments from continuing. This included a number of those under development through cooperative credit schemes in Jambi Sejati, schemes that later faced problems in their own right.

Under the cooperative credit scheme in 1995 the Garuda Group had formed a company, PT Petani Makmur, to collaborate with a partner – the local cooperative, Koperasi Unit Desa (KUD) Jambi Sejati Ulu. Subsequently, under this partner arrangement village land was freed up for the oil palm development. Prior to the economic crisis PT Petani Makmur had obtained land from three villages in Peluang, with 70 per cent of the land to be allotted to customary landholders in the cooperative and 30 per cent to be allocated to the company's Nucleus Estate. Accordingly, 2075 villagers were listed as 'candidate smallholders'.

PT Petani Makmur began to clear village rubber gardens for oil palm in 1996. The first of the oil palm began to be productive in 1999 in the village located alongside the trans-Sumatran highway. Living along this major highway, these villagers were already well integrated into the wider economy. Many had participated in a Nuclear Estate development in the neighbouring transmigration site, and PT Petani Makmur had developed the oil palm in this area before the economic crisis. In contrast, poor villagers in the two neighbouring and more remote villages lacked direct experience of oil palm developments and were comparatively unprepared for the transition. Some subdistrict officials alleged that during the economic crisis PT Petani Makmur's finances dried up and the company discontinued the development of 'satellite' oil palm gardens there. However, many villagers suspected more underhand motives, alleging that particular people – either in the cooperative or the company – were conspiring to extend their landholdings at the expense of villagers. In any case, in these two villages PT Petani Makmur mismanaged the oil palm gardens, for instance failing to apply the specified amount of fertilizer, allowing trees to die, and not planting the specified number of oil palm trees per hectare.

In 2003, some five to seven years after they had given over areas of rubber garden to PT Petani Makmur, villagers still saw their land out of production. At this point brokers went around the village offering to buy up the cooperative membership cards that entitled villagers to 2 hectares of oil palm smallholdings. In two of the remote villages where PT Petani Makmur was active, the desperately poor majority of villagers sold their membership cards, assuming that the cooperative credit venture had failed. Brokers working on behalf of the company, local entrepreneurs and even a manager in the local cooperative (KUD), told villagers that they would be better off selling up. In this way they snapped up the cooperative membership entitlements at bargain prices, with one subdistrict official buying eight 2 hectare areas and a prominent local entrepreneur gaining control of 1,000 hectares of land planted with oil palm.[10] Then, despite the story of the company's economic failure which had been circulated, shortly after the purchase PT Petani Makmur began to cultivate the areas that had been left unplanted. Those villagers that had managed to retain their satellite entitlements and finally gained possession of their plots began to benefit from productive oil palm smallholdings. However, the villagers who had sold their cards bitterly looked on as their village's former swidden areas were converted into plantation smallholdings by outsiders. With even less land available for cultivation within their customary village territory, they were in a dire position. In the words of one village leader, the development left 90 per cent of villagers 'naked'. Having given over large rubber gardens, and without sufficient land remaining to subsist, they were left 'living like chickens scavenging for whatever grain they could find'.[11]

Although cooperatives are – at least normatively – a form of collective organization, interviews with Melayu villagers indicated that they had never been able to hold cooperative leaders accountable. During the New Order period the government had set up a national programme to supply rice on credit to poor villagers, through a network of village cooperatives. In Peluang the leadership of this vertically integrated cooperative network (KUD Jambi Sejati Ulu) and its subordinate village level units was chosen during 'consultations' (*musyawarah*) in the subdistrict township by selected villagers invited to the meeting. In the early 1990s, under the weight of bad debts, this cooperative, like many others across Indonesia, became inactive due to corruption and poor management. In 1995 prominent subdistrict actors reconstituted the KUD Jambi Sejati Ulu as a vehicle for partnering oil palm companies in developing cooperative credit 'satellite' projects. With its headquarters in the subdistrict capital and a membership of over 2,000 spread across the subdistrict, the cooperative effectively remained under the control of these actors from the beginning.

When asked about the cooperative, many Melayu villagers made cynical comments. According to one disaffected villager, only villagers wishing to go along with the managers joined the cooperative: those who didn't agree quickly dropped out. Members were unaware of how the cooperative leadership were chosen, how cooperative funds were spent, and the basis of compulsory credit repayments extracted from the oil palm produced on their oil palm smallholdings. They were also unsure why, even though oil palm becomes productive after three years, it took seven years for the cooperative to turn over oil palm plots to the cooperative members with entitlements in the 'satellite' development. An observer resident in the subdistrict township said that 'the profits were only enjoyed by the managers' of the cooperative, describing how, after becoming a manager, a local schoolteacher bought a new house and a luxury car. Meanwhile, in the absence of effective mechanisms for collective action at the village level, villagers merely vented complaints outside cooperative meetings.

State regulations stipulate that local smallholders form cooperatives when they engage in business activities – such as timber or oil palm developments – to obtain permits and to participate in a fashion that distributes the benefits of development more widely. De Soto controversially has argued that, for poverty alleviation, property rights operating outside the state system of land tenure need to be brought under a formal unified legal property system in order to ensure that they can serve as a means of generating capital for development among the poor (de Soto 2000). In a similar vein, a community leader noted the problem that, without formal certification, village customary property rights remained outside of the state recognized system of entitlements. The cost and other difficulties of obtaining formal tenurial rights made it difficult for villagers to obtain credit from a bank.[12] As an alternative, villagers could form a cooperative with the status of a 'legal body' (*badan hukum*), he argued. This would enable them to operate in the formal economy. For instance, if they sought to collectively obtain legal certificates recognizing their land tenure under state law, they could do so at a much lower cost for each individual involved. With the legal status of a cooperative and with legal tenure, they could then obtain credit from a bank.

The failures experienced in Jambi Sejati were not inevitable. While cooperatives in PIR-Trans schemes were not always successful, interviews in neighbouring transmigrant communities revealed that the problem was less severe. With village cooperatives under the leadership of 'poor' transmigrants, these cooperatives tended to be less vertically differentiated and horizontal networks operated more effectively. Consequently, villagers were better able to coordinate and this heightened their ability to hold the leadership of their cooperatives (elected among themselves) to account. Consequently, cooperatives in many transmigrant villages were more successful and better able to secure outcomes for the collective good of their members.

In contrast, in Melayu villages longstanding resource endowments could only be translated into legal entitlements via the state established cooperative mechanism – which in these contexts constituted an externally imposed and bureaucratized mode of resource control. In the absence of effective recognition of village or *adat* property and resource management rights, commercial development of these resources could be mediated only via cooperatives. Yet these cooperatives came under the control of subdistrict elites that worked in collusion with the oil palm estates. Consequently, cooperatives in the Melayu villages were highly differentiated, and lacked the normative controls and horizontal networks of engagement that might advance effective institutional outcomes (see Putnam 1993).

Hence the two or three managers who controlled the cooperative supervisory committees also controlled the means by which community members could gain legal rights to exploit timber or obtain oil palm smallholdings within 'satellite' oil palm developments. In the absence of effective collective accountability mechanisms, all too often those who manage the cooperatives end up controlling the benefits of resources held under its name (ILO 2004). Once again, those occupying key positions in the local bureaucracy or otherwise enjoying access to capital, knowledge and permits, captured the benefits from oil palm and forest policies under regional autonomy.

Emerging problems after regional autonomy

Despite the endemic disputes associated with oil palm developments, district and provincial governments have remained enthusiastic about developing oil palm plantations in the regional autonomy period. Given that revenues from the timber sector have now fallen away – due to mismanagement during the Suharto era and uncontrolled logging in the immediate aftermath – district and provincial governments had particular incentives for promoting oil palm development to raise revenue from land (PBB) and other taxes that could be levied on plantations. While district powers over the forestry estate remained limited under the regional autonomy legislation, districts had initially gained some authority over the granting of plantation licences (*izin lokasi* and *izin prinsip*) within their domains.[13] Accordingly, in 2001, the Governor of Jambi announced plans to develop a million hectares of oil palm in the province.[14]

Earlier, the 1992 Spatial Planning Law (UU 24/1992) had set out a framework for zoning land use according to social and environmental criteria. It also provided for community consultations, as well as a framework for bureaucratic regulation and sanctions to ensure that developments accorded with these plans. In theory the law allowed for district, provincial and national governments to develop their own spatial plans which would be brought together to achieve coordinated and coherent land management across sectors and levels of government while incorporating the results of community consultations. However, despite the best intentions of this law, as we will see, outcomes were shaped by the competition between different agencies seeking to maximize their jurisdictional control over resources and opportunities to capture rent in this area.

The Jambi Sejati district administration faced significant problems in facilitating new plantation developments. An earlier provincial spatial plan had excised large areas from the forestry estate in 1993. This spatial plan was reconciled with the forest planning maps, leading to an integrated map (Padurasi) for the province in 1999. This process had recategorized some 40 per cent of the forestry estate within Jambi as 'areas for other uses' (APL) and made it available for plantation development. However, by 1999, all large areas of this district listed as APL had already been allocated to plantations. This meant that, if the district wanted to facilitate further plantation development, it would need to take new areas out of the forestry estate. While the district land agency (BPN) now had the power to grant a temporary 'location permit' for a plantation development if it fell outside official forest boundaries, the Ministry of Forestry retained its authority over land use decision-making, including for timber concession permits and conversion to other uses within the forestry estate. The district could only recommend to the governor that an area be excised from the forestry estate for plantation development. In 2001 Jambi Sejati district wished to develop a new district spatial plan that would further reduce the forestry estate by rezoning significant forestry areas as 'areas for other uses'. After this rezoning, the district would be able to make forestry areas available for timber extraction under district timber licences and for conversion to palm oil plantations. If the district succeeded in its objectives, land use decision-making might be more accommodative of village aspirations to open land in areas subject to customary claims but currently mapped inside the national forestry estate.

Until the district spatial plan was finalized and reconciled with the province and forestry agencies concerned, the final status of many areas was unclear. According to one source, the district delayed this final day of reckoning by failing to provide a revised spatial plan to the province. Making use of this lack of certainty, the district administration could still issue 'location permits' as well as small-scale timber exploitation licences in areas it said had been reclassified as 'areas for other uses' in order to collect revenue on these activities. By the time the forestry agencies found out about these changes, the areas would already be *de facto* plantations, and the forestry agency would be forced to accept this *fait accompli* and redraw its maps. However, this uncertain status quo clearly threatened the interests of the Ministry of Forestry and local forestry offices (*dinas kehutanan*) alike: any *de facto* reduction in the forestry estate would amount to a reduction in the land area over which it had authority. These problems were to come to a head in a complex dispute over timber plantation developments in Jambi Sejati.

The dispute over ex-timber concession lands in Peluang

In 2002 some 46 per cent of Jambi Sejati district still remained mapped as forest estate. The Ministry of Forestry, however, had failed to ensure the sustainable management of production forests in these areas. While large areas of logged over secondary forest remained, these natural forests no longer offered sources of commercial timber for concessionaires. With the district government angling to rezone areas of unproductive forest for oil palm development, and with villagers opening gardens in many 'forest areas', forest agencies faced the dismal prospect of losing *de facto* control over their resource base. The development of timber plantations (HTI) gave forestry agencies a means of maintaining their power over the forestry estate. The ministry acted quickly, issuing new tree plantation concessions in Jambi Sejati's national forestry estate.

In the 1990s PT Pulp and Paper, a daughter company of a major national conglomerate, had already obtained control over extensive former concession areas in Jambi, including in Jambi Sejati. These areas provided the raw material for a mill run by the giant paper and pulp conglomerate. By 2001 this company controlled some 115,000 hectares of Jambi Sejati. It was hardly surprising that PT Pulp and Paper soon found out about the district's plan to rezone areas of the forestry estate within district boundaries. If the spatial plan was allowed to stand uncontested, PT Pulp and Paper faced the loss of significant areas of its timber plantations to oil palm plantations. By this time PT Pulp and Paper had a Rp 600 billion investment in Jambi. The parent conglomerate also had significant 'political investments' both within Jambi and at higher levels that could be used to shape how decisions were made. In the face of PT Pulp and Paper power, the district eventually withdrew: in 2004 the district spatial plan was revised, dropping the proposed reduction of forestry estate in the district by 40 per cent. The district later also acquiesced in the decision to extend PT Pulp and Paper's plantation holdings in the district under, what some alleged, was a behind the scenes *quid pro quo* deal. In early 2004, the Ministry of Forestry issued new timber plantation permits to PT Pulp and Paper in Jambi, including permits over three former timber concession areas within the district. This brought the total area subject to PT Pulp and Paper to well over 200,000 hectares, mostly within Jambi Sejati, leaving PT Pulp and Paper in control of the majority of the forestry estate found in Jambi Sejati.[15] In 2004 some of these areas had been or were being clear felled and replanted with *Acacia mangium*.

Consequently, apart from the remaining forest found in an adjacent national park and a former concession bordering the park, which was also in the process of being allocated as a timber plantation, this policy practically sounded the death knell of the remanent natural secondary forest areas found in the district.[16]

Given the extent of these holdings, now most of the forestry related business in Jambi Sejati consisted of PT Pulp and Paper activities: PT Pulp and Paper had become the primary patron of the forestry agency in the district. According to a local official interviewed during 2004, PT Pulp and Paper made significant 'contributions' to a range of officials in the district. At least with respect to forestry, the district capital had become a PT Pulp and Paper company town.

Many of the above dynamics have come together in a dispute over 6,000 hectares of 'forestry estate' land in Peluang. This area had originally formed a part of a logging concession, after which effective management of the area had ceased. During deliberations over its spatial plan, the district government wanted to rezone this as 'areas for other uses'. In the meantime, PT Pulp and Paper found out about this plan and successfully lobbied in Jakarta to obtain an HTI timber plantation concession in this area.

Historically this area fell within the village boundaries of the corridor of villages now sandwiched between PT Minyak Manis on one side and the PT Banyak Sawit/Nucleus Estate oil palm development on the other. Given the land shortages facing many farmers described earlier, the land found in the ex-timber concession area was particularly attractive: it was adjacent to a main road, apparently without a landlord, and ideal for land pioneers wishing to open oil palm gardens. For this reason, the area attracted the attention of small landholders wishing to extend their holdings and speculators who calculated on obtaining compensation if they were forced to move. Many villagers genuinely in need of land aspired to cultivate this former concession area. In interviews during 2004, some villagers pointed out that the companies had already taken over most of the former village territories, including most of their swidden agricultural areas. In their view, before the permit for the timber plantation was issued, the local government should have worked out how many families lacked land, so that those without land had somewhere to go. Villagers were passionate about the issue. As one villager noted: 'So ordinary people face obstacles while the big people are helped? Where is justice?'.[17] A number of sources within the district intimated that some district government decision-makers wished to accommodate the villages. Some may have also wanted to see the area of APL expanded, and either encouraged, tacitly allowed for, or (according to one forestry official) provoked land pioneering in this area, as a means of contesting forestry agency claims.

In 2003–2004 villagers moved into the area to open gardens. Deploying *adat* as a framework for establishing land rights claims, they worked in groups to open land according to *adat* principles: villagers felt entitled to open land wherever water drained into the watershed where the village asserted *adat* entitlements.[18] Despite fear of later eviction, those villagers who took part decided to act first to establish claims on the ground. As one pioneer said, 'we go first, we don't want to watch on. We are native to this area. PT Pulp and Paper can go ahead, but please don't disturb the community.'[19] Some more wealthy actors also paid poor villagers to cultivate land here on their behalf. This set the scene for a complex dispute.

Usui and Alisjahbana (2003) have noted that an effective planning practice which responds to 'people's demands, priorities, and preferences, is one of the indispensable prerequisites for successful decentralization' (2003: 6). However no community consultations were undertaken before the decision was made to give industrial plantation concessions to PT Pulp and Paper.[20] Nevertheless, at least one village proposed that the district divide up the former timber concession area among farmers without sufficient land. According to the head of the village council (BPD), nothing ever came of this proposal. When some villagers made a formal approach to the subdistrict government regarding the availability of the land, the subdistrict forestry official sent a letter to the district forest agency. Predictably the forestry agency took the side of PT Pulp and Paper, stating that the area could not be cultivated because it was forestry land. PT Pulp and Paper personnel and forestry officials from both the provincial and district offices visited the area, advising the land pioneers to abandon their efforts.[21]

Behind the scenes the position of the district government remained somewhat more ambiguous. Under Ministry of Forestry regulations, the district head (*bupati*) was required to provide a letter of recommendation before the ministry would issue a timber plantation concession. During 2004 the district head of Jambi Sejati made two critical decisions that indicated the extent of PT Pulp and Paper's political investment and influence in the district. In March 2004, the district head wrote a letter of support for PT Pulp and Paper to obtain a plantation timber concession over the 8,685 hectare ex-HPH logging concession area within the district. The letter stated that the area 'very much caused concern due to the illegal logging and community land clearance', especially because the previous concession holder had been unable to 'secure the area concerned'.[22] In April the district head wrote a second letter once again agreeing in principle to PT Pulp and Paper, obtaining further control of two other former concession areas extending over 10,195 hectares.[23] Spatial planning and environmental impact assessment laws provide for public consultations before environmentally significant decisions are made. Moreover, spatial planning legislation provides procedures to ensure that sufficient land is set aside for community agriculture. If these provisions had been brought into force, perhaps a case could have been made for excluding the pioneer gardens from the timber plantation, in this way avoiding a dispute. However, the district head took these decisions without any public consultation with the villages that were making *adat* claims within these territories.

During this controversy district government was caught in the middle. As a cooperative leader noted: 'the local government faces regulations from the centre, but on the other side they want to help the community. The community's aspiration is to convert production forest into areas for other uses ... The production forest is very large, so why shouldn't a small section be taken out for the little people. PT Pulp and Paper controls 17,000 hectares in the subdistrict, so why not let 2,000 hectares be worked by the community? The community needs to live.'[24]

In mid-2004, PT Pulp and Paper organized a meeting in Peluang subdistrict to 'socialize' their activities, inviting provincial, district and village officials, *adat* leaders and other prominent villagers to attend. PT Pulp and Paper argued that the villagers who had cleared land along the road had no right to work this land, and indicated that it was prepared to use the police or the army to enforce its will.[25] However, a spokesman for local people warned the company 'to watch out' if they pushed villagers off the land.[26] According to press reports, by June 2004 PT Pulp and Paper was seeking to solve the problem through a 'cooperative partnership' that would to some extent accommodate village grievances (*Jambi Ekspress*, 7 June 2002). In a follow-up meeting in the district capital during July, officials representing the district government argued that PT Pulp and Paper should accommodate local people and let these areas be excised, especially if PT Pulp and Paper 'wanted to secure its future in the area'.[27] This accommodation would appease village grievances without affecting the structural position of PT Pulp and Paper or substantially touching on its large landholding in the district. In this way the district government sought to secure its position with powerful actors above while ensuring its esteem with aggrieved actors below.

This was particularly important in the run-up to the election of a new governor that occurred in November 2004, an election in which the district head was to compete as a candidate. Clearly there were no votes in being seen to support the company. Given that it was an open secret that PT Pulp and Paper secured its interests by 'investing' in the district apparatus, during this period, the district head wished to avoid being seen as working in collusion with the company. During 2004, according to village observers, the district administration hovered somewhere in the middle on the issue. As one villager noted, 'it does not forbid, but it does not grant'.[28] Another observed that, taking note of community aspirations, the deputy district head had reportedly said that the land occupation was not forbidden, 'after all, it is the community's government'.[29] Land pioneers working the area said that some officials told them 'to just manage what they had already, but not to add any new areas'.[30] According to an official in the subdistrict office, local officials, many residing in the villages involved, recognized that the '90 per cent' of villagers opening land in the former concession area were poor farmers, and generally preferred to see them hold out there.[31] This suggested the possibility of an informal accommodation.

However, during the Megawati period (2001–2004) the pendulum had already begun to swing back towards centralized power in Jakarta and the Ministry of Forestry. With the passing of a new forestry regulation (PP 34/2004), the Ministry of Forestry regained control over significant permits and licences in the forestry sector (Barr *et al.* 2004). In March 2005 the new president, Yudhoyono, issued a Presidential Decree (No 4/2005) requiring the whole government apparatus, including all governors and all district heads, to make a priority of fighting illegal logging and to act strongly against it. The instruction also urged government agencies to harmonize and streamline regulations and procedures in order to eliminate contradictory rules. The new policy climate worked against land pioneering that could be construed as illegal forestry activities.

In May 2005, bulldozers backed up by mobile armed police brigades began to remove the land pioneers out of the disputed timber plantation area. PT Pulp and Paper bulldozers began pushing over oil palm and rubber gardens that the villagers had planted in the area. Intimidated by the security forces, many villagers withdrew. When four villagers sleeping in their huts were arrested in June, this sparked demonstrations in front of the PT Pulp and Paper office. After villagers threatened to burn down the PT Pulp and Paper office, the four men were released. In a further altercation, a villager was arrested for hitting a company worker.

According to one man who faced losing his land here, the year before they had believed that the district government had supported their position, but now villagers felt let down by the district government. He raged about the lack of consistency in decision-making, accusing local government officials of 'playing a role in a soap opera' – performing one part to the conglomerate and another to the local community. At some moment, he said, 'this problem could explode', and the villagers could burn down the company's office. He now faced eviction after investing his own efforts and limited resources in developing the land.

Conclusion

With due consideration to how the changing matrix of laws and policy formulations affected resource entitlements, I wish to draw some conclusions regarding the key factors determining outcomes for ordinary Melayu villagers. Here I consider how local demands were addressed in planning and decision-making; what factors affected processes of bargaining and negotiation; and to what extent the reforms enhanced the capacity to integrate across and mediate among the differences kindling conflicts in this district. Finally, what conclusions can we draw regarding the possibility for justice and sustainability – in ecological, social and economic terms – within the new dispensation?

In the past villagers had the capacity to command resource bundles under what might be broadly considered *adat* entitlements. During the New Order the state system of resource entitlements to varying degrees transformed local people's ability to access forest and land resources. These changes were part of a structural transformation of long-standing extensive systems of agriculture that rested on a diverse ecosystem. Following the enclosure of village 'commons' by timber concessions, oil palm and timber plantations, the original inhabitants of the area lost their access to common pool resources previously availed in *adat* territories. With the passing of time Melayu villagers became increasingly aware of the affluence of oil palm estates and the comparative prosperity of transmigrant farmers who had gained productive oil palm smallholdings under nucleus estate developments. At the same time they were left locked into shrinking areas of less productive 'jungle rubber' land holdings. The end of the Suharto period occurred as Melayu villagers were becoming more acutely aware of their comparative poverty. With the opening political situation, the range of possible strategies available to farmers changed.

During the New Order some local landholders had joined the transmigrants in the PIR-Trans schemes or the cooperative smallholder (KKPA) schemes.

Indigenous Melayu farmers were clearly disadvantaged by the terms under which they were integrated in these 'satellite' schemes. Despite this, in the face of grinding rural poverty, there is evidence that Melayu villagers participating in these schemes did improve their economic situation compared to others now unable to pursue the diverse portfolio of activities that used to support village livelihoods since large areas of community land have been enclosed by plantations (Susila 2004; Zen *et al.* 2005; Papenfus 2002: 13).[32] Interviews carried out in the course of this study with the limited number of villagers who either obtained small plots from the redistribution of PT Banyak Sawit's area or KKPA plots in the PT Petani Makmur satellite area indeed found that landowners who gained productive areas of high-yielding oil palm expressed satisfaction regarding their changed economic situation.

As elsewhere, during this rural transformation poorer peasant farmers risked being locked into their old crop patterns and farming practices or otherwise undertaking transition under conditions that further their marginalization (Kay 2002). The rolling out of oil plantations over the landscape, the carving up of large areas for timber plantations, and the subdivision of family smallholdings with each generation have made land increasingly scarce. As elsewhere, the enclosure and commoditization of customary (*adat*) lands has been coupled with a process of agrarian differentiation due to uneven accumulation of land (Li 2002). Unable to pursue their previous livelihood strategies, poorer villagers face the challenge of making smaller and smaller areas of rubber gardens more productive by intensifying farming practices.

Without access to credit, superior planting material and fertilizers, however, those who attempt to plant oil palm on their own obtained yields that tend to be 50 per cent lower than those participating in the various schemes (Zen *et al.* 2005). Further, when independent smallholders sell their lower quality products to local mills, they obtain low prices. Consequently, those individuals who successfully enter into oil palm production on their own initiative tend to be staff and workers from estates, well-endowed local businessmen, village elites, and successful migrants in search of additional land. These are individuals who are able to buy land from poor local landowners and have the social networks, knowledge and capital to plant high-yielding varieties and engage with the market on favourable terms.

The questions of social sustainability raised by this agrarian transformation are compounded by questions of ecological sustainability. With rapidly growing markets for vegetable oils, now motored by accelerating demand for biofuel, and with Indonesia the most cost efficient producer of palm oil in the upstream production chain, oil palm offers quick profits.[33] Indonesia is in the midst of a palm oil boom: analysts have forecast that the total area of oil palm plantations in Indonesia is set to increase to 11.2 million hectares by 2020 (Wakker 2005). It is estimated that two-thirds of oil palm plantations are developed from the conversion of forest areas, mostly in endangered lowland evergreen tropical rainforest. This conversion is associated with forest fires, carbon emissions, and biodiversity loss (Pagiola 2000; WRM Bulletin 2004). While in neighbouring areas oil palm development is coupled with forest conversion on the frontier, the area subject to this study lies along the trans-Sumatra highway. Here, the conversion of forest to oil palm estates occurred earlier – in the early 1990s. Now, the most critical ecological transformation involves the conversion of remnant forests into acacia monoculture. This is occuring alongside the ongoing expansion of oil palm monoculture smallholdings, replacing 'jungle rubber', which is considered an environmentally desirable land use because it has a complex ecological structure and 'supports biodiversity in a similar fashion to a secondary forest in its mature phase' (Papenfus 2002: 3).

The ecological deterioration associated with these changes is intensified by the other pressing environmental issues associated with the plantation sector, air pollution and the accumulation of effluent in rivers. For instance, palm oil mills produce enormous amounts of liquid waste. A small factory with a capacity of 30 ton Fresh Fruit Bunch (FFB) per hour will produce 600 cubic metres/day liquid waste. The Biological Oxygen Demand (BOD) associated with this level of liquid waste is equivalent to the sewerage produced by a city of 75,000 people (Zen and McCarthy 2008). Given that the fresh fruit bunches need to be processed within two days of harvesting, these highly polluting mills are nested along the rivers in between villages. The factories deposit most of this waste into rivers and small streams, which poor villagers depend upon for washing, bathing and drinking water. These problems are intensified by the high demand for water required for growing oil palm trees and the impact of large fertilizer and pesticide runoff on ground and river water (Ginoga *et al.* 1999). This raises questions regarding the desirability of spatially intensive oil palm cultivation over the long term. The oil palm boom may demand too high a price in terms of ecological values and human health. Further, with the transition to oil palm transforming whole landscapes, villagers and regional economies risk becoming overly dependent on a limited range of commodities and livelihood options, leaving themselves open to threats from pests, diseases, fire and unpredictable fluctuations in market demand (Potter and Lee 1998).

Nonetheless, the economic opportunities associated with this crop together with the political openings of this period presented new opportunities for those well positioned to take advantage of them. The wider political economy of oil palm clearly opened a field of activity for members of the local elite and affected district government decision-making in their interests. For instance, people with knowledge, connections or capital might obtain a small-scale timber concession from the district government and pay villagers to work over areas of forest for a timber mill. These activities could generate the capital to buy land and to move into oil palm cultivation. With small amounts of capital they could buy up village smallholdings and cooperative memberships from villagers holding entitlements to 'satellite' oil palm holdings.

Districts have an interest in facilitating the exploitation of forests and their conversion into plantations to raise revenues from licences and permits, through formal and informal payments, and taxes levied on timber extraction and plantation lands. While the reforms left district governments without effective formal control over areas mapped as state forest land, districts had discretionary powers to facilitate new plantation areas and small-scale logging concessions, particularly in the early period of *reformasi*. This interest in maximizing the areas outside the forestry estate, together with the relative weakness of the Ministry of Forestry and the security apparatus in the early years of the Reform Era, meant that district administrations more or less turned a blind eye to villagers opening gardens within areas classified as 'forestry estate'.

At the same time, Indonesia's decentralization process was meant to produce local development plans based on more genuine public participation compared to the centrally dominated ones of the past (Usui and Alisjahbana 2003: 5). The reformed governance system provided processes that were meant to facilitate local participation in decision-making, including the formulation of laws, district budgets, and spatial zoning plans. However, as a number of cases discussed in this paper demonstrated, village demands failed to significantly affect decision-making. Indeed, the district head made decisions regarding plantation concessions that directly contradicted fundamental village aspirations without community consultation.[34] As one study of other areas across Indonesia noted, the problem remains that 'local society is not active in the bargaining process' that leads to decisions and the formation of policy. 'Instead, they are subject to the policies imposed upon them' (Asia Foundation 2004: 27). As in other cases, the affected groups were only involved at the point when decisions were 'socialized' among affected communities. Community consultations tended to conform to a minimal, instrumental model of token local involvement that did not lead to serious policy innovations in response to demands from below.

A key problem here was what political scientists call 'interest articulation'. The political system does not yet effectively process political demands from below (The Asia Foundation 2004). Local district assemblies (DPRDs) and district administrations were only poorly accountable and unresponsive to village demands. One issue, as an interview with a candidate who stood for election to the district assembly in 2004 revealed, was that centrally directed party organizations controlled the levers of political representation. Although popular local candidates who were critical of the system could stand, their position on the party ticket depended upon their financial and political relations to party apparatchiks. Votes attracted by a particular party were aggregated and seats in the assembly allocated to candidates according to their position on the party ticket. Consequently, candidates that were listed too far down the ticket were unlikely to be allocated the number of votes required to take a seat in the district assembly regardless of the number of primary votes they had attracted.

As in the past, established district level networks of accommodation and exchange continued to overshadow these channels for popular participation (McCarthy 2004). In Jambi Sejati, as in other districts, the need to increase district revenue and to amass political and economic capital affected the strategies of politicians and actors in the district administration and district assembly. Although this is difficult to document, interviews in the district pointed to the way investments by corporate actors – particularly the large timber conglomerate – influenced significant outcomes and decision-making processes.

At the same time villagers tended to lack forms of organization that are 'politically and economically efficacious' and that could act as countervailing powers to ensure state accountability.[35] In rural Jambi villagers act at some distance from wider popular movements, such as those organized by urban NGOs. Yet, the efforts of local NGOs are critical in disseminating information, outlining possible strategies for action and linking communities with outside sources of assistance. In seeking to find redress for their grievances and channel their aspirations, villagers have very limited means and correspondingly restricted aims. Even when NGOs attempt to organize, community solidarity tends to be fragile. Those involved in organizing village resistance have too often been coopted, and this tends to be corrosive of community trust. Even when NGOs are present, this crisis of confidence works against collective action.

The political reforms of the post-Suharto period have so far failed to provide the legal means for recognizing *adat* entitlements, particularly in areas under plantation concessions or mapped in the forestry estate. The Ministry of Forestry had successfully maintained its *de jure* control over state forest land, and an institutional interest in defending and maximizing that control. This left *adat* property claims legally invisible. Even if they had wanted to, the district government lacked the authority to 'free up' areas for Melayu farmers. Given the poor position of *adat* claims under the law, in the course of a dispute company representatives did not fear litigation. Companies did not negotiate in 'the shadow of the law', and this weakened the position of villagers in informal negotiations. Furthermore, in most cases informal dispute resolution processes tended to occur where there were wide imbalances in power between villagers and their company adversaries. This has also had a detrimental effect on village attempts to seek redress (World Bank 2004). The only incentive for companies to accommodate villagers lay with the expense, inconvenience and risk created by continuing village resistance. In disputes with plantations, degrees of success were only possible when the constellation of local interests – particularly local elites and the district government – were in alignment. This occurred when there were congruent interests working across these different scales, and when state law could be applied. For instance, in the PT Banyak Sawit case, this worked in favour of outcomes that accommodated (to some degree) ordinary villagers.

From before *reformasi*, and continuing into the regional autonomy period, policy favoured the role of cooperatives as the vehicle for popular participation in the economy. These cooperatives in themselves constituted a new form of entitlement – in the sense of legitimate effective command over alternative resource bundles (Leach *et al.* 1999: 233). In the absence of effective recognition of village or *adat* collective rights or of village management rights, commercial development of village common pool resources can only be mediated via cooperatives. This policy effectively forces villages to forego other long-established indigenous forms of collective resource management and *adat* conceptions of property. These cooperatives were constituted by state regulations stipulating how they are registered and operate. Cooperatives opened possibilities for obtaining oil palm smallholdings, but individual farmers had to give up a significant portion of their land and become subject to forms of organization open to manipulation by local elites and their corporate partners.

To be sure, transmigrants interviewed during the course of this study had improved their situation by working oil palm smallholdings in transmigrant settlements under well-functioning cooperatives. In some villages a proportion of indigenous Melayu farmers – especially better established farmers and village elites who had joined nuclear estate developments and finally obtained productive oil palm smallholdings – had also improved their livelihoods. However, organized by local government, Melayu village cooperatives became dominated by unaccountable subdistrict elites. As they operated without the social cohesion that might once have been provided by Melayu *adat*, they lacked the kind of normative controls and horizontal networks of engagement that might have provided better outcomes for poor Melayu villagers. Despite the rhetoric of community mutuality in which the concept of cooperative organization is rooted, in mediating village opportunities to participate in commercial timber and plantation enterprises on *adat* lands, the *modus operandi* of cooperatives tended to truncate the possibilities that might have been opened by genuine participation.

The cooperative model also served an additional function. Under the wing of a plantation company, villagers in a dispute with a company could be offered participation in new 'satellite' developments, in the process defusing the conflict while also expanding the area producing oil palm for company mills. Where vocal leaders were recruited into cooperative management, they could be readily turned into company partners who cooperated in this process. In the absence of effective forms of accountability, access to resources via cooperatives typically failed to work in the common interests of ordinary villagers in whose name they operated.

During the years immediately after 1998, the state legal system became ever more heterogeneous. At this time Indonesia's competing national and district regulatory regimes were coming to terms with increasingly assertive customary orders in turn interacting with political networks of patronage, exchange and accommodation in the regions now also growing in strength. During this time actors could invoke rules associated with these diverse but shifting national, district, and localized orders to find support in struggles over power, meaning and the control of resources. The interconnections between these coexisting legal and normative orders remained precarious, and the institutional arrangements determining access and use of local resources were shifting and uncertain (McCarthy 2004).

We have seen these dynamics in the Jambi context. The district government had an interest in using its discretionary powers to facilitate the exploitation and conversion of forests. At the same time the Ministry of Forestry wished to maximize its control over the national forestry estate. However, due to the weakness of the central government through much of this period, the ministry lacked the political means of doing so: it faced the danger of losing control as local government and local people established facts on the ground. Echoing the allegory of the tragedy of the commons, these dynamics led to a race to control and divide up lands and resources irrespective of the impact on native forests, biodiversity and future livelihood needs of surrounding communities. Finally, the ministry responded to this threat by allocating a large number of timber plantation concessions to a single corporate actor, a policy direction that left almost 40 per cent of this district in the hands of one large conglomerate. The increased power of this corporate giant clearly also affected decision-making processes. In the course of this study respondents suggested that this actor had increased its 'political investments' in the district administration, effectively turning the district capital into a company town. In the short term the volatile configuration of the years after 1998 appeared to have been resolved in favour of a corporate actor able to draw on relationships inside local and national government to secure resource control.

A number of other factors affected the outcomes of the disputes studied in this case. Meetings, consultations, the participatory mechanisms of local government and demonstrations did provide fora for local people to vent their feelings and space for dialogue and protest. In the course of these discussions and consultations, accommodations might be made, and the heat of the conflict attenuated. As other cases suggest, decentralization may have provided a means for the state to extend its reach, enabling it to defuse volatile issues, but without resolving the underlying grievances.

Here I do not wish to overdraw this gloomy conclusion. With new spaces being opened for contestation, even these incremental steps towards negotiation and the instrumental forms of participation that have emerged under regional autonomy represent an improvement on the asymmetrical relations of the New Order period. Yet key problems remain. These include the absence of appropriate political vehicles required to give the reforms real meaning. There is a requirement for strong farmer organization and associations with knowledge of the decision-making processes and legal framework, and with bargaining power vis-à-vis decision makers and politicians. The absence of such forms of organization in Jambi and the lack of effective legal openings for redressing village grievances leaves farmers dependent on local protests and land occupations. 'Guerrilla' style acts of vandalism and outright violence represent one of the few options available to address grievances (Barron and Madden 2004). However, in most cases such actions only have a limited effect at best. At the point of implementing decisions that would lead to significant conflict, such as the PT Pulp and Paper decision, there may be occasions for venting local opinion. To be sure in one or two cases, such as the PT Banyak Sawit case, there were limited victories. Yet this hardly allowed the majority of Jambi villagers to find their place at the table of collective welfare, still less a position sustainable over the long term.

This points to the persisting unresolved problems facing Indonesia as a political community expected to act and resolve differences in the 'common interest' of all its citizens. As Cohen (1989) has pointed out, community is 'an aspiration to common interest which is all too often missing in reality'.[36] Yet, if Indonesia is to maintain any confidence in being a 'national community' of shared interest where conflicts are mediated through institutional arrangements rather than by violence, it will seek to accommodate the aspirations and address the grievances of its most disaffected citizens in a more encompassing fashion. As a former *adat* leader pointed out, otherwise Jambi will end up 'just like Aceh' – a place riven by violent, unresolved conflict.

Acknowledgements

Research for the paper was undertaken under an Australian Research Council fellowship at Murdoch University, in conjunction with a Centre for International Forestry Research (CIFOR) project, 'Making Decentralization Work for Forests and the Poor: Policy Research to Promote Sustainable Forest Management, Equitable Economic Development, and Secure Local Livelihoods in Indonesia' funded by the Australian Centre for International Agricultural Research (ACIAR). The field research was undertaken together with Zulkifli Lubis and the author is particularly grateful for his insights and collaboration in the field.

Notes

1 Forestry and wood-based industry in Indonesia. Industry overview. http://goliath.ecnext. com/coms2/gi_0199–6039884/Forestry-and-wood-based-industry.html. Reportedly, by 2006 less than 3 million hectares of these HTI plantations had actually been planted. 'No chip mill without wood', *Down to Earth* No. 71, November 2006. In 2006 the Ministry of Forestry announced plans to extend the area of HTI by another 9 million. 'Indonesia to offer 17 million hectares of forest concessions', AFP Jakarta 28 September 2006.

2 The *Oxford Dictionary* defines an endowment as a 'gift, power, capacity, or other advantage with which a person is endowed by nature or fortune'. In contrast, 'entitlement' is derived from 'entitle'('to lay claim to'), in the sense of a right (especially a right to benefits) granted by state law. Hence Leach *et al.* (1999) distinguish between endowments, the 'rights and resources that actors actually have' and entitlements, the 'legitimate effective command over alternative resource bundles'.

3 The place-names in this chapter are pseudonyms.

4 Interview with village head, 27 June 2004.

5 WALHI Jambi is the local chapter of WALHI- Friends of the Earth Indonesia, a network of NGOs who define their mission as working to build a 'social transformation movement, community sovereignty and sustainable livelihoods', http://www.walhi. or.id/ttgkami/. WALHI Jambi sees itself as supporting efforts to improve village livelihood conditions and resolving conflicts between farmers and plantation owners and has been involved in advocacy against oil palm and timber plantation developments that are seen to work against the sustainable livelihoods for communities living near the forests.

6 The PIR-Trans scheme (*Perkebunan Inti Rakyat–Transmigrasi*, or Nucleus Estate and Smallholder Transmigration Scheme) involved the state funded infrastructure development, land acquisition, smallholder plantings, initial living expenses, and housing for smallholders, mostly transmigrants participating in a 'satellite' (*plasma*) development and the provision of credit at concessionary rates for companies to develop a 'nucleus' (*inti*) estate, plant oil palm and establish crushing facilities (see Cassons 2001: 50).

7 See Henley (2007).

8 Sharma (1997: 74–82). See also Article 3, Law No. 25/1992 concerning cooperatives.

9 Ketetapan Majelis Permusyawaratan Rakyat Republik Indonesia Nomor XV/Mpr/1998 Pasal 3.

10 Interview, 31 May 2004.

11 Ibid.

12 As other research has noted, there are considerable obstacles for people wishing to

convert their unregistered land rights into official certificates of title in light of formal conveyancing requirements (Fitzpatrick 2007).

13 'Pembangunan Perkebunan di Jambi Lamban', *Jambi Express*, 24 March 2004.

14 'Oil palm investments opposed', *Down to Earth* 49, May 2001. Under a new policy, local administrations required 'an initial stake of 10–20 per cent in every natural resources investment project in the area, with an option of increasing this to 51 per cent'

15 These were set out in Forest Ministerial Decisions during 2004. As most of the PT Pulp and Paper concessions overlap with those in neighbouring districts, it is difficult to give an exact figure for PT Pulp and Paper holdings within the district.

16 According to sources within the district, the ministry will soon release a HTI license over a neighbouring area.

17 Interview, 7 August 2004.

18 Ibid.

19 Interview, 6 August 2004.

20 While top-down decision-making has long dominated development planning in Indonesia, the planning framework also provides for a 'bottom up' process. Under the district planning process the 'bottom up' process begins with village development meetings (*Musbangdes*) that are chaired by the village head and which involve village councils (BPD) and ordinary villagers. However, while Musbangdes process allows for discussion of development plans, it does not work as a vehicle for eliciting community inputs into land use decision-making.

21 Interview, 7 August 2004.

22 Surat Bupati Jambi Sejati/Dinhutbun/2004.

23 Ibid.

24 Interview, 12 August 2004.

25 Interview, 9 August 2004.

26 Interview, 6 August 2004.

27 Interview, 12 August 2004.

28 Interview, 28 May 2004.

29 Interview, 28 May 2004.

30 Ibid.

31 Interview, 9 August 2004.

32 According to Susilo's (2004) study, most participants in these schemes have an income of more than Rp 10 million per month, with average incomes approximately 50 per cent above the poverty line in the areas surveyed.

33 *Financial Times*, 8 October 1997, cited in Gelder 2000:16.

34 Further, when villagers requested government assistance through the *Musbangdes* forum in order to make the transition to productive rubber or oil palm, outcomes were similarly disappointing. According to members of the district department of agriculture, the regional legislature (DPRD) and the district government did not provide sufficient resources for an effective program to support farmers to make the transition, usually because of competing priorities.

35 Evans quoted in Sundar (2000); see also Agrawal and Ribot (1999).

36 Cohen 1989, quoted in Mosse (1998).

Bibliography

Agrawal, A. and J. C. Ribot (1999) 'Accountability in Decentralization: A Framework with South Asian and African Cases', *Journal of Developing Areas* 33: 473–502.

Asia Foundation (2004) Indonesia Rapid Decentralization Appraisal (IRDA) 5th Report, Jakarta: The Asia Foundation, Decentralization and Local Governance Team.

Barr, C., Resosudarmo, I. A. P. *et al.* (2004) 'Decentralization of Forest Administration in Indonesia: Implications for Forest Sustainability, Community Livelihoods, and Economic Development', unpublished report. Bogor: CIFOR.

Barron, P. and D. Madden (2004) 'Violence and Conflict Resolution in "Non-Conflict" Regions: The Case of Lampung, Indonesia', *Indonesia Social Development Paper 2.* Jakarta: World Bank.

Casson, A. (2001) 'Oil Palm and Resistance: The Political Ecology of the Indonesian Oil-palm Sub-sector in an Era of Turbulent Change', PhD Thesis, Australian National University.

de Soto, H. (2000) *The Mystery of Capital: Why Capitalism Triumphs in the West and Fails Everywhere Else.* New York: Basic Books.

Dove, M. R. (1993) 'Smallholder Rubber and Swidden Agriculture in Borneo: A Sustainable Adaptation to the Ecology and Economy of the Tropical Forest', *Economic Botany* 47 (2): 136–47.

Dove, M. R. and D. M. Kammen (2001) 'Vernacular Models of Development: An Analysis of Indonesia under the New Order', *World Development* 29: 619–39.

Fitzpatrick, D. (2007) 'Land, custom and the state in post-Suharto Indonesia: A foreign lawyer's perspective', in J.S.Davidson and D. Henley (eds.), *The Revival of Tradition in Indonesian Politics. The Deployment of Adat from Colonialism to Indigenism*, London: Routledge, pp. 130–48.

Gelder, van J. W. (2000) 'US Corporate Actors and the Indonesian Oil Palm Production Chain', research paper prepared for WWF-US, http://www.panda.org/downloads/forests/.

Ginoga, Kirsfianti, Oscar Cachob, Erwidodoc, Mega Luginaa and Deden Djaenudinaa (1999) 'Economic Performance of Common Agroforestry Systems in Southern Sumatra: Implications for Carbon Sequestration Services', Working paper CC03. ACIAR project ASEM 1999/093, http://www.une.edu.au/febl/Econ/carbon/.

Henley, D. (2007). 'Custom and *koperasi*: the cooperative ideal in Indonesia', in J.S.Davidson and D. Henley (eds.), *The Revival of Tradition in Indonesian Politics: The Deployment of Adat from Colonialism to Indigenism*, London: Routledge, pp. 87 – 112.

ILO (2004) *Indonesia: Working out of Poverty: Job Creation and Enterprise Development (SMEs and Local Development).* Jakarta: International Labour Organization.

Kay, C. (2002) 'Chile's Neoliberal Agrarian Transformation and the Peasantry', *Journal of Agrarian Change* 2 (4): 464–501.

Leach, M., Mearns, R. *et al.* (1999) 'Environmental Entitlements: Dynamics and Institutions in Community-Based Natural Resource Management', *World Development* 27 (2): 225–47.

Li, T. M. (2002) 'Local Histories, Global Markets: Cocoa and Class in Upland Sulawesi', *Development and Change* 33 (3): 415–37.

McCarthy, J. F. (2000) 'The Changing Regime: Forest Property and Reformasi in Indonesia', *Development and Change* 31 (1): 91–129.

McCarthy, J. F. (2004) 'Changing to Gray: Decentralization and the Emergence of Volatile Socio-legal Configurations in Central Kalimantan, Indonesia', *World Development* 32 (7): 1199–223.

McCarthy, J. F. (2006) *The Fourth Circle: A Political Ecology of Sumatra's Rainforest Frontier.* Stanford: Stanford University Press.

McCarthy, J. and Zen, Z. (2008), *Managing the Oil Palm Boom: Assessing the Effectiveness of Environmental Governance Approaches to Agro-industrial Pollution in Indonesia.* Paper presented at the International Association for Impact Assessment, Perth, May 2008.

Mosse, D. (1998) *Making and Misconceiving Community in South Indian Tank Irrigation.* Crossing Boundaries, Conference of the International Association for the Study of Common Property, Vancouver, http://dlc.dlib.indiana.edu/archive/00000099/.

Mubyarto (1992) *Desa dan perhutanan sosial: kajian sosial-antropologis di Prop. Jambi.* Yogyakarta: Aditya Media.

Pagiola, S. (2000) 'Land Use Change in Indonesia, Environment Department', World Bank, http://econwpa.wustl.edu/eps/othr/papers/0405/0405007.pdf.

Papenfus, M. M. (2002) Investing in Oil Palm: An Analysis of Independent Smallholder

Oil Palm Adoption in Sumatra', Southeast Asia Policy Research Working Paper, No. 15. Bogor: ICRAF.

Potter, L. and Lee, J. (1998) 'Tree Planting in Indonesia: Trends, Impacts and Directions', Bogor: CIFOR.

Sharma, G. (1997) Cooperative Laws in Asia and the Pacific. Delhi: ICA Publications.

Sundar, N. (2000) 'Unpacking the 'Joint' in Joint Forest Management', *Development and Change* 31: 255–79.

Susila, W. R. (2004) Contribution of Oil Palm Industry to Economic Growth and Poverty Alleviation in Indonesia', *Jurnal Penelitian dan Pengembangan Pertanian*, 23 (3) 107–13.

Usui, N. and Alisjahbana, A. (2003) *Local Development Planning and Budgeting in Decentralized Indonesia: Update*, International Symposium on Indonesia's Decentralization Policy: Problems and Policy Directions, Shangri-La Hotel, Jakarta, http://www.gtzsfdm.or.id/documents/dec_ind/o_pa_doc/UsuiArmida_PlanningBudgeting_Jan2003.pdf.

Wakker, E. (2005) *Greasy Palms: The Social and Ecological Impacts of Large-scale Oil Palm Plantation Development in Southeast Asia*, AIDEnvironment in collaboration with Sawit Watch Indonesia and Friends of the Earth, http://www.foe.co.uk/resource/reports/greasy_palms_impacts.pdf.

Wilson, G. A. and Rigg, J. (2003) 'Post Productivist Agricultural Regimes and the South: Discordant Concepts?' *Progress in Human Geography* 27 (6): 181–207.

World Bank (2004) 'Village Justice in Indonesia: Case Studies on Access to Justice, Village Democracy and Governance', Jakarta, World Bank.

WRM Bulletin (2004). 'Indonesia: Palming the forest', 85 (August).

Zen, Z., Barlow, C. *et al.* (2005) Oil Palm in Indonesian Socio-Economic Improvement: A Review of Options. *Working Papers in Trade and Development*. Canberra: Research School of Pacific and Asian Studies, ANU.

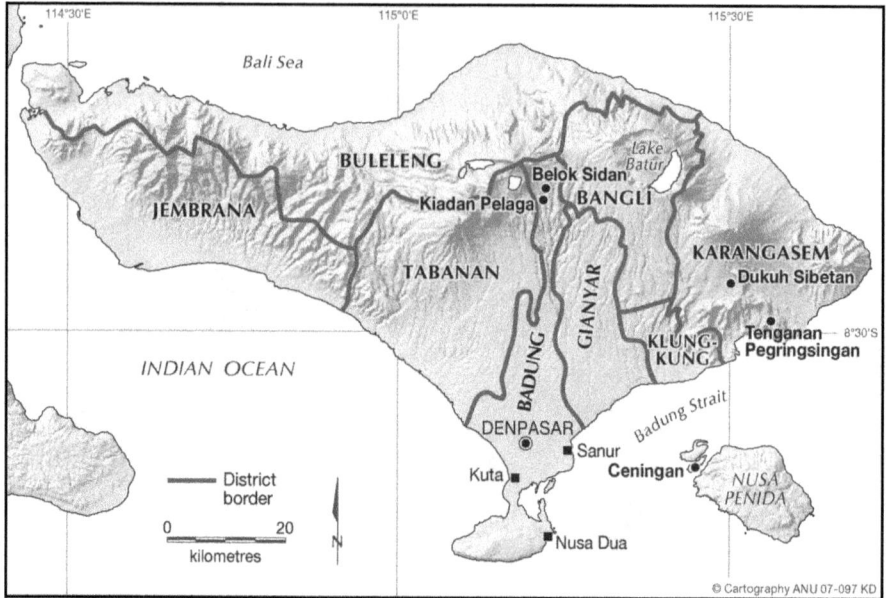

Map 8 Bali Province.

8 Off the market?

Elusive links in community-based sustainable development initiatives in Bali

Carol Warren

In Indonesia's Reform Era, NGO interventions in local communities throughout the country have been aimed at reversing asymmetrical institutional relationships between state and local authority and enhancing local capacity to plan sustainable and equitable alternative development paths. Many of these projects suffer from short timeframes, under-resourcing, and unrealistic assumptions about local needs and desires, as well as the limited ability of community-based enterprises and their NGO facilitators to marry social and environmental goals with the market mechanisms of a neoliberal globalizing economy in practical ways.

This case study tracks the efforts of several remote upland communities involved in an NGO-sponsored participatory mapping project aimed at planning for sustainable development in Bali. The philosophy behind the project was that strong customary (*adat*) community institutions could be harnessed to provide alternative options to patterns of overdevelopment in the tourism sector, perceived as threatening Bali's culture and environment. Pollution, erosion, land alienation, water and electricity shortages, as well as pressures on the social and cultural resources that have sustained Bali's way of life for generations, contributed to intense introspection and critical policy debate since the late New Order.[1]

The control of land sale and land use is now widely regarded as crucial to reversing negative trends in the direction of economic development, and the commodification of culture and environment. But the sustainable and equitable livelihoods imagined by these projects depend critically upon strengthening the terms of trade for agriculture and providing prospects for enhancing rural incomes through the development of sustainable down-stream processing and other complementary industries. This chapter explores efforts to use village level *adat* institutions towards these collectively defined ends. It examines the challenges to the pursuit of equity, sustainability and the 'common good' posed by the state, the market and competing interests within and between these communities and their NGO counterparts.

Confronting displacements – land use and resource management

Throughout Indonesia during the New Order period, as we have seen, local control over resources was marginalized or overridden in the face of centrally driven and increasingly capital-intensive development policies that served elite political and economic interests. In Bali, as elsewhere, local management practices were undermined by national legislation and New Order developmentalist policies, which cannot, of course, be separated from the global process of capital accumulation that accelerated dramatically in this period. In particular, the mega-project developments of the 1990s dramatically altered the character of the Balinese economy, and increasingly politicized Balinese society.

A disturbing aspect of the new pattern of development that emerged in this period was the establishment of a luxury residential property market. The fusion of tourism development and a voracious new real estate industry that was part of a global model of resort living for the wealthy was a significant step beyond the already controversial mass tourism oriented development model. The latter was theoretically limited to the Nusa Dua peninsula on the southern tip of the island under the 1972 master plan for tourism on Bali, but these regulations were largely ignored by government in the process of issuing development project permits. Consuming hundreds of hectares for a single project, the new mega-developments stimulated speculative acquisition of land and promised limited long-term economic spin-offs for the local economy. Throughout the 1990s land was diverted from agriculture for new projects and accompanying infrastructure at rates averaging a thousand hectares annually, often through forced expropriation. In consequence, land alienation became one of the pivotal issues in post-New Order debates about planning and development on the island.

The land issue was intimately bound up with questions of local control over Bali's cultural and economic resources and its environmental security. Controversies concerning the relationship between land alienation, the inflow of external capital, unchecked in-migration from other parts of Indonesia and the integrity of Balinese culture, which was becoming increasingly divorced from its agrarian roots, have pervaded local anxieties about the perils of globalization for at least the last decade. Escalating pollution, strains on the island's water resources, forests and marine environment have been directly or indirectly driven by the rapid expansion of tourism and the luxury property market since the early 1990s. Heated debates over planning and resource control had emerged with the announcement of the first resort and real estate development at Tanah Lot during the late New Order investment boom. At that time unprecedented protest demonstrations opposing the project on cultural and environmental grounds had forced an eight-month moratorium, but ultimately failed to stop the watershed project that shifted the direction of development in Bali so dramatically (see Warren 1998; Suasta and Connor 1999).

Public criticism of poor government planning, overemphasis on tourism, and neglect of the agricultural sector[2] was fuelled by the serial buffeting of the tourism industry as a result of the terrorist nightclub bombings at Kuta in 2002, the SARS epidemic in 2003, and a second bombing attack in 2005. Along with an acute new awareness of the insecurity of employment, revelations of the scale of outside ownership of the tourism industry fanned discontent with development policy. Claims that 85 per cent of the Rp 150 trillion (approximately $US 16 billion) in tourism assets is owned by investors from outside the island stimulated a series of critical responses in the columns of the local newspaper. The original article which launched this controversy reported an Indonesian economist's assertion that Bali had become a 'waste basket' (*keranjang sampah*), carrying the infrastructure burden, as well as the social and environmental costs of tourism development, while profits went elsewhere (*Bali Post*, 22 November 2004). This startling claim, and the economist's argument for a revision of tax distributions and planning regulations to deal with the problem, prompted the newspaper to devote its special '*Giliran Anda*' (Your Turn) column to public opinion on the subject.

Popular responses in the column, which ran for several weeks, were similar to those levelled at the Javanese dominated central government in the same paper during the heated controversy over the Tanah Lot development a decade earlier. Now under decentralized governance, however, Balinese themselves would carry a great deal more responsibility for policy outcomes than had been the case under the New Order. Contributors to the *Giliran Anda* column in 2004 expressed anxieties about the loss of control over the island's destiny, and saw the land question and local *adat* institutions as keys to restoring local control over future development. The following excerpts reflect widely held opinions on these subjects.

We need to be conscious that the time has come to do something to protect Bali ... It is taken for granted that the control of the economy is directly connected to the use of capital that is aimed only at profit .... With respect to the disposition of land, for example, we have to understand that control over the fabric of the community resulting from displacing even as little as one handspan measure of land now will have impacts on efforts concerning landholding and land use in the village territory for the foreseeable future. (DWA, Jimbaran, *Bali Post*, 1 December 2004)

Balinese as individuals are not able to invest large amounts of capital. Balinese communities (*Krama Bali*) are more used to carrying things out collectively through *banjar, subak, tempekan, sekaa*,[3] etc. If the *banjar* [customary hamlet] system that includes thousands of communities were activated, that would certainly become an extraordinary asset ... If this could be made a reality and a significant part of Bali's tourist assets could be owned by *banjar*, that could turn Bali from a 'waste basket' into a 'pearl palace' .... (KS, Denpasar, *Bali Post* 03 December 2004)

Paralleling alarm over the loss of control over Bali's economy have been concerns about the capacity of regional government to confront Bali's acute environmental issues. Because regional autonomy reforms have been focused at district (*kabupaten*) rather than provincial level, decentralization has not contributed to coherent treatment of the island province's environmental problems. Bali's eight district governments are now required to find the funds to cover most of their own budgets. Adopting some of the same rent-seeking practices for which the central government became notorious under the New Order, they have not generally proved willing or able to tackle environment and development problems within their own boundaries, let alone cooperate with other districts in island-wide planning. Professor Johan Silas of the Surabaya Institute of Technology, an advisor to Indonesia's Minister of the Environment, argued that Bali's environment cannot be managed in piecemeal fashion, and that the current processes of official planning are inappropriate for an island ecosystem. Like many local critics, he sees constraint of land sale and land use conversion as an urgent step toward resolving Bali's most serious problems:

> The displacement of agricultural land need not continue if Balinese themselves are solidary (*kompak*) … Farmers sell land because their costs of production are not matched by harvest incomes, because the youth are not attracted to become farmers, and because government policy doesn't protect farmers …
> (J. Silas, quoted in *Bali Post*, 30 November 2004).

Land taxes, for example, are officially based on market value instead of current use.[4] This introduces extreme distortions where resort or real estate development drives land values beyond levels which low-income producing agricultural users could afford. Silas proposes state zoning and tax policies that would make changes in land use function difficult and expensive.

But since the beginning of the Reform Era, there have been few signs that national, provincial or district level governments are seriously prepared to tackle Indonesia's troubled land issues or its rapidly deteriorating environmental 'commons'. While farmers are rhetorically urged not to sell their land, virtually no policy initiatives through official governance channels have been introduced which might improve the conditions of rural smallholders. Indeed, new policies reducing subsidies on agricultural inputs and fuel, and opening up markets to agricultural imports under pressure from the IMF and World Bank, added another crippling blow to rural farmers, exacerbating the rising costs of fuel and agricultural inputs.[5]

The question of whether the political system under reform could deliver on measures to regulate future development policy in the face of a now more differentiated economic, social, cultural and political landscape continues to be critically engaged by the public through the highly populist local radio and press media.[6] Given the inadequacies of official government planning and regulation, if there were any prospect that Silas' assertion that 'if Balinese themselves were compact', they might be in a position to resolve their own problems, it seemed logical to look to local customary community institutions – the banjar and desa *adat* – for ground-up approaches to planning and development. Corporatist principle and collective action practices are well established in these institutions, and were thought to offer real possibilities of alternative development strategies, as suggested in so many of the letters published in the local press and audience feedback on talk-back radio.

There is of course intense scholarly debate about the 'myths' surrounding the twin concepts of '*adat*' and 'community'[7] in the anthropological literature. Insofar as elements of the Dutch scholar administrator's vision of the 'village republic' (see Korn 1984) in Bali ever existed, they have been very much altered by Suharto's New Order, not to mention the colonial and pre-colonial states which also influenced the evolution and development of the famously corporatist orientation of Balinese *adat* communities. Even the New Order state, however, never fully accomplished the subordination of local forms of citizenship or notions of the local as a site for constructing culturally specific concepts of the 'common good', or 'commonweal'.

The turn to local *adat* institutions for dealing with contemporary problems, including cultural protection, environmental degradation and land alienation, has been a prominent feature of Reform Era politics. This revival was already gathering momentum in Bali[8] before the collapse of the Suharto regime in the face of the economic crisis of 1997–1998. Since then, *adat* communities across the island have been reasserting claims over land and resources, sometimes with novel interpretations of their collective property rights, as for example in Gianyar where members of one *adat* community across the gorge from a luxury hotel were able to demand payment of 'rent' for the view of their exquisite rice terraces. Claiming authority over resources and social relations within their customary domains (*hak ulayat*), communities in Bali are now assuming management of tourist sites, developing new industries, regulating peddling, forest access and road use. They are also negotiating terms with hotels and other outside interests regarding levels of local employment and the funding of community projects and rituals. With newfound momentum *adat* communities have been reframing their relationships to government and private actors who previously ignored customary domain rights. In the process, they have also had to confront divergent views on the character and reach of those collective regulations over themselves, as individual members of the *banjar* or *desa* community.

Reclaiming place – community mapping

As elsewhere in Indonesia, experiments with participatory community mapping were taken up in Bali as part of local efforts to reclaim a greater role in defining the character and direction of social change and economic development in this period of governance reform. One of the most systematic of these community mapping projects to be carried out in Bali in the Reform Era involved five upland communities, working with a local NGO, Yayasan Wisnu, and supported by funding from the national foundation, Yayasan Kehati. The latter in turn received its funding from overseas donors, including USAID. With the primary objective of promoting participatory, sustainable and locally negotiated alternatives to the top-down models of planning and resource management that had characterized New Order practice, the programme focused on remote communities, still primarily dependent upon agriculture. Although not directly affected to date by the extreme displacements transforming many lowland villages, all had experienced varying degrees of conflict with state and private interests over local resource issues. Ceningan, a small island off the east coast, had been targeted by its district government to become the site of a major tourism development. Belok Sidan and Kiadan Pelaga were in the process of being linked via a new road and bridge to mass tourist sightseeing destinations through a project funded by the district government of Badung for this distant mountain periphery. As significant watershed catchment areas they were in contention with regional government agencies over green zone restrictions and the diversion of their water sources to coastal tourist centres. Banjar Dukuh in Sibetan was seeking alternatives to the erratic and increasingly poor terms of trade for its primary product, salak fruit, through downstream processing and small-scale community-based tourism. Tenganan Pegringsingan, already a tourist 'object' because of its distinctive culture and history, was concerned to gain greater control over the benefits and impacts of day-trip tourism, to date largely managed from outside the village.

All five villages were struggling at the same time with internal differences over the direction of social change and the degree of collective control that might be exercised over it. Many were concerned in one way or another to fend off the dislocations they saw in the overdeveloped parts of the island. Protection of the local culture and environment were explicit objectives of the mapping project participants, both NGO and local volunteers, although the relative strength of these commitments – when the restrictions they often imply conflict with new economic opportunities – has been a matter of periodic contestation since the project's inception in 2000. At the heart of local debates were concerns over the shifting base of Bali's economy, the pull of modern consumer lifestyles, and the extent to which customary institutional authority over land and resources might be a practicable basis for experiments with alternative directions for future local development.

The regulation of land use and sale was one of the key issues to be confronted in the early stages of the project. With the exception of Tenganan – where all land is tightly regulated by customary law, and sale is prohibited – the other project villages have historically evolved varying degrees of direct and residual control over land within their customary domains. These typically involve a mix of communal *adat* land with various customary entailments, and private land, certifiable under national land law and therefore subject to sale. Most of the agricultural land in the other four communities is privately held (whether or not officially certified), as is the case in most of the densely populated lowland areas of Bali. This is not to say that land under private (*hak milik*) title is free of customary controls, however. While the formal status of communal *adat* lands and rights of *adat* institutions to at least residual control over what they regard as customary domain remain ambiguous in national law,[9] other means of imposing collective control could still be employed by these communities. Tightening *adat* codes represented the most obvious planning and land use management tool that lay within the authority of local institutions. But this was evidently not possible without control over in-migrants not subject to local *adat*, and without seriously circumscribing the options of members who held state certified private land rights.

Effective regulation would have to be part of a scheme that could persuade broad segments of these communities that restriction was a matter of 'common interest' in which costs and benefits would be shared. The experiment with community mapping was intended to enable these issues to be addressed by the participating villages on their own terms, and in a context that would permit self-regulation to be considered in tandem with prospectively oriented planning for sustainable development. It was aimed at articulating a common vision for future conservation and development within each particular locality, and at community empowerment by reinforcing internal solidarities and strengthening local capacities to deal with external power-holders and to make their own land use and development plans.

The hamlet of Dukuh, a *banjar adat* in the mountain village of Sibetan in eastern Bali, has to date taken the self-regulatory objectives of the planning programme furthest.[10] It used the mapping project not only to reinforce restrictions on sale of unregistered *adat* lands,[11] but extended this authority to privately held land already certified under the national land titling system. As one of the Dukuh participants explained,

> Outsiders cannot buy *adat* land or residential land. Now private lands have been returned by the people, so that this land also isn't permitted to be sold ... because until now we have been just [passive] observers [in its management] ... We want to bequeath it to our children, so that they aren't left with only a story (interview with MS, 23 November 2001).

The experiences of participants from this community will serve to illustrate some of the successes and limitations of this NGO intervention in achieving its broad planning and institution-strengthening goals toward sustainable development. Dukuh had become involved initially with Yayasan Kehati at the suggestion of an academic from the agriculture department at the local university. He had been researching salak fruit, the main crop in this village where 75 per cent of the population are farmers. The development of downstream processing options for Dukuh's farmers was one objective of the training programmes tied to the community mapping project. Another was the explicit use of highly structured local customary institutions, the *banjar* and *desa adat*, as development planning and land use management and monitoring agents.

Achieving a common position on the regulation of rights to sell land has been among the most contentious of issues confronted during the project, as is clear from the account of debates surrounding these issues by several of the participants from Banjar Dukuh (group interview, 23 November 2001):

> NS (*banjar* dinas official) ... [T]he debate [on rights to sell land] was extremely sharp. Sometimes we would run out of time. So we would bring it to the next meeting. Again the debate was long. In the end we asked, 'So how does the community stand on this? Are we agreed?' ....

> ND (farmer): We needed space to think, to focus .... The debate went on, we took our thoughts home and asked questions. What are the good and bad effects that need careful weighing up? Sometimes there was someone who would say: 'I can't give agreement yet. My son thinks like this ...' So we would call his son and explain our rationale.

WS (farmer): The debate was really intense; From one side: 'This is my land, why can't I develop it? Why is the *banjar* prohibiting sale?' 'The reason is like this – for the environment, for the next generation ....' 'Well, if that's the reason!' 'Do you agree?' 'Agreed!'

Deliberations in the course of the mapping programme brought to a head fundamental conflicts between land as a symbol and cultural resource on the one hand, and its pivotal role as a commodity in neoliberal economic development frameworks on the other. The diversion of communal land to private control and the question of collective authority over the disposition of privately owned (*hak milik*) land are key issues throughout the island. Indeed, conflicts over land have been primarily responsible for the dramatic rise of internal *adat* disputes in Balinese communities in the post-Suharto period.

Land alienation poses serious challenges to the concept of collective authority over people and resources that customary community relationships in these villages theoretically presuppose. *Adat* land in particular is tied to village service obligations (*ayahan*),[12] and there is a residual sense in which even land within village boundaries that is regarded as privately held (the largest proportion of agricultural lands in most villages), is nevertheless bound up with the customary rights and responsibilities of local citizenship.[13] Despite the remoteness of the five villages, several had already found themselves in the situation where small parcels of land had been sold or contracted to outsiders through the national system of certification and sale. The rights to control the disposition of land having been appropriated by the central government under the Basic Agrarian Law of 1960, official land transfer under the law does not require formal approval through *adat* institutional channels. What communities could theoretically control through *adat* sanctions (expulsion, refusal of access to temples or cemeteries, etc.) were members' actions that transgress collectively determined rules.

The attempt to deploy customary obligations as a broad regulatory mechanism by remote, more or less culturally homogeneous communities such as these has serious practical implications, of course. Using *adat* institutions and sanction systems as a framework for planning and collective decision-making on community development ties notions of local citizenship to adherence to local customary law, and precludes 'outsiders' (whether immigrants from other ethnic groups, non-local Balinese, or for that matter members of the *krama banjar/desa* itself who fail to uphold these obligations and risk becoming outsiders through expulsion) from full participation in village affairs. This raises dilemmas analogous to those surrounding protectionist policies aimed at assuring priority for national interests and values in the world economy. At the same time as such regulatory regimes are intended to constrain exploitation by more powerful external holders of capital, they are also mechanisms of social exclusion. Economists would argue such restrictions would also have the unintended effect of restricting efficient allocation of resources and of inhibiting economic development. At the local level this dilemma is faced most acutely with respect to the question of land sale.

Rapidly rising land values and the escalating cost of living in Bali, both driven by tourism and real estate development, represent a relentless combination of pressure and enticement to bring land onto the open market. Ordinary people in villages like these face a difficult choice when windfalls from land speculators offer the prospect of a dramatic increase in life chances and living standards, however limited and short term the window of opportunity this may present to those with land in strategic locations. The possibility of turning windfall income from land sale into a longer-term private benefit through investment in a business, the education of the next generation or a bribe payment to obtain a civil service position is a consideration not to be dismissed lightly. These highly conditional options may bias real life decision-making against collective solutions to social and environmental issues. They set economic against socio-cultural values and individual short-term opportunities against potential long-term collective interests and alternatives, in the classic 'tragedy of the commons' paradigm.

The underlying question of the ultimate social and cultural embededness of economic interests embodied in the land question poses important questions for De Soto's (2001) controversial thesis that formalization of property rights through legal registration is the single most important means of eliminating poverty in developing countries. De Soto does not give serious attention to the social and cultural bonds to land that go beyond its money capital-generating commodity aspect. His failure to address the critical issue of alienability is a sleight of hand that allows him to evade the ultimate implication of market-oriented titling systems. Property has to be 'freely' transferable to maximize its capital value on a 'free' market. Both Marx and Weber recognized that without the capacity to separate property from its various social and cultural encumbrances, its 'mysterious' role in the capitalist system is circumscribed. Lauding the advantages of fungible conversion of property assets permitted through a formalized property system (2001: 55–7), De Soto's path to reform through a property career for a larger proportion of developing country populations disingenuously circumvents this paradox. The term 'alienable' does not appear in the book's index, and the wider implication of selling land bound up with cultural and social ties to place is unexamined. Even in strictly commodified asset terms, the loss of this basic resource represents uncalculated risk to the wellbeing of the next generation. The challenge he has set 'of making a transition to a market-based capitalist system' (which he considers 'the only game in town') that 'respects people's desires and beliefs' (2001: 242) founders on this crucial blindspot.

Balinese communities typically display a strong sense of local identity, underscored by powerful socio-religious beliefs and ancestral obligations. Their concerns include the protection of 'assets' that are not only of a material kind, but also aesthetic, social and religious values, which had not previously faced the meaning and evaluative transpositions posed by capitalist markets. Like all forms of property, land rights are about social, and in the Balinese case often spiritual, as well as economic relationships.[14] Harnessing this social and cultural 'capital' for the solidary response to local environment and development issues that Silas prescribed (see above), was precisely the challenge taken up by these villages through the NGO sponsored community participatory mapping programme.

Participants credited the mapping process with giving them a broader and more balanced perspective on the cultural and material resources available in their communities, as well as a more acute sense of the vulnerabilities and potential conflicts presented by different patterns of resource use and different development options. In Dukuh, in addition to regulations on the ownership and sale of land, a range of regulations on building and development in the village were instituted. By local determination, no hotels could be built without *adat* community approval. Homestay style tourist accommodation would be permitted, but only within the residential area of the *banjar*, and with limits placed on the number of rooms that could be converted for these purposes.

One particular area of privately held and certified land known as the *pemukuran*, with spectacular views from the mountains to the sea, will eventually pose the test of Banjar Dukuh's resolve to place community development interests first. Tracts of land at this location had already been sold or contracted to Balinese living outside the village. One contracting landholder sought permission in 1999 to build a homestay on this land. That proposal was rejected at the time by *adat* leaders on grounds that it would cut off the view to the sea which they believed important to keep under public control in the common interest (interviews NS, 17 March 2005 and 30 March 2005). The *banjar* head described the impediments posed by the national land certification process which gives no more than notional acknowledgement to *adat* claims and processes in the official land titling regime:

> Because land that is certified can be sold through a notary, or only with the knowledge of the head of the official (*dinas*) village; that is how [*adat* regulations] are frequently circumvented .... That land [at the *pemukuran*] is said to be intended for a guesthouse that they say is for meditation. So it isn't built yet, and is still at the planning and 'issue' stage. But, if he wants to build he has to request permission from Dukuh in whose territory it belongs, even though it has already become his property. At that point we can give a judgement based on the decisions and regulations which apply in the territory of Dukuh. (interview, NS, 17 March 2005)

At the instigation of Banjar Dukuh the wider *adat* village of Sibetan eventually formalized local regulations regarding land sale to outsiders in its *adat* codes (*awig-awig*). These were officially recognised by the district head (*bupati*) of Karangasem on 28 October 2004. They require the carrying out of *adat* service duties (*ayahan*) by all those occupying land in the village, and provide that any sale, loan, contract, pawn or gift of land requires approval of *desa adat* authorities (section 23, *Awig-awig Desa Sibetan*).

If Banjar Dukuh retains this commitment to restrict land sale and thereby external investment and the potential economic windfall[15] that an open market in land and unrestricted development might offer to a few strategically located landholders, then other avenues for economic improvement would need to be on the horizon. The NGO project was meant to assist in capacity building and exploration of alternative economic development options through the mapping exercise and training programmes. The project explicitly aimed at building on the social and economic capital of the community in collective terms.

With government and NGO assistance Banjar Dukuh experimented with production of wine from surplus salak fruit to diversify income sources, value add, and reduce market price fluctuation. In the third year of the project it joined with the three remaining communities[16] and the local NGO in establishing an inter-village trading cooperative and a community based ecotourism network (Jaringan Ekowisata Desa [JED]), which they hoped would offer alternative paths for local development.

Imagining alternative futures was a positive and in the view of most participants an empowering exercise.[17] But translating the vision produced through the mapping programme into a practical alternative development strategy proved another challenge entirely.

Fragile partnerships and elusive markets

Critical to any prospect of success for this innovative local planning and development programme is the problem of confronting the weak position of rural products and small-scale NGO- and community-sponsored industries in their efforts to deal with markets. Although diversification of the economies of the participating communities was one of the project's ultimate goals, and despite significant innovation and experimentation, such as with salak wine production in Dukuh, and quality coffee selection and packaging in Kiadan, the niche markets envisaged by the project have been slow to materialize.

In Dukuh, the focus on value adding to their primary agricultural product, salak fruit, resulted in the development of an unusual organic wine which should have a relatively accessible local market in Bali's tourist centres. A community-based wine industry could offer the prospect of additional labour opportunities within the village, as well as addressing the problem of the low market price for salak at the height of the harvest season. The intent was to find a commercial use for salak of uncommercial size on the fresh fruit market and to maintain a higher price for the fruit at the height of harvest season by diverting otherwise perishable surpluses into wine production.[18] The initiative therefore ought to have a serious prospect of directly enhancing the wellbeing of all members of the community.

At its high point, 18 people from Dukuh joined the production team for which they were paid labourer's wages. About 1,250 kg of salak was required to produce 250 litres of wine in a basic fermentation process using plastic gerrycans, stored in the backroom of the *banjar* meeting house. Small 300 ml bottles with recycled paper labels produced by Wisnu's recycling business sold for Rp 20,000 per bottle (US$2.20) at a small profit of about Rp 3,000 per bottle to the producers, and Rp 5,000 to the JED cooperative as distributors.[19] This was, however, very expensive compared with local rice beer and palm wine, which sold for around Rp 8,000. Even Indonesian-produced European products such as Guinness beer retailed at Rp 11,000 in supermarkets and small general stores in the tourist areas. Breaking into the higher value wine market, which retailed at four times the price of beer in the restaurant trade, meant obtaining licences for the higher alcohol content, meeting health standards, and establishing a reputation and reliable distribution arrangement. But despite an initially positive response from a few outlets sympathetic to the project,[20] Sibetan's wine industry foundered on a combination of licensing and marketing problems, which point to the pivotal role of intermediaries between producers and markets that are critical to realizing the commercial potential of community-based enterprises.

Forstner (2004) presents a useful summary of the strengths and weaknesses of different intermediary agents acting to facilitate commercial arrangements between communities and markets. She argues that these agents need to combine their strengths to overcome the limitations which any one of the potential partnerships presents for the development of viable community based enterprises. Although Forstner's study is primarily concerned with community-based tourism, this sector is treated in the broader context of alternative approaches to enhancing the disadvantaged position of rural producers in developing countries generally. Remote agricultural communities everywhere face similar difficulties of information and market access, economies of scale, quality assurance and standards compliance, transport costs and promotional reach. Parallel issues arise across rural economic sectors in the increasingly globalized marketing system, where competitive pressures tend to depress prices for primary products lacking value-added characteristics, and where large-scale global producers are typically able to penetrate remote markets more easily than local producers can make their mark in global circulations (Stiglitz 2002; Levi and Linton 2003; Dunkley 2004; Tudge 2004; Daviron and Ponte 2005).

The private sector clearly has the resources and experience to provide market access, but its profit maximizing bottom line disadvantages the sourcing of products in remote locations, which present lower profit margins and higher risks. Where low cost, high value local products from remote communities are attractive, private sector intermediaries tend to move towards increased control and eventually monopolization of production and distribution. Experience suggests the private sector also tends to withdraw from experiments with partnerships in the face of the 'slow pace of capacity building at community level' (Forstner 2004: 510), which are essential to any broadly based and participatory approach to market engagement.

The public sector has the capacity to use regulatory frameworks to link sustainability and poverty reduction objectives to marketing instruments. It is also in a position to bring together resources to invest in infrastructure and promotion. Because the state is expected to have some welfare obligations towards its citizens, and can operate on a macro-level that could overcome the problems of economies of scale in promotion and vertical integration, the public sector should have the motivation and means to take on such a role. On the other hand, public sector officials are actors with multiple agendas, and may have competing interests in the much greater tax and rent collecting opportunities presented by large-scale, capital intensive industries. Limited staff and resources result in neglect of the difficult work of assisting small producers in favour of the advantages of serving the big end of town. Even with good intentions, the upper levels of government lack the grass roots local knowledge and trust that NGOs seem better able to tap.

NGOs tend to be closer to community level, more flexible, and more readily able to commit to capacity building and liaison work between public, private and community based organizations. But NGO activists often lack the professionalism, incentives or the expertise of the private sector in marketing and management. They rarely have the substantial resources at the disposal of both private and public sector actors that would enable them to see a venture through to a self-sustaining stage. The ethos of many NGOs may also mean that ideological and political commitments conflict with local priorities (Li 2002) or efficient economic allocation of resources.[21]

Associations of community-based enterprises/cooperatives offer possibilities for greater economies of scale than individual communities or local cooperatives may be able to achieve on their own, while retaining the local knowledge essential for effective programming and building local commitment to projects. But these associations tend to suffer similar problems to those experienced by NGOs and the cooperative movement generally. In particular, they face practical difficulties in providing substantial and equivalent benefits to all members, and may be crippled by internal disputes over resource distributions. They also face the difficulty of balancing the demands of socially oriented as opposed to economically 'rational' goals in decision-making regarding resource allocation and income distribution versus reinvestment. Like NGOs they may be disabled because of social philosophy, conflicting interests, and constituents' other priorities (ILO 2001).

Forstner (2004) argues that it is unlikely any of these intermediaries alone could assure success for community-based ventures, and advocates integrating approaches where these separate parties might complement each other's strengths. To date the JED cooperative has been focused on institutionalizing links between the NGO and the component communities through a formal cooperative structure in which the four villages and local NGO hold equal shares.[22] While efforts toward collaborative arrangements with the private and public sectors were at best lukewarm and at worst actively spurned in the early years, carefully targeted and negotiated engagements with private and public sector agents eventually proved necessary to regenerate the momentum of the early stages of the project and place it on a viable commercial footing.

The state bureaucracy has been both a facilitator and an obstacle in Banjar Dukuh's efforts to turn its wine production venture into a viable enterprise. Karangasem district provided for the original training of the village wine makers, who then experimented further to develop the special characteristics of this salak wine with the help of Agriculture Department officials and an academic researcher. But in order to capitalize on its unique product commercially, the village must obtain a licence through the national alcohol agency. Had their ability to satisfy formal standards and testing requirements been the only issue, the extension services of either the Department of Small Industries or the Department of Agriculture should have been able to assist farmers to get their product to the point where it could meet these standards. But villagers are convinced that this impediment had been manipulated because they were not in a position to pay the huge bribes expected by officials for obtaining a licence in this lucrative industry.[23]

Complicating their situation was the uncertainty introduced among permit-granting agencies by decentralization and restructuring in the wake of new regional autonomy legislation. Among the bureaucratic impediments[24] the fledging industry experienced when attempting to obtain the several permits necessary for production and marketing of alcoholic beverages was the fact that no regional government regulations had been issued which set out new standards and procedures. As regional government stands to earn an estimated Rp 7 billion from imposts on alcohol (*Bali Post*, 7 March 2005), it should have a powerful incentive to ensure that local products find their way into this market. This is particularly the case for kabupaten Karangasem, which is one of the poorest districts in Bali.[25]

As of 2006, salak wine production had halted, because the Dukuh producers were unable to clear their stored stock through the one small NGO outlet that would still risk distribution of unlicensed alcohol. Its production was too costly to compete with rice and palm wine on the local market, and the cooperative lacked a means of breaking into the higher priced restaurant or export market without selling out to powerful producers and distributors or finding a way to satisfy bureaucratic licensing demands.

The private sector had shown interest in the new wine from the outset. Overtures came from urban-based rice and palm wine producers to buy out processing and distribution rights to the villagers' product.[26] However, such a development would have meant ultimately losing community-based control of the fledgling industry. A high production volume made possible with investment from an outside private company with sufficient capital might raise salak prices in the short term and increase employment prospects, but ultimately the capacity of private interests to source fruit and process the product elsewhere could eventually undermine Dukuh's comparative advantage.

The economic viability of the ecotourism venture proved equally disappointing in the initial years. Launched just before the Bali bombing, with the slogan 'It's now safe to turn off mass tourism', the JED cooperative ventured into an industry in crisis.[27] What market there might have been for this alternative brand of tourist experience immediately evaporated. The impact of the crisis on mass tourism might have offered the opportunity to alter the character and direction of development in the sector. But regional governments seeking tax revenue for their budgets continue to put their resources into promoting the capital-intensive and mostly foreign owned sector of the industry, and have so far done little to turn the rhetoric of a 'people's economy' (*ekonomi kerakyatan*) into serious policies aimed at supporting equity and sustainability.

For their part, the community participants and their NGO partners in the JED ecotourism network lacked the skills and experience necessary to single-handedly attract the attention of a highly specialized nature- and culture-based tourism market.[28] There was no website for three years after the official project launch, and publishers of alternative travel guidebooks that could have carried information on the project had not been contacted until 2005. Other than a small number of tours run for like-minded NGO groups in the Southeast Asian region, the JED Community Based Ecotourism Network, one of the most important outcomes of the community mapping project, languished in the face of promotional difficulties, as well as internal problems of organization, infrastructure support,[29] and lapses in transparency.[30] It is difficult to gauge to what extent these problems can be attributed to the loss of momentum and sense of direction under adverse political and economic circumstances following the Bali bombing. Unquestionably more focus, experience and external support would be needed to overcome the range of problems they faced.

Responsibility needs to be laid also at the feet of the national and international donors who failed to carry through with support for the critical final implementation stage of the project. The anticipated third year of funding for developing programme plans was prematurely aborted. This was partly because shifts in USAID priorities in the wake of the September 11, 2001 terrorist attack in the United States affected Yayasan Kehati's own funding capability. Kehati funding criteria in any case precluded grants other than for capacity building and training programmes. By this stage the project participants felt they were more in need of investment in infrastructure than further training. Their application for a Rp 499-million (US$60,000) grant toward the construction of information centres and other start-up requirements was rejected by Kehati as inappropriate to this stage of the project's development and outside the partner organization's funding guidelines. The local partners parted ways with Kehati, with some misgivings on the part of participants (interviews with NS, 10 November 2003 and NK, 1 August 2004; see also Ambarwati 2004: 67ff). The four remaining villages and local NGO made the decision to launch their ambitious programme in February 2002 in the expectation that income from their inter-village trading cooperative and ecotourism enterprise would subsidize further expansion of the project. Funds were to be pooled between the four villages and the local NGO as equal shareholders in order to subsidize the venture.[31] But as we have seen, the markets they desired did not readily materialize on a viable scale.

The local NGO, which used its offices and equipment as the base for the venture, was constantly struggling to find the resources to cover overheads and provide basic salaries[32] to its own small staff. It sought out new grants for other purposes in the face of inadequate spin offs from the fledgling cooperative, with the consequence that it did not give marketing difficulties sufficient attention. Despite exceptional enthusiasm, and a great deal of creative endeavour by participants, the JED cooperative's ventures had not been profitable enough to provide reasonable wages to its workers or benefits to the core cooperative's member communities until 2008.[33] Inadequate resourcing and time frames are widely reported as negatively affecting outcomes in other studies of NGO development project effectiveness (Hilhorst 2003; Salafsky *et al*. 1999).[34] The power of funding agencies to unilaterally define and revise the conditions of assistance, the inflexibility of project timelines, and the fickle nature of development agency agendas (often driven by political interests of donor and recipient nations) work against the prospects of alternative development options.

Under pressure to fund its own infrastructure (which had now become the base for the JED trading and ecotourism network) and without the anticipated start-up capital to tide them over, the local NGO leadership turned its attention to links with a larger Java-based umbrella NGO that could provide some continuity in support for the local one. But this was inevitably at the expense of spreading their commitments thin and diverting attention from the needs of the fledgling inter-village cooperative programme.[35] Village-based cooperatives linked to the JED network were more or less left to their own devices as far as promoting their value added products, and were on occasion discouraged from taking independent initiatives. In one instance involving coffee marketing from Kiadan, the community found itself in competition with the JED network itself, which insisted on handling the packaging of the upmarket branded product, even though the village cooperative had purchased equipment to do this themselves. At that point the Kiadan coffee farmers were offered higher prices for their best quality coffee beans from a nearby commercial processor. Temporarily at least, these farmers found themselves better served by the private sector than the cooperative network of which they were joint owners (interview WJ, 23 October 2003; AK, personal communication, 9 June 2005).

Tenganan continued to receive a steady flow of tourists brought by outside guides and tour companies. By comparison with the 28,000 tourists arriving by these other means in 2003, JED brought only a trickle of visitors, mostly from sympathetic regional NGO groups.[36] Tenganan farmers also maintained existing private arrangements for sale of their rice crops before harvest, rather than selling to JED for distribution to other villages through the cooperative. There was a general perception that the cooperative could not compete with private sector middlemen who had historically monopolized the market, and certainly the physical distance between the member communities added appreciably to overhead costs.

Banjar Dukuh participants and NGO staff expressed mutual disappointment at failures to carry through on lines of action that each felt the other ought to have pursued to help develop the customer base for their wine. Community participants had hoped for greater involvement of the NGO in negotiating with regional government and seeking out potential markets. From the NGO point of view, local leaders' failure to call meetings and take their own initiative were at least as important as obstacles presented by the bureaucracy or their lack of capital in accounting for the limited development of the wine industry. The production stalemate in Dukuh was partly blamed on cash flow problems. JED owed Banjar Dukuh Rp 5 million for wine it had distributed. It had used these funds for operational expenses in the absence of the remaining unpaid share commitments from member villages (interview ND, 31 August 2002; AK, personal communication, 28 February 2005 and 30 March 2005).

The local NGO itself suffered from internal problems in this period, when structural changes were initiated to cement its new partnership with the Java based NGO, and as a result of lack of serious oversight from its Board. Conflicts of personality and principle emerged among staff, virtually immobilizing the organization for much of 2003 and 2004. Nevertheless, relationships were maintained throughout this period between the community participants and the

NGO, thanks to the dedication of its director and field staff. Six-monthly meetings continued to take place between NGO and village representatives to the JED cooperative, but no serious steps were taken to break the deadlock which inhibited the development of the JED industries until financial pressures and two separate transparency issues forced all participants to take stock.[37] In December 2004, the cooperative and NGO organizational structures were again revised following conflict resolution meetings, facilitated by their new national NGO partner. An agreement was forged to set up its several activities as independent units under a new board of management.

The minutes of the meeting called to confront this crisis cite differences in perspective, personal conflicts and weak accountability in managing the group's resources for future investment among the key problems that had to be addressed. They explicitly articulate the need to take account of the 'social capital' at stake in deliberations over the restructuring to solve these problems.[38] Nevertheless, in March 2005, the network made the pragmatic decision to cease trading in agricultural products and to focus its attention on ecotourism, abandoning for the time being at least one of the major objectives of the original programme – the balanced development of the formerly neglected agricultural sector.

Notable throughout the project has been the high level of reflexivity among participants – NGO and village based alike. During the three-day evaluation meeting in 2005[39] to assess project outcomes and reconsider its future direction, participants were remarkably frank about its inadequacies. Dukuh participants, for example, reported that there was still suspicion in their community over money issues, and that there was an 'aid project' mentality, where 'participation' was often for appearance. The JED initiative had taken serious steps backwards marked by declining enthusiasm, and lack of professionalism, direction and community support. 'The sudden end to [Kehati's financial] support for the programme mid-stream had negative impacts, making it difficult now to convince the community [of its viability]' (summary report from Dukuh participants presented by NK, meeting transcript 5 May 2005). Other criticisms from participants which emerged during discussion focused on contradictions in the role of their NGO partner:

> Actually the people desire to be accompanied [by the NGO partner], but often this is under pressure, like the formation of the JED and primary [village level] cooperatives. The permit had to be gotten first, even though at the grass roots there were still discussions going on – so it is just like the government. In the end, JED just lengthens the marketing chain, because everything has to be bought through JED (ND, Banjar Dukuh, meeting transcript, 6 May 2005).
>
> JED also looks for profit, so the market chain gets increasingly long. [The NGO director] is always saying, 'even though the cooperative is more expensive by 50 *rupiah*, it belongs to us'. We can understand that, but it doesn't make sense to others (KJ, Ceningan, meeting transcript, 6 May 2005).

Community participants also asked for simplification of different accounting systems: 'I feel colonized; we have to report finances to the government, Wisnu and the community' (KJ, Ceningan, meeting transcript 6 May 2005). Others expressed concern that they had been too optimistic, and having been encouraged to 'follow this dream', they became caught in the trap of their own making, with heavy responsibilities on which they had not been able to deliver (PS, Ceningan, and ND, Dukuh, meeting transcript, 6 May 2005).

On the other hand, the NGO–community partnership was framed positively in terms of what local people might recognize as linking and bridging forms of social capital (Woolcock 1998) that support solidarities needed for an active civil society and collective action in the interests of the 'commonweal'.

> I think it isn't only because of the funds, but since working together with the NGO the solidarity (*kekompakan*) among the community is increasingly evident and there are a lot of things that can be done. … [T]he people of Kiadan are now more outspoken in expressing their opinions for example in monthly village meetings. Before only the leaders would talk. [Now] there are always villagers that ask questions and we are more tight-knit. (WJ, Kiadan, meeting transcript 5 May 2005).

Notwithstanding the disappointments of the early years, the project participants continued to experiment, with new applications for funding agency grant support, the arrival of an overseas volunteer to assist with tourism promotion, and renewed efforts to tap into the fair trade market. [40]

Communities and markets in global transformations

Globalizing forces pose significant opportunities as well as challenges for the efforts of communities such as these to deal with change in collective terms. The problems experienced by the JED cooperative that emerged from the community mapping and participatory development project cannot be separated from fundamental problems facing rural smallholder producers and cooperatives in the global arena generally.[41] What degrees of freedom do socially oriented systems of production and distribution have to adapt the narrowly economic 'rationality' of prevailing neoliberal policies and market practices to their wider collective purposes?[42] Governance issues become central in efforts to bridge this divide, and regulatory regimes which seriously incorporate social and environmental values represent a crucial 'collective good' still largely missing in the global arena, as evidenced in the failure of the international community and its agents to establish more than marginal mechanisms which could effectively submit the global processes of capital penetration, accumulation and economic transformation, to widely touted triple-bottom-line (social and environmental as well as economic) principles (Stiglitz 2002: 247ff; Dunkley 2004: 215ff; Hadiwinata and Pakpahan 2004; Hawkins 2006; Hopkins 2006).[43]

Neither the local NGO, nor the communities involved in this project on their own possessed sufficient capital or skills to reach markets on a scale that would enable their efforts in ecotourism and value-added organic product marketing to be economically viable under current conditions. At the same time, alliances with outside agents or capital providers posed the risk of losing community control and therefore the social capital that these endeavours contribute to and are dependent upon. Banjar Dukuh not only lacked the financial capital to produce its unique organic wine on an economic scale, it also lacked the symbolic capital that public relations firms and advertising agents 'produce' to value add and carve out a strategic place in the market. Those potential private industry partners offering to fill the capital gap have proposed to do so at the price of privatizing common property. This perception of risk to the broader common good principle similarly prevented the JED cooperative network from exploring potential partnerships with travel operators and tour agents in the early years.

Li argues against tendencies in common property studies to represent 'the market' as a hostile external force located physically and conceptually beyond village boundaries (Li 2001: 172). Certainly, contentions emerging within and between these Balinese communities and their intermediaries over the course of this project reveal competing values regarding private versus collective interests and social versus market driven values that were at least as palpable as those positive opportunities and hostile interests they perceived as 'external' in the sense of operating beyond the reach of village rules and reciprocities. On the other hand, it is clear that deliberations on the use of *adat* mechanisms to regulate land sale and to build local community-based industries were primarily about constructing distinctions of right and responsibility between insiders and outsiders which would support specific, local and collective determinations of value in steering the direction of development and change. Their regulatory agenda did not indicate a self-imposed moral economy of subsistence, or a backward-looking peasant ethos of limited good (cf. Li 2001: 159–67), but rather an understanding that some such bounded rule-making regimes would be necessary to place the engagement process on terms which could offer the possibility of benefiting some broader construction of local economic interests, and situating these in relation to social, cultural and environmental values.

Such determinations are, however, contentious and provisional. For example, although in the course of the community mapping exercise Ceningan villagers rejected the regional government's proposed tourist resort development in favour of retaining control over the island's coastal resources for sea-grass farming, they have yet to come to consensus on rules regulating the sale of land on the island, with the likelihood of a slow incremental loss of control over the resources that had initially led them to reject district government development plans. More generally, the question of the extent to which customary community regulation can most fairly and feasibly be applied to the management of land and resources in the interest of a local construction of the 'commonweal' remains at issue in most of the project villages.[44]

For Dukuh, which did introduce regulations to control land sale, it is not a matter of sealing out the market, but of whether selective engagement on local terms – with agricultural and ecotourist instead of land markets – will prove viable and provide some leverage in what they see as a David and Goliath relationship. For the time being this *adat* community appears determined to resist external market penetration of their land regulation system, seeing this as pivotal to the protection of local values and heritage. On the other hand, the kind of market relationship that Dukuh villagers would like to pursue through value adding to their local farm products and through small-scale community based tourism that could be integrated into locally owned facilities and a diversified agricultural economy – precisely the kind of balanced, culturally and environmentally sensitive approach to Bali's future development that is being called for in the press and among the general public – appears only a fragile possibility in the absence of sustained interventions.

Whether the alternative economic experiments tied to the Bali project eventually accomplish their objective of providing practical and sustainable alternative development strategies remains to be seen. The JED cooperative is still struggling to build the institutionalized alliances and links to promotion and distribution networks that would allow them to engage on fair trade terms under the 'free market' model of globalization's promise with a locally defined agenda. Without these links, and without a different model of the global playing field, such alternative development options appear little more than a vaguely glimpsed mirage.

Experiments with agricultural cooperatives, fair-trade promotion schemes, or small-scale community based ecotourism need to be analysed comparatively with mainstream systems of production and distribution in terms of sustainability and equity criteria, instead of the narrow measures of economic efficiency and profit maximization currently privileged. This would at least put community-based enterprises on a fairer basis for comparison in evaluations of their relative costs and benefits, and in assessing the viability and appropriate strategies for externally fostered interventions and partnership arrangements. Ultimately the level playing field also depends upon a global regulatory regime that imposes social and environmental ground rules on an overarching footing vis-à-vis the currently dominant principles of private profit maximization. In other words, the key question for those who are serious about equity and sustainability principles is how to recuperate the role of the economy as a means rather than an end in itself in the global trading system of a post-industrial age.[45]

Building communities of shared vision and interest, with organizational capacity and extensive extra-local networks is pre-requisite to enabling more balanced engagements with state and market forces, and between popular and elite interests. As we have seen, alternative development strategies face seemingly intractable difficulties when dealing with bureaucratic gatekeepers (as Dukuh has with its salak wine licence), and the enormous economies of scale and vertical integration in the global marketplace (as the JED network faces in establishing a foothold in the travel industry),[46] against which they cannot easily compete without greater resources, wider networks and principled intermediaries. Niche marketing and downstream processing of organic, fair trade, value added farm products linked to small-scale industries and carefully planned culture- and nature-based tourism offer appealing options for rural communities seeking to resist the seemingly monolithic pressures of dominant market players. But the roles of intermediaries and regulatory mechanisms in bolstering more equitable and sustainable initiatives are interlinked and need to be given more serious attention to make this possible.

Parnwell (2005) and Li (2001) point to the importance of including market issues within the community development agenda. Clearly, this is the arena in which governments, communities, academics and NGOs have so far proved less than effective in bringing social and environmental objectives into successful accommodations with economic improvements. In Dorothea Hilhorst's (2003: 136–7) study of NGOs in the Philippines, for example, marketing proved the weak link undermining the efforts of an NGO-sponsored women's income-generating project, as marketing proved the most serious problem for the development of the newly formed JED cooperative's plans for promoting their coffee, wine and ecotourism industries.

Hilhorst's (2003) detailed study of 'the real world of NGOs' provides a critical assessment of the misfit between formal rational accounting procedures that dominate capitalist markets as well as donor agency policy and the broader notions of moral accountability that inform the 'real world' of local communities and the NGOs which engage with them directly. Li (2001: 163, 175) frames the argument differently. The 'real' world is an already globalized one in which locals share many of the same desires – for convenience, accumulation, status – as their romanticizing others in the academic and NGO worlds.[47] She argues that communities such as these cannot be left to bear the burden of creating sustainable alternatives on their own.[48] From both perspectives long-term partnerships between government, NGOs, local groups, and interdisciplinary 'action research' teams in search of replicable alternative models of market engagement are a crucially important direction to pursue.[49] At this critical juncture, sustained experiments with partnerships of this kind have to be part of a concerted ongoing exploration of new forms of governance for land use and development in a rapidly globalizing world order. These must be linked to significant reforms of national and global structures that expand and give teeth to international social and environmental conventions – unless we intend to leave the increasingly inequitable and unsustainable 'free market' driven set of structures as 'the only game in town'.

Epilogue

By mid-2008, persistence on the part of the communities and NGO partners in the JED network and more effective collaborative relationships with government, business and non-profit international agencies had begun to bear fruit. Tourist visitor numbers through JED in the first seven months of 2008 had already exceeded the annual total for all previous years at 222 visitors. For the first time the JED cooperative was able to cover all its staffing and infrastructure costs and meet its commitments to contribute to the social and environment funds of member villages. The upturn primarily resulted from partnerships forged with French and English alternative travel agencies and a new upbeat website[50] which enabled prospective clients to book online. With its small-scale ecotourism venture finally taking off, the JED cooperative members agreed at their bi-annual meeting in March 2008 to resume efforts toward diversification of agricultural products and allied environmental experiments. Dukuh has obtained some of the twelve required wine production licenses with the assistance of its new district head and has been granted loans and equipment from regional government departments responsible for cooperatives and small industry to upgrade production processes (notably without the unofficial payments formerly expected). They planned to produce a thousand litres of salak wine following the August 2008 harvest. Ceningan's cooperative was in the process of building a cottage to accommodate overnight stays on the island by JED visitors. Pelaga had built an impressive new open pavilion, built entirely from members' labour and contributions, which served as meeting hall for the *subak* irrigation society and doubled as a space for JED group activities and communal meals when tourists visited the village. Tenganan was about to start up its own rice mill powered by micro-hydro energy enabled by a grant the village obtained from the Global Environment Facility. Wisnu was meanwhile branching out into new experiments with organic farming and biogas production from animal waste. Plans to expand involvement of youth in the four villages were also underway. Though still fragile and small scale, this community-based development program's earlier halting efforts to source outside expertise and resources had begun to come together with the energy and enthusiasm of core participants to produce the kinds of synergies of which villagers and NGO alike had previously 'dreamed' (ND Dukuh, 6 May 2005).

Acknowledgements

The fieldwork for this research project was made possible through funding from the Australian Research Council and the Asia Research Centre at Murdoch University. I am grateful to the villagers and NGOs involved in the community mapping and cooperative projects for their contributions to this collaborative research, in particular I Made Suarnatha, Ambarwati Kurnianingsih, Ni Made Puriati from Yayasan Wisnu, and villagers from Belok Sidan, Ceningan, Kiadan (Pelaga), Dukuh (Sibetan), and Tenganan Pegringsingan. Some of the material in this chapter related to the early stages of the community mapping project was previously published in *Development and Change* (Warren 2005).

Notes

1   An unusual number of articles tracking environmental conditions was published in the *Bali Post* in 2006, reflecting widespread debates on future development strategies. Some 23 per cent of Bali's coral reef system is reported to have been completely destroyed, and 69 per cent of the remaining 50 sq km of reef is in seriously damaged condition (*Bali Post*, 24 August 2006). Inventories of forests, mangroves, and beaches were showing similarly serious impacts of human activity (*Bali Post*, 16 August 2006). An estimated 25 per cent of Bali's forests had been diverted to other functions and the remainder are severely degraded (*Bali Post*, 11 August 2006). Drought and declining water levels in mountain lakes dried up most of the rivers in the northern district of Buleleng, resulting in withdrawal of 741 hectares from rice production due to lack of irrigation water (*Bali Post*, 23 August 2006). Rivers in the Denpasar area are highly polluted as a result of unregulated dumping of chemicals and dyes from the 200 factories producing textiles

(*Bali Post*, 23 August 2006). A particularly pointed editorial, titled 'The environment first, then the economy', summarized the serious state of affairs and urged a radical change of policy ('Alam dulu, baru ekonomi', *Bali Post*, 9 August 2006).

2 See, for example, the *Bali Post* end of year report (29 December 2005) decrying the marginalization of Bali's farming sector, which it attributes to the exclusive focus of policymakers on tourism development, until the two bombings 'raised it from the grave'. The article calls for subsidies, appropriate technology and better planning to solve the problem.

3 These refer to highly corporatist local institutions in Bali: *banjar* (hamlet), *subak* (irrigation association), *tempekan* (neighbourhood subdivision of the *banjar*, which is in turn a subunit of the adat village – *desa adat*), *sekaa* (clubs and work groups). For discussions of these local organizations, see Geertz (1959), Warren (1993) and Lansing (2006).

4 An editorial in the regional press asks whether the use of market value based taxation is not simply a 'polite' (*halus*) way of forcing farmers off their land (*Bali Post*, 16 January 2006).

5 See for examples of public complaints regarding loss of subsidies and low agricultural prices: 'Petani Bali menjerit – air susah, pupuk mahal' [Balinese farmers cry out – water short, fertilizer dear], (*Bali Post*, 30 August 2006).

6 Some recent examples of these critical themes in the populist Balinese media include: 'Jangan mudah beri izin investor – laksanakan konsep desa pakraman' [Don't grant easy permits to investors – Carry out the concept of the desa pakraman (*adat*)] (*Bali Post*, 22 March 2005), 'Tingkatkan peran desa pakraman menjaga tata ruang Bali' [Raise the role of the desa pakraman in protecting Bali's spatial plan] (*Bali Post*, 5 April 2005), and especially reports published in the regular *Bali Post* 'global coffee shop' column summarizing talk back radio commentary: 'Cegah krisis air bersih – Tolak proyek yang berpotensi rusak lingkungan' [Reject projects that could damage the environment] (*Bali Post*, 29 August 2006); 'Beri kemudahan perizinan – Pemerintah sumber kerusakan lingkungan' [Granting easy permits – Government is the source of environmental destruction] (*Bali Post*, 24 August 2006); 'Cegah abrasi, jangan izinkan membangun di pantai [Prevent erosion, don't issue building permits on the beach] (*Bali Post*, 8 August 2006). Commentators in the December 2004 'Your Turn' column recommend severe restriction on in-migration to limit population growth; state tax and *adat* service relief for those maintaining green zone reserves; and strengthening the rule of law by incorporating *adat* sanctions.

7 See Warren (1993: 7–28, 68–89) on myths and metaphors embedded in the *adat* and village community concepts. For critical discussion of the 'community' concept in relation to natural resource management, see essays in Agrawal and Gibson (2001) and Brosius *et al.* (2005).

8 See Warren (2007: 171, 191) for an account of the 1997 successful resistance by the *adat* community of Kesiman to a hotel development at Padanggalak in which it threatened *adat* sanctions against the Governor of Bali. The case reasserted a broad interpretation of the collective authority of the *adat* village over its community customary domain defined in territorial, social and religio-symbolic terms.

9 See Lucas and Warren (2003) for a detailed account of the contested process of revising the 1960 Basic Agrarian Law, both honoured for its socialist premises and legal commitment to land reform, and reviled because of its privileging of certified (*hak milik*) over only nominally recognized customary *adat* rights, and of 'national interest' claims over local entitlements.

10 For a more detailed account of the community mapping project and the other villages involved, see Warren (2005).

11 The complexity of variations of the 'communal land' concept is illustrated by the fact that Sibetan has six named categories of land which could be classified as communal *adat* land, each of which entail obligations of a different nature to various local *adat* institutions: family held agricultural and residential lands are tied to labour service obligations to the village (*tanah ayahan desa*) and *banjar adat* (*karang desa*); communal lands are set aside for temple support (*laba pura*), as well as for cultivating

specific plants which are used for ceremonies (*tanah swaka*); and include lands which compensate the 10 village leadership positions that have religious and organizational responsibilities (*tanah bukti keliang* and *bukti pemangku*).

12 Since these customary conditions of land tenure imply both civic and ritual responsibilities, they represent an obstacle to outside investors and in-migrants. Even ethnic Balinese from other villages may not subscribe to all beliefs and practices stipulated by local *adat*, and efforts to impose collective social control are sometimes a source of serious internal conflict. Willingness to abide by collectively determined common *adat* rules and obligations remains an important criterion of local citizenship, and the exercise of severe *adat* sanctions has become a more open feature of local politics in Reform Era Bali. Members of *adat* communities themselves may lose rights to reside in their own community through serious transgressions of *adat* (Warren 1993, 2007).

13 Complex understandings of place and ancestral relationship are mutually bound up with Balinese notions of local citizenship and responsibility for collective spiritual and material wellbeing (see Warren, 1993). For Balinese, rights to certain categories of land entail heavy communal labour service obligations for ritual purposes.

14 See the impressive collection of essays in Verdery and Humphrey (2004) on the myriad new and old forms of property, which restores richness and cultural depth to a literature that continues to be dominated by models founded on a mythically abstracted notion of pure calculable value, and by commodified, privatized and unencumbered notions of ownership and rights of disposal. As Benda Beckmann's cover comment describes it, property is 'one of the most powerful devices of exclusion and hierarchy'. Its double-edged dimensions may be both liberating and dominating. Property is fundamentally about relationships and no property right is ultimately unencumbered, unqualified, or entirely 'private'. Even at the most individualised freehold end of the spectrum of rights, some collectively legitimated governance regime is ultimately required for ongoing recognition of what would otherwise be force-based claims.

15 Differentials in land values depend upon a number of factors, including fertility, accessibility, infrastructure, distance from existing tourist enclaves, scenery, etc. Between locations with and without these value enhancements, variations in land prices can be of the order of several thousand per cent – at present from as little as Rp 3 million to as much as Rp 130 million (US$350 to $14,000) per are (100 square metres) according to various informants in 2005. Even at the low end of the spectrum, the sale of one hectare of land could bring income equal to many years' earnings for the average farming household.

16 Belok Sidan left the project after the first year, because of tensions over local accountability processes and scepticism about the potential for economic benefit to any significant proportion of the membership of this large village. There was also some friction between NGO, state and local actors concerning the relatively confrontational character of NGO engagement with government agencies adopted in the course of the project.

17 With respect to the relative power position of these communities vis-à-vis state and private interests, this activist intervention went some way toward reversing asymmetrical institutional relationships between official and customary, central and local authorities. Villagers from Ceningan Island rejected government plans for a major US$ 200 million tourism development on the island, when the community mapping exercise enabled them to weigh up potential costs alongside benefits, and to imagine alternative options. The community eventually also succeeded in negotiating a percentage of the income from the district government licensed private concession on a lucrative bird's nest site on the island.

18 In 2004, the price of salak at the height of fruiting season dropped to as little as Rp 500 per kg (US$.06), but reached Rp 3,000 per kg (US $.33) for the small volumes of fruit produced in the off season. A kilogram of rice at this time cost around Rp 2,500. The Dukuh work group produced some 1,000 litres of wine in 2001–2002 (ND interview, 31 August 2002).

19 The proportion taken by JED as distributor/promoter was questioned by villagers, especially given the NGOs failure to secure markets for their product. See below.

20 Initially, only Wisnu's roadside café at the front of their peri-urban office and recycling complex, a small hotel in Sanur owned by another environmental NGO, and a sympathetic retailer at the airport were prepared to sell the wine. The airport retailer eventually stopped carrying the wine because of the risks of trading an unlicensed product in such a highly visible outlet. Wisnu's café eventually closed because the location was too far from the main business district to acquire a steady clientele, and the NGO hotel only sporadically received delivery of the wine as the initial level of activity surrounding the project began to wane.

21 The distance between the participating villages in the JED cooperative imposed heavy transport costs, while the self-help ideology and the focus on the social capital bonds built through their NGO-sponsored programme inhibited inclusion of other communities that might have enabled them to achieve economies of scale.

22 The communities were unable to raise the full Rp 25 million (US $2,700) share, which each had agreed to contribute over a four-year period to cover the cooperative's start-up costs. To date only the NGO has contributed its full share of the Rp 125 million originally estimated as necessary to capitalize the project.

23 See the *Bali Post*, 7 March 2005 on the general issue of corruption and lack of transparency in the alcohol licensing system.

24 A regional government officer commented on the familiar bureaucratic wrangle which has long characterized the stifling model of top-down governance in Indonesia's civil service: 'We here would really like to help our enterprising friends, but … we can't do anything without an instruction from the head of department, and the head of department can't do anything without an instruction from the *bupati* [district head], and so on. If we actually get a recommendation from the bupati to the department head and the department head gives us directions, we are ready to process it! (interview with departmental officer, Karangasem district, 28 February 2005).

25 According to census statistics prior to the impact of the terrorist attack in 2002, Karangasem had the lowest per capita income of any of the regions in Bali at Rp 3.1 million per annum (approximately US $390), compared with Rp 10.1 million for the wealthiest regency, Badung, and a Bali-wide mean of Rp 5 million per capita (BPS 2001).

26 There is an interesting intellectual property question here. Only two villagers actually know the precise process for producing this wine. NS and ND were among the small group of farmers from several *banjar* in Sibetan intially taken into the training programme organized by the regional agriculture department. The two continued experimenting to get the brew to the right level of alcohol and to achieve its unusual piquant tart-sweet character. As the 'recipe doctors' (*doktor resep*) they are sole holders of knowledge of all aspects of the process. ND has been approached by private sector interests to sell this information, but commented he 'wouldn't dare' since it was regarded as common property by the *banjar*, notwithstanding their individual roles in the development of the process (interviews with WS, 23 November 2001, ND, 31 August 2002).

27 An estimated 29 per cent of the workforce in Bali lost their jobs in the months following the bombing, with per capita income reported as declining by an average of 43 per cent (UNDP et al. 2003: iii).

28 See Ambarwati (2004). The author, who has been actively involved in the Wisnu project since its inception, describes the difficulties faced by the ecotourism component of the programme with a particular focus on Desa Tenganan's experience.

29 The Biodiversity Conservation Network provides data for one of its funded projects (Cradle Mountain in PNG) in which conservatively estimated management costs accounted for 46 per cent of total costs. These and other fixed costs had been unreported in project accounts, leaving the impression that this forest research tourism venture was profitable, when in fact the project sustained substantial losses relative to income.

Despite careful criteria in the selection of projects in a number of countries with a 'high likelihood of success', the BCN review ranks only six of the 45 enterprises in 39 communities that it funded as likely to remain financially successful over the medium term (Salafsky *et al.* 1999: 20).

30 Three incidents involving misappropriation or private 'borrowing' of cooperative and donor funds arose over the period 2001–2005. All took some time to be dealt with because of the social sensitivities involved. Close relationships between participants make it difficult to confront the failures in oversight both within the villages and by the NGO in open meetings, despite considerable emphasis on this issue in the early stages of the project (MD, interview 26 July 2004). On the other hand, the fact that these cases were over time addressed, despite the lack of external monitoring since involvement with the national NGO Kehati ended in 2002, does indicate that transparency is being taken seriously.

31 None of the member villages had yet fully paid up its Rp 25 million share in the cooperative. In 2008 it was decided to reduce shares to Rp 12 million and use profits from the eco-tourism venture to finance remaining contributions.

32 Some NGO staff were receiving monthly salaries as low as Rp 300,000 ($US 35), at one point several months in arrears (MD, interview 26 July 2004). Most of the volunteers within the local cooperatives received no wages beyond a small honorarium or participation subsidy for sponsored workshop and training activities or an occasional bonus from the small income generated from JED activities.

33 Several of the component community cooperatives on the other hand, proved more viable because they were more pragmatic, flexible and adapted their activities to local needs and priorities, in comparison with the umbrella JED cooperative.

34 It is worth noting, however, that the Biodiversity Conservation Network study of 45 enterprises it funded in connection with conservation projects showed a surprising lack of correlation between the amount of funding provided to the project, or the level of cash as opposed to non-cash benefits to participants, and the degree of conservation success, measured as positive outcomes in terms of reduction of threat to biodiversity (Salafsy *et al.* 1999: 26–36).

35 Villagers as well as several NGO fieldstaff criticized the invention of a '*banjar* school' programme tailored to the needs of their new Jogjakarta based partner NGO, which incorporated the Wisnu network into one of its Indonesia-wide governance projects. Participants from the four villages hadn't been consulted in the development of proposals for the new project, and expressed irritation at the implication that their robust local *banjar* organization needed either revitalization or tutelage in democratic principles (interviews MD, 26 July 2004; WJ, 31 July 2004; and NS, 3 August 2004).

36 JED statistics list a total of 168 visitors to member villages arranged through it for the year 2002 in the nine months before the October bombing at Kuta. Visitor numbers dropped to 27 in 2003 and began to rise with the general increase in tourism in 2004 when it catered to 108 visitors (JED trading statistics 02–05; Ambarwati 2004: 81 and 83, Table 6). A second terrorist bombing again decimated the tourist trade in 2005.

37 The two separate transparency incidents, involving misuse of funds by an NGO and community cooperative staff member respectively, created a crisis, which was both fiscal and one of trust. Both issues were eventually resolved through negotiated repayment arrangements. But the process was slow and painful because of the personal relationships involved.

38 JED Activity Report, December 2004: 2.

39 Transcript JED bi-annual meeting, 5–7 May 2005.

40 The volunteer assisted in listing JED with responsible tourism networks in Europe. JED member communities also agreed in 2006 to contribute Rp 3 million each to support the salary of a hired externally advertised coordinator for the ecotourism programme. They applied successfully for small grants for alternative energy and eco-tourism marketing development.

41 An ILO report (2001) concerned with revival of the cooperative movement stresses their potential in the age of structural adjustment and privatization. It suggests the employment and social service displacing effects of these policies offer an 'opportunity' for cooperatives and other self-help enterprises. In addition to costs of providing social and advisory services to members, however, high reporting and transaction costs may disadvantage cooperatives and other socially embedded ('encumbered' from an orthodox economics perspective) enterprises.

42 A remarkable experiment with community owned industries in the village of Sanur in the early 1970s (see Warren 1993), involving the establishment of restaurants, a service station, laundry and a bank, had initially been extremely successful. Sanur was able to support health and education services as well as providing a significant number of jobs to villagers with benefits comparable to the national civil service scheme. But the Sanur village owned enterprises began to stall in the late 1980s in the face of competition from external investors and heavy social claims on their surpluses. Indicative of these tensions was the controversy surrounding the decision to separate the village bank from the other village-owned industries. In the deregulatory environment of the late 1980s, bank profits outstripped those of the more labour-intensive village industries, and bank staff demanded wages and incentives comparable to those in the private sector. Eventually, the bank was allowed to establish its own salary structure, while continuing to contribute a proportion of its profits to the Sanur Village Foundation. The bank achieved far greater autonomy in its financial decision-making, disembedding itself from and arguably undermining some of the corporate community principles under which it had been established. As Geertz (1963) notes in describing similar ventures in the 1950s, this is the point at which economic rationality faces the limiting implications of its embeddedness in non-economic social structures and cultural values, and at which the logics of Weberian substantive (socially constituted) and formal (market economic) rationality meet head on. See also Parnwell (2005: 19) for an analogous example from his research in rural Thailand.

43 See also Watts (2000: 45ff), who returns to Polanyi's thesis regarding the fundamental importance of re-embedding the economic domain so that its formal 'economic rationalist' principles are subordinated to the social welfare and environmental sustainability purposes that the economy ought to serve.

44 One among several reasons the fifth village, Belok Sidan, withdrew from the community mapping project was the belief that there was no way in which the development of a small-scale ecotourism industry, even with significant government and NGO assistance, could possibly fulfil the unrealistic expectations that had been generated by the project for enhancing rural incomes for its large village population.

45 Daviron and Ponte (2005) give an excellent overview of the problems and prospects of old and new forms of regulation, and an explicit agenda for reversing the decline in commodity trade disadvantage in the global South. Their study focuses on the coffee industry, but applies generally across the smallholder producer sector, and is notable for confronting the paradoxes of the power relations tied to the production of symbolic quality attributes, including the value of 'place' in product development and labelling, which needs to be factored into serious efforts to revamp the global asymmetries in commodity trading. Relevant to Dukuh and other cases studied here, they propose linking a regulatory regime revolving around the promotion of IGO (indication of geographic origin) labelling and promotion to intellectual property rights (2005: 264ff).

46 How, for example, in fair trade terms could the JED network compete with a package tour from Perth to Bali —including airfare, three days' accommodation in a star hotel, breakfast and bonus barbecue and tour, costing A$695 – while a return airfare alone on the same airline for the same period cost A$900 (Flight Centre quotes 11 May 2005). The preferential arrangements afforded large commercial operators through wholesale discount arrangements between airlines, hotels, and mass-market travel agencies mean

that the community-based ecotourism network could not compete on price terms, even if they charged nothing for their services.

47 Clifford Geertz, pursuing a Weberian-informed modernization logic, warns even more strongly (before the spectre of environmental catastrophe put this teleological trajectory into question): 'To urge this sort of "grass-roots and small-industry" policy on Indonesia, whether in the name of libertarianism or on the basis of a sentimental regard for the vitality of local enterprise, is to condemn her to wander in the no man's land of transition indefinitely' (1963: 156).

48 Similar positions on the practical rationale for institutionalising subsidies and other regulatory mechanisms to redress these triple bottom line imbalances can be found in Stiglitz (2002), Tudge (2004), Dunkley (2004), Salafsky et al. (1999). See also McCay and Jentoff (1998), whose position on this arises from the classic in principle opposition between socially embedded reciprocal and redistributive exchange and the 'negative reciprocity' characterizing the supply–demand, cost–benefit calculus of the market, as substantivist economic anthropologists and historians have long argued.

49 See Bebbington et al. (2006) for a discussion of the critical role of intermediaries in the World Bank funded 'social capital' projects. They found that in their rural study communities in Java and Sumatra, where agriculture was no longer able to sustain livelihoods, the 'need to establish links with other parties has been critical in attempts to confront problems that derive from government policy, market structure, and nature' (2006: 1966).

50 See http://www.jed.or.id accessed 14 August, 2008.

Bibliography

Agrawal, A. and Gibson, C. (eds) (2001) *Communities and the Environment: Ethnicity, Gender and the State in Community-Based Conservation*. New Brunswick, NJ: Rutgers University Press.

Ambarwati K. (2004) 'Jaringan Ekowisata Desa: Tradisionalisasi Diri Orang Bali di Tengah Modernisasi', unpublished MA dissertation, Yogyakarta: Program Pasca Sarjana, Universitas Gadjah Mada.

Bebbington, A., Dharmawan, L., Fahmi, E., and Guggenheim, S. (2006) 'Local capacity, village governance, and the political economy of rural development in Indonesia', *World Development* 34 (11): 1958–76.

BPS (2001) *Bali Dalam Angka 2000*. Denpasar: Badan Pusat Statistik Propinsi Bali.

BPS (2004) *Bali Dalam Angka 2003*. Denpasar: Badan Pusat Statistik Propinsi Bali.

Brosius, P., Tsing, A., and Zerner, C. (eds) (2005) *Communities and Conservation: Histories and Politics of Community Based Natural Resource Management*.New York: Altamira Press.

Davidson, J. S. and Henley, D. (2007) *The Revival of Tradition in Indonesian Politics: The Deployment of Adat from Colonalism to Indigenism*. London: Routledge.

Daviron, B. and Ponte, S. (2005) *The Coffee Paradox: Global Markets, Commodity Tade and the Elusive Promise of Development*. London: Zed books.

De Soto, H. 2001 *The Mystery of Capital: Why Capitalism Triumphs in the West and Fails Everywhere Else*. London: Black Swan.

Dunkley, G. (2004) *Free Trade: Myth, Reality, and Alternative*. London: Zed Books.

Forstner, K. (2004) 'Community ventures and access to markets: the role of intermediaries in marketing rural tourism products', *Development Policy Review* 22 (5): 497–514.

Geertz, C. (1959) 'Form and variation in Balinese village structure', *American Anthropologist* 61: 991–1012.

Geertz, C. (1963) *Peddlers and Princes: Social Change and Economic Modernization in Two Indonesian Towns*. Chicago: University of Chicago Press.

Hadiwinata, B. S. and Pakpahan, A. K. (2004) *Fair Trade: Gerakan Perdagangan Alternatif*. Yogyakarta: Pustaka Pelajar, and Oxfam.

Hawkins, D. E. (2006) *Corporate Social Responsibility: Balancing Tomorrow's Sustainability and Today's Profitability*. New York: Palgrave Macmillan.

Henley, D. (2007) 'Custom and *koperasi*: the co-operative ideal in Indonesia', in J. S. Davidson and D. Henley (eds), *The Revival of Tradition in Indonesian Politics*. London: Routledge, pp. 87–112.

Hilhorst, D. (2003) *The Real World of NGOs: Discourses, Diversity and Development*. London: Zed Books.

Hopkins, M. (2006) *Corporate Social Responsibility and International Development: Is Business the Solution?* London: Earthscan.

ILO (2001) Report V (1) on the Promotion of Cooperatives, for the 89th International Labour Conference. Geneva: International Labor Organization, http://www.ilo.org/public/ english/standards/relm/ilc/ilc89/rep-v-1.htm (accessed 6 September 2005).

Korn, V. E. (1984 [1933]) 'The village republic of Tenganan Pegeringsingan', in J. L. Swellengrebel (ed.), *Bali: Studies in Life, Thought and Ritual*. Dordrecht: KITLV/Foris, pp. 301–68.

Lansing, J. S. (2006) *Perfect Order: Recognizing Complexity in Bali*. Princeton, NJ: Princeton University Press.

Levi, M. and Linton, A. 2003 'Fair trade: a cup at a time?', *Politics and Society* 31 (3): 407–42.

Linton, A. (2005) 'Partnering and sustainability: business-NGO alliances in the coffee industry', *Development in Practice* 15 (3): 600–14.

Li, Tania M. (2001) 'Boundary work: community, market, and state reconsidered', in A. Agrawal and C. Gibson (eds), *Communities and the Environment: Ethnicity, Gender and the State in Community-based Conservation*. New Brunswick, NJ: Rutgers University Press, pp. 157–79.

Li, Tania M. (2002) 'Engaging simplifications: community based resource management, market processes and state agendas in upland Southeast Asia', *World Development* 30 (2): 265–83.

Lucas, A. and Warren, C. (2003) 'The state, the people, and their mediators: the struggle over agrarian law reform in post-New Order Indonesia', *Indonesia*, 76:87–126.

McCay, B. and Jentoft, S. (1998) 'Market or community failure? Critical perspectives on common property research', *Human Organization* 57 (1): 21–30.

Parnwell, M. (2005) 'The power to change: rebuilding sustainable livelihoods in north-east Thailand', *Journal of Interdisciplinary Environmental Studies* 4 (2):1–23, <25–34.http:// www.journal-tes.dk/.

Salafsky, N., Cordes, B., Parks, J., and Hochman, C. (1999) *Evaluating Linkages Between Business, the Environment, and Local Communities: Final Analytical Results from the Biodiversity Conservation Network*. Washington, DC: Biodiversity Support Program.

Scott, J. C. (1998) *Seeing like a State: How Certain Schemes to Improve the Human Condition Have Failed*. New Haven, Yale University Press.

Stiglitz, J. (2002) *Globalization and Its Discontents*. London: Penguin.

Suasta, P. and Connor, L. (1999) 'Democratic mobilization and political authoritarianism: tourism developments in Bali', in L. Connor and R. Rubinstein (eds), *Staying Local in the Global Village*. Honolulu: University of Hawai'i Press.

Tudge, C. (2004) *So Shall We Reap: What's Gone Wrong With The World's Food – And How To Fix It*. London: Penguin.

UNDP, USAID and World Bank (2003) 'Bali Beyond the Tragedy: Impact and Challenges for Tourism-led Development in Indonesia'. Denpasar, Indonesia: Unpublished Report, released 19//20/2003.

Verdery, K. and Humphrey, C. (2004) *Property in Question: Value Transformation in the Global Economy*. Oxford: Berg.

Warren, C. (1993) *Adat and Dinas: Balinese Communities in the Indonesian State*. Kuala Lumpur: Oxford.

Warren, C. (1998) 'Tanah Lot: The cultural and environmental politics of resort development in Bali', in P. Hirsch and C. Warren (eds), *The Politics of Environment in Southeast Asia: Resources and Resistance*. London: Routledge, pp. 229–61.

Warren, C. (2005) 'Mapping common futures: customary communities, NGOs and the state in Indonesia's Reform Era', *Development and Change* 36: 49–73.

Warren, C. (2007) 'Balinese *adat* in discourse and practise', in J. Davidson and D. Henley (eds), *The Revival of Tradition in Indonesian Politics*. London: Routledge Curzon, pp. 170 -202.

Warren, C. and McCarthy, J. (2002) 'Customary regimes and collective goods in Indonesia's changing political constellation', in S. Sargeson (ed.), *Collective Goods, Collective Futures in Asia*. London, Routledge, pp. 75–101.

Watts, M. J. (2000) ' Contested communities, malignant markets, and gilded governance: justice, resource extraction and conservation in the tropics', in C. Zerner (ed.) *People, Plants and Justice: The Politics of Nature Conservation*. New York: Columbia University Press, pp. 21–51.

Woolcock, M. (1998) 'Social capital and economic development: toward a theoretical synthesis and policy framework,' *Theory and Society* 27 (2): 151–208.

Zerner, C., (ed.) (2000) *People, Plants and Justice: The Politics of Nature Conservation*. New York: Columbia University Press.

9   Locating the commonweal

Carol Warren and John F. McCarthy

In introducing the themes of this book, we argued for a broad integrated approach to understanding the complexity of ecological, socio-cultural and economic dynamics affecting local resource governance. This position reflects an emerging consensus in the social sciences, arising from a realistic recognition of the difficulties and contradictions faced by decentralized governance and sustainable development policy goals.[1] We have seen in the Indonesian case how economic and political crises became inextricably linked to the country's environmental calamities as the excesses of an authoritarian regime brought about a serious deterioration of the nation's natural wealth and extraordinary biodiversity. But we have also seen that democratization and decentralization in the post-Suharto 'Reform Era' did not automatically reverse this decline. The extended case studies presented in this volume show evidence that political reforms have stimulated revival of old and experimentation with new forms of local resource management, in some cases offering the prospect of a more sustainable and equitable future. In other cases, these political transformations have been accompanied by the acceleration of environmental decline and social disparity. In this final chapter we examine patterns across the case studies, considering the practical and theoretical problems exposed, as well as the enabling and constraining factors affecting social and environmental outcomes. We consider the links between social equity, environmental sustainability and resource governance raised in the context of debates surrounding 'commons' management, and reflect on the lessons suggested by our cases for prospects of constructing a broadly framed commonweal.

Equity, sustainability and resource governance regimes

The themes of environmentally sustainable and socially equitable resource governance with which the case studies presented in this volume grapple, require us to address problems as conceptually and practically complex as they are urgent. 'Sustainability' and 'equity' are of course widely deployed concepts in academic and policy-making circles; but their ideal-typic definitions founder with attempts to apply them in concrete terms with any consistency. Still, we can broadly identify those policies and practices that have proved *in*equitable or *un*sustainable,[2] and begin to assess what governance measures and conditions are more likely to contribute towards improving social and environmental security for the general public.

In the Indonesian context the economic, political and environmental crises that marked the end of the Suharto regime remained so raw and palpable that even small increments in livelihood benefits through improved access to resources represented improvements for ordinary people. But some of the measurable advances that have been achieved by direct action, legitimated if not accomplished by reformist efforts toward fairer distribution through decentralized governance, appear less promising when considered in the longer timeframes required by concepts of sustainability and intergenerational equity. This is apparent in the enthusiasm of policymakers as well as local people for conversion of forests to oil palm developments discussed in McCarthy's chapter on Jambi,[3] the accelerating deforestation and elite capture of benefits through community cooperative timber concessions reported in the Hidayat, Ballard and Kanowski chapter on Papua, the over-exploitation of Javanese coastal fisheries, and the bleak prospects for Jepara's once prosperous teak furniture industry described in the chapters by Lucas, Schiller and Fauzan.

The long time-frame through which these studies trace current struggles draws out the mutual implications of sustainability and equity considerations. If the gross injustices of the New Order period contributed to the rapid liquidation of resources during the crisis of 1997–1998, these case studies alongside recent work by a number of other researchers[4] show that inequities continue to drive another version of the 'tragedy of the commons' scenario in the era of decentralization and democratic reform. In the realignment of interests between strategically placed elites at various levels of governance in Indonesia, patronage and collusion still characterize official decision-making at the expense of a wider 'common good'. Plantation, timber and mining interests work with one or another arm of the state to entrench and extend the expropriation of local domains that took place under the New Order. In some instances local 'community' actors collude with private interests and state agents in stripping natural assets for mutual but short-term advantage. Meanwhile, environmental degradation and resource depletion have been fuelling vertical and horizontal resource struggles among local groups across the country.

Some of the most striking and urgent examples of the multi-dimensional and multi-scaled character of resource competition come from the fishing sector. Here large populations depend upon once seemingly limitless marine resources for cash incomes and basic protein.[5] Coral reefs, sea grass beds and spawning aggregation sites have become locations of intense conflict between subsistence and commercial fishers, local and outsider interests, and now also between the entire fishing industry and conservation agencies concerned with the looming crisis in the world's fish stocks (Butcher 2004).[6] As yields decline with increasing extractive efficiency, and in the absence of adequate regulation and protection of coastal fisheries, local artisanal fishers resort to destructive fishing techniques using mini-trawls, blasting, cyanide poison and other means of maximizing their takes in a losing battle to maintain their livelihoods.

The intense internal and inter-village conflicts over resource rights described in the fisheries studies in this volume are paralleled in accounts from the forested 'outer islands' (beyond Java), where inter-ethnic conflicts are associated with spontaneous and government-sponsored migration spanning decades. In many of the outer island areas of the archipelago, resource competition and entitlement claims frame the identity politics that are driving regional administrative reconfigurations (see Resosudarmo 2005; Erb *et al.* 2005; Schulte Nordholt and van Klinken 2007; Davidson and Henley 2007). Competing inter-ethnic identities and interests are the main focus of Acciaioli's study of resource contestation among variously defined indigenous groups and mainly Bugis immigrants at Lake Lindu in Sulawesi (Chapter 4). Similarly, inter-group resource competition fuelled by state sponsored inter-island transmigration (in this case mainly Javanese) forms the backdrop to the environmental politics of the Papuan case study (Alhamid, Ballard and Kanowski, Chapter 6). Meanwhile, protracted and sometimes violent land conflicts between customary landholders and the plantation and timber companies that obtained concessions over large areas of customary community land during the New Order continue, as in the Jambi (McCarthy, Chapter 7) case.

Yet the story is not one sided. The integrity of indigenous cultural identity and negotiative inter-community relations in the study communities of Kalimantan (Bakker, Chapter 5), Bali (Warren, Chapter 8), Sulawesi (Acciaioli, Chapter 4) and north Jepara (Schiller and Fauzan, Chapter 2) – in some cases facilitated by NGO intermediaries – indicate the potential for more promising local management under favourable conditions. These examples illustrate the importance of shared values, consensual decision-making, articulate and responsive leadership, and the experience of successful cooperation, which researchers attempting to specify contexts most likely to support sustainable commons management have stressed. (Ostrom 1990; Agrawal 2001, 2007).[7]

Clearly the dynamics of resource contestation and accommodation in the contemporary period demand analysis in contexts that extend beyond the local and national. Contradictory global demands for both expanding market access *and* more stringent environmental protection impinge directly on local resource struggles in the most remote parts of the archipelago. Moreover, declining biodiversity on a global scale simultaneously increases both the economic and the conservation value of forests and coral reefs; and local dependence on still intact ecosystems intensifies at the very point their conservation becomes of urgent importance on global environmental agendas.[8] In consequence, the problem of 'locating' the commonweal is an increasingly complex one, requiring greater integration across scales and disciplinary frameworks for analysis and policy development. Not only must efforts towards serving a broader common good take account of triple bottom line (ecological, socio-cultural and economic) dimensions of development processes, but practical efforts to grapple with building governance institutions for more equitable and sustainable futures require us to simultaneously consider the global, national and local articulations which could make the commonweal possible, or conversely undo the best intended efforts.

'Community' 'state' and 'market' – old concepts and new articulations

The three institutional spheres that have most exercised the commons debates to date – the community, the state, and the market – remain at the heart of increasingly pressing questions concerning distribution of resources and conservation of the environment.[9] Our focus on the articulation of these sites of engagement in the 'commonweal' responds to the appeal of McCay and Jentoft (1998:27) for a more 'expansive construct of community, one that would stretch from homesteads to townships to seats of central government and on to loose alliances among environmentalists or business leaders, the fragile institutions of international relations, the more robust institutions of global commerce, and even to "epistemic communities" (Haas 1990) of scientists and others engaged in trying to cope with common pool environmental problems.'

The cross scale dynamics and differential outcomes produced in the diverse cases presented here indicate the need to reformulate the relationships between 'community', 'state', and 'market' in order to support new articulations that would enable something approaching a true commonweal to be realized.

Community and indigeneity constructs

The role of 'community' is as problematic as it is pivotal to the development of effective local resource management. Since the 'tragedy of the commons' became the key metaphor for policy debates surrounding environmental decline, revisionists have looked to community regulatory regimes to demonstrate that it is possible to maintain an environmental 'commons', under certain conditions more effectively or equitably than state regulatory or market-driven private property regimes.[10] At the same time, a critical stream within anthropology has challenged tendencies to homogenize and reify our understanding of community structures and the moral economies that have been asserted to underpin them (Agrawal and Gibson 2001; Greenough and Tsing 2003; Brosius *et al.* 2005).

The case studies in this book attempt to draw together both strands in the literature, recognizing the variety of 'communities' of identity and interest and the diverse character of responses pursued by local groups to the socio-economic and environmental challenges they face. The Mului community of Bakker's study (Chapter 5) is taking a strong conservation position towards preserving its forests for the long term through customary *adat* mechanisms, while the north Jepara fishers of Schiller's case (Chapter 2) have attempted to prevent depletion of marine resources through the formation of the North Jepara Fishers' Forum and NGO facilitated negotiations to regulate fishing technologies used in their coastal waters.

In both the Lindu and Mului cases (Chapters 4 and 5) the protected status of national parks, superimposed on their *adat* territories by the state legal regime, reinforced local resistance to logging and dam projects that have been promoted by outside interests and by other arms of the state apparatus itself. Conversely, Bakker's other study community, the entrepreneurial village of Kepala Telake, backed a timber company that had its licence revoked by the central government's forest department for logging transgressions within protected forest. Both Mului and Kepala Telake exhibit high levels of local solidarity, shared values, and responsive leadership, all virtues prominent on the list of conditions held to connect community social capital with constructive collective action. But while Mului solidarities are tied to culturally and environmentally protectionist goals, Kepala Telake employs the same community attributes towards a developmentalist agenda pioneered by its elected official (*dinas*) village head and businessman.

In this vein, the concept of 'indigeneity' is, like 'community', the subject of critical treatment in the literature on the politics of Indonesian resource management (Li 2000; Sakai 2002; Brosius *et al*. 2005; Davidson and Henley 2007). More complex understandings of community dynamics have sensitized us to examine uncritical assumptions: that a conservation ethos implicitly underlies indigenous knowledge systems; that distributive equity is central to indigenous social values; or that whatever environmental protection and social redistributive practices may have operated in these societies previously have maintained their integrity in the face of neo-colonialism, cultural imperialism and other contemporary processes of modern global transformation (Ellen *et al*. 2000; Zerner 2000; Brosius *et al*. 2005; Ellen 2007).

The critique of oversimplified community and indigeneity constructs must lead us to interrogate any automatic privileging of sustainability and equity claims asserted in these terms. Yet the correspondence between the bio-diverse regions of the global south and the parallel diversity of the cultural minorities that inhabit these environments is hardly coincidental. Based on research across Southeast Asia showing how high levels of biodiversity are actively sustained in many traditional agricultural regimes, Dove *et al*. (2005) argue that such carefully managed human systems can actually enhance environmental integrity. We need to understand what conditions make these regimes possible, and cannot afford to underrate the potential offered by local institutions and the culturally embedded knowledges, meanings and values which remain of palpable significance across the regional cultures of the archipelago.[11] Nor should we permit their legitimate claims to precedence and entitlement on the basis of long histories of prior use or demonstrably effective management to be written off.

Adat identity is a significant consideration in all of the outer island case studies presented in this volume from Sumatra, Bali, Sulawesi, Papua and Kalimantan; although it is difficult, and perhaps disingenuous, to artificially distinguish between the intrinsic meaning and instrumental deployment of *adat* constructs in analysing resource contestation in these cases. *Adat* social and symbolic sensibilities were central to Balinese imaginings of future prosperity (Chapter 8).) But while *adat* served as the basis for territorial claims in the Melayu villages of Jambi (Chapter 7), it lacked the resilience to provide an effective alternative to state institutional arrangements and did not translate into concerted collective action that might achieve common aims. The To Lindu of Sulawesi (Chapter 4) embraced the NGO-sponsored *adat* peoples' movement, which enhanced their bargaining power with the state and in-migrant settlers. For the indigenous communities of Mului and Kepala Telake in Kalimantan (Chapter 5), *adat* was also the basis of claims to forest resources, but formal '*masyarakat adat*' identity – tied as it has become to stereotypes promoted by outside NGOs – faced mixed reception. The concept was useful to Mului efforts to position themselves as indigenous stewards of the forest, but proved constricting to Kepala Telake's developmentalist ambitions. For all the failings of Special Autonomy in Papua (Chapter 6) and of the quasi-*adat* institutions upon which 'reformed' local resource development is predicated, retreat from *adat* foundations under current circumstances is regarded by the authors of that study as politically unthinkable.

The resurgence of local customary (*adat*) institutions (Davidson and Henley 2007) disempowered under Suharto's New Order regime poses particular challenges for constructing common interest solutions where migration and social change in recent decades have resulted in complex local ethnic configurations, as at Lake Lindu and in Jambi and Papua. In McCarthy's Jambi (Chapter 7) study, New Order policy coupled the expansion of oil palm estates over customary domains with the development of state-sponsored transmigration settlements. The establishment of formerly impoverished Javanese transmigrants, now prospering from oil palm cultivation on what had been Melayu customary lands, affected the lifeworld of Melayu villagers. It also altered the interests and alignments of Melayu elites, thereby constricting the plausible options for accommodation or resistance to state policies that might be pursued by Melayu communities. Even in Warren's case study (Chapter 8) of ethnically homogeneous Balinese highland communities, the perceived threat to local environment and culture stimulated village efforts to control land alienation and tourism development by restricting outsiders' access to local resources.[12] In contrast, in-migration was actively encouraged in the remote Kalimantan community of Kepala Telake, which was intent on expanding its economic prospects. In-migrants under these circumstances were prepared to respect the local *adat* regime in return for incorporation into a community that offered them promising livelihood options. Important for conservation in the Lindu area of Sulawesi (Chapter 4), Bugis in-migrants began to accept the application of Lindu *adat* regulations on fishing and forest felling that they had previously ignored after experiencing the collapse of fish stocks due to over-exploitation of the Lake Lindu fishery and acute water shortage for their ricefields resulting from deforestation in watershed areas. They were less sanguine, however, about the imposition of limits on agricultural landholdings proposed by To Lindu *adat* leaders, which would impinge on their expansive coffee and cacao gardens.

In analysing the extent to which the resource rights of previously disen-franchized local groups can find institutionalized expression within emerging patterns of local governance, we need also to take account of the new forms of marginalization which may be produced by current reforms. How can the local construction of a 'commonweal' bridge historically and culturally constructed differences to establish new kinds of identities and institutional arrangements that better accommodate equity and sustainability? Is this possible without further disabling the positive forms of social and symbolic 'capital' available to the *adat* communities of Indonesia described in these cases?[13] Effective *adat* regimes and other enduring local solidarities supporting collective action represent valuable collective assets, which ecologists have come to consider a missing link in the conservation agenda (Pretty and Ward 2001; Pretty and Smith 2004).[14] But these models presuppose a 'community' of interest among local stakeholders that is likely to be tenuous and vulnerable to contestation. All of our case studies demonstrate the permeable character of the local, and the crucial issue of how internal structures and identities in villages and urban wards articulate with wider spheres. The importance of a conducive governance context that respects both human rights and environmental integrity, and encourages positive relationships to wider community, government agencies and NGO networks cannot be underestimated.[15] Even those local regimes seriously grappling with the question of how to equitably and sustainably manage resources are unlikely to withstand the steady incremental pressures of more powerful competing interests from within and without in the absence of a supportive governance framework at wider levels of decision-making.

State regulatory frameworks and governance reform

Like the 'community', the concept of the 'state' is contested ground, implying the presence of a unitary institution able to achieve coherent policy outcomes, but describing institutions fragmented in practice by bureaucratic and jurisdictional divisions, as well as by the diverging interests of the actors who comprise them. Conflicts of interest arise not only among sectoral departments and across scales and levels of governance, but perhaps more importantly between the unofficial private interests of state authorities and the responsibility in their official roles to act for the 'common good'. Yet the effectiveness of modern governance still depends upon the engagement of state institutions able to deliver policy outcomes in coordination with other actors. Within a governance framework, the capacity of the state is understood to be contingent on its ability to mobilize social actors to its purposes, and for this it requires institutional coherence and organizational capacity. From a governance perspective, for instance, the state needs to be able to establish 'patterns of long term relationships between mutually interdependent actors, formed around policy issues or clusters of resources, and to provide the administrative and institutional context for efficient and effective service delivery' (Jervis and Richards 1997:13). Most germane to the cases in this book, the state needs to be capable of establishing the regulatory framework for environmental management. Yet governance outcomes continue to be undermined by conflicts of interest that arise within the state. Some of the problems emerging from our case studies, such as the mismatch between jurisdictional responsibilities applied to fishing zones managed by central and district governments described by Lucas (Chapter 3), or the lack of an appropriate framework at district level for product licensing in the Bali case (Chapter 8) may be considered matters of legal harmonization and coordination that could be ironed out in time. Others are more fundamental and pervasive.

Corruption is the core component of the vicious cycle that divorces governance practice from legal and policy principle, derailing public interest and formal process in favour of private and informal dealings, and eroding trust in institutional relationships that are needed to support common-interest solutions. Not least important with respect to the Indonesian state's role in law enforcement and resource management are endemic patterns of patronage, bribery and misappropriation that pervade every level of governance. These practices have particularly corrosive effects within those institutions that have the responsibility to make the law work – the police, the military, the courts and the permit-granting gatekeepers in the bureaucracy. Rent-seeking begins at the point of recruitment and is reproduced as actors, who paid large bribes to obtain an official position, seek a return on their 'investment'. The pattern is reinforced as agents of the state and party functionaries become accustomed to lifestyles that require substantial supplementation of their meagre government salaries.

The processes that have rolled out under the name of decentralization compound this vicious cycle pattern, creating the need to raise official revenue at lower levels of government while proliferating opportunities for rent-seeking and patronage.[16] Regional autonomy led to a rearrangement of pre-existing patterns of accommodation and intensified competition among vested interests. Decentralization processes replicated many of the corrupt rent-seeking practices associated with patronage politics of the centre in the now more autonomous regions; and collusion between regional government functionaries and private interests, in many cases now joined by aggrieved local groups, contributed to further despoliation of natural resources. Fox *et al.* (2005: 105) observe: '[I]t is by no means certain that better laws with greater harmonisation of legislation along with improved demarcation of jurisdiction will result in more efficient management of Indonesia's natural resources. Decentralisation is essentially a political process involving competition among ... vested interests.'

Furthermore, the high cost of attaining and holding competitive political office in Indonesia heightens the incentives for corruption. This has a serious effect on law enforcement and therefore weakens the potential for an effective state-based regulatory framework that could avert future tragedies of the commons. Expectations of bribes or risk of intimidation affected decision-making at one point or another in most of our cases. In the context of pervasive corruption, regulatory regimes often serve to promote the illegal practices that enrich agents of the state and the private interests that co-opt them.[17]

The objective of democratization and decentralization reform is to bring decision-making responsibilities closer to the people and to facilitate public scrutiny and accountability. However, success depends heavily upon responsive local institutions and an active and organized civil society capable of holding their representatives to account. As our research and other studies on 'participation' (see below) show, there is a great deal of variation in the ability of local communities and resource user groups to control their leaders,[18] particularly where relationships with state authorities or commercial interests conflict with their collective responsibilities. Decentralization reforms have enhanced the discretionary power of local government, allowing for more openness and flexibility within a very broad framework set by the centre. Yet, when decentralization proceeds in the absence of robust local forms of representation and accountability, as in several of our cases, local elites tend to become the primary beneficiaries of the redistribution of power and resources.

A decade of 'governance reform', while allowing for more localized patterns of resource management, and an expanded role for activist NGOs and the media, has to date failed to build the 'nested' governance regimes that could use checks and balances within and across scales to promote the common good in a consistent and predictable way. It is the shadow side of official governance structures, where public purpose is diverted to serve private opportunity, that prevents effective problem solving which should otherwise be possible given good communication, effective representation and appropriate monitoring mechanisms and sanctions (cf. Marifa in Resosudarmo 2005: 256–8).

While most of the studies in this volume predictably found evidence of more inclusive and responsive approaches to decision-making by government authorities in this period of democratic and decentralization reforms, only in rare cases when there was a coincidence of economic or political interest did authorities adequately respond to local resource management conflicts and concerns. When local solidarities could be consolidated around particular issues, district governments would often broker negotiations, particularly where the potential for violent conflict reared its head or where electoral kudos might result. However, as we saw in the disputes over oil palm and timber plantations in the Jambi case (Chapter 7), given the chasm between those entitlements granted in law and those claimed by customary landowners, such *ad hoc* local accommodations were rarely legally inscribed. Almost invariably authorities retreated from formal commitments when the political and economic stakes militated against action that might alienate powerful patrons or large segments of their constituencies. Too often underlying grievances remained unresolved, as in the Jepara and Tegal fisheries cases, and in the protracted handling of local claims over the feldspar mine in Jepara (Chapters 2 and 3).[19]

Interestingly, in the case of the dispute between north and central Jepara fishers over the use of mini-trawl *cotok* nets, neighbouring signatories appear to be respecting the negotiated exclusion of this damaging technology from north Jepara waters despite the retreat of the district government from commitments to promulgate a formal ban on their use. The combination of local solidarity, NGO-facilitated negotiations and the risk of direct action by the North Jepara Fishers' Forum, has apparently been sufficient for the time being to enforce the agreement informally. But the outcome of a similar conflict along the Java coast in Tegal was more in line with the 'tragedy of the commons' scenario. There the government's failure to enforce zoning regulations against inshore use of mini-trawl technologies led fishers who had initially resisted their introduction to eventually follow suit, adopting the mini-trawl technology themselves in a short-term survival strategy. As the 'tragedy of the commons' allegory would suggest, without the effective regulatory regime required to prevent long-term degradation by all parties, there is no economically 'rational' incentive for any one party to restrain its own resource use.

Hardin's use of the 'tragedy of the commons' allegory, however, missed the primary insight of the political ecology literature that asymmetries of power and wealth systematically permit a small number of well-placed actors to stack the decision-making process in their favour through informal as well as formal structural means. This blind spot is also a weakness of the more optimistic neo-institutionalist and social ecology approaches, when they 'romanticize' the community management model.[20] Political economy and political ecology approaches, on the other hand, tend to underrate the potential for cultural values and social organizational mechanisms to check abuse and to forge cooperative solutions (Uphoff and Langholz 1998). Deliberative strategies appear to have brought 'common good' solutions to competing inter-village resource claims in Mului (Chapter 5), and at least provisionally (with the aid of NGO intermediaries) in the north Jepara fishers' altercation (Chapter 2), despite the absence of effective overarching authority. Among the Mului, longstanding deliberative processes in the context of shared *adat* have so far been capable of bridging inter-community differences on resource use despite the relative informality of these processes. In the north Jepara case, the agreement excluding mini-trawl fishing seems to have survived even the incapacity of the North Jepara Fishers' Forum to maintain operations of the boat it purchased for enforcing the agreement.

As noted in the introductory chapter, the emphasis on community resource governance reflects and reinforces an earlier move towards promoting local participation as both an instrument and a goal in conservation and development planning. The subject of 'participation' is of course also one of the many contested fields reflecting deeply polarized positions in the conservation and development literature (Henley 2007; Li 2007). Sceptical assessments contrast the rhetoric promising empowerment and appropriate development through public participation with what actually happens in the real world of unequal power relations within communities, and between them and the various agencies of the state, the corporate world or the advocacy-oriented NGOs that are increasingly involved in mediating between these three spheres. Critics dismiss participation as more rhetoric than substance, sometimes disguising the manipulation of communities and groups by social change agents pursuing their own agendas under a veneer of community consent. This view calls for reassessment of a set of practices which are at best naive about questions of power and at worst serve to systemically reinforce, rather than ameliorate, existing inequalities. But other views contest this position, asserting that rather than a vacuous and depoliticizing concept, participation can – as always, given certain conditions – be linked to genuinely transformative processes with positive outcomes for marginalized communities and groups.[21]

But while transparency, accountability and public participation are pre-requisite to good governance, they do not automatically assure priority for equity or sustainability principles. Since neither nature nor future generations can vote (Dutton 2005), there are practical limitations in any case to the capacity of democratic political institutions to represent all 'stakeholders' implicated by broad understandings of these concepts. Furthermore, the complexity of these issues adds to the difficulties of developing far-sighted public policy. In this respect activist non-government organizations have come to play an important role in emerging governance regimes.

International NGOs have become the vocal proponents of human rights and environmental protection causes, and act as the unofficial voice of an amorphous but increasingly influential global civil society, whose priorities do not always match those of local counterparts, however. Questions of resourcing, competence, accountability and representation also pose challenges to the capacities and legitimacy of the loosely allied and highly diverse NGO movement to carry out this task.[22] Local participants are often critical of what they perceive as the *ad hoc* and instrumental approach sometimes exhibited by NGO partners. NGO personnel, like their bureaucratic counterparts, are often urban middle-class 'outsiders' with their own objectives to pursue and infrastructures to resource. Nonetheless, NGOs and global multilateral agencies represent both complementary and alternative channels for articulating new cross-scale and cross-level relationships and for pursuing critical engagement across spheres of interest, while their diverse commitments and sources of funding provide 'forum shopping' options for local groups.

The case studies in this volume show that at the very least participatory processes, expanded as a consequence of formal democratization and decentralization reforms in Indonesia and fostered by NGO facilitated programmes, have created political space opening up greater possibilities for collective action, for giving voice to local needs and concerns, and for bringing these into wider arenas of engagement. Through participatory mapping, communities in Sulawesi and Bali (Chapters 4 and 8) achieved notable changes in the degree of local empowerment they could exercise, and some better conservation outcomes. But the failure of efforts towards more inclusive approaches to conservation issues in the Karang Jeruk Marine Sanctuary in Tegal (Chapter 3) and the remnant nature reserve in Jepara (Chapter 2) are a sobering reminder that participation is no magic bullet. In these examples, the importance of the wider governance context to individual and collective action contributed critically to these failures. Individual behaviour and collective decision-making will be strongly influenced by uncertain estimations of others' behaviour and expectations (Ostrom 2005: 48–9). Small differences in actors' perceptions, plausible choices, or institutional setting may shift outcomes in one direction or another; although opportunities for feedback and adaptive learning or political realignment may alter those always provisional outcomes further down the track.

In the Papua and Jambi cases (Chapters 6 and 7) what passes for participation and autonomy has so far in practice meant little more than the co-option of local elites to share in the profits primarily flowing to state and private interests. By most accounts, state mandated participation in local resource management through government established cooperatives has yet to fulfil its promise to contribute to a more equitable 'people's economy' (*ekonomi kerakyatan*) or to more sustainable development of the plantation or timber industries in Indonesia's outer island regions. Yet cooperatives are regarded as among the few mechanisms for empowering primary producers in the face of market disadvantage. McCarthy suggests that with good facilitation, cooperatives appear to have been more successful among transmigrants than was the case among indigenous Melayu he studied in Jambi (Chapter 7); and there is some evidence that the *adat*-based cooperative movement in Bali is proving more effective than the earlier KUD network established under auspices of the state.[23]

Neo-institutionalists concerned to avert commons tragedies (see Chapter 1) argue that the strength of social networks and intensity of interaction among user groups – through high levels of participation in the processes of identifying and investigating environmental problems, rule-making, monitoring and enforcement – are essential components of effective local environmental governance. It is tempting, following this line of argument, to attribute the 'good news' stories among our cases to features of group cohesion, resilience of customary institutions or capacity building support from outside NGOs. But we do not find any simple, formulaic or predictive explanation as to why the Mului chose a culturally conservative and conservationist strategy in decision-making regarding their forest resources, while their equally remote neighbours with similar *adat* backgrounds in Kepala Telake and other surrounding Paser communities have not. Nor is it self-evident why the cooperatives of the transplanted Javanese transmigrants in Jambi have facilitated better economic (if not ecological) outcomes for their members, while those established among indigenous Melayu came to serve primarily the interests of indigenous elites who were able to take private advantage of strategic positions mediating resource access and use. Similar problems cripple the Kopermas cooperatives grafted on to politically compromised *adat* institutions in Papua (Chapter 6). The uneven outcomes reported for NGO sponsored efforts at boosting the local economies in the villages involved in the Bali project (Chapter 8; see also Warren 2005, 2007) also complicate arguments that might be made concerning social capital accumulation and transfer for conservation and development in the context of strong *adat* institutions.[24]

Various interpretations would lead us to look at these differential outcomes in terms of degrees of internal cohesion and the social and ecological resilience it can engender (Putnam 1993; Berkes *et al*. 2003); the effects of engagement with agents of the state and market (McCay and Jentoft 1998; Dove *et al*. 2005); and/or the subversion of common good solutions by powerful interests and instrumental deployment of symbolic and social capital by actors in privileged positions (Bourdieu 1990; Li 2007).[25] These factors are largely interdependent, and analysis hinges on matters of emphasis and degree. But policy models require the provision of narratives that serve as a means for navigating complexity and planning future interventions.

At the same time, the high degree of complexity of both social and natural components of ecosystems has a number of implications that confound simple policy solutions. The weight of evidence indicates that the scientific knowledge upon which environmental policy must be based will always be confronted with substantial uncertainties; that transaction costs in time and resources required for creating and managing the institutions necessary for decision-making, monitoring and evaluation are high; and that the vested interests of the economically and politically powerful can normally be expected to hobble implementation of 'common good' solutions to equity and sustainability problems (see Berkhout *et al.* 2003; Visser 2004).

The consequence is that these real world complexities escape prescriptive preferences of policymakers, at least as they apply to individual cases, which is not to say that policy cannot create conducive frameworks to improve outcomes in general terms. Policymakers need to understand the specificity of particular contexts, taking seriously the conception of 'institutional bricolage' which recognizes that institutions are engaged in socially embedded processes that are 'multi-functional, semi-opaque and contingent' (Cleaver and Franks 2005: 5). This perspective underscores the need to comprehend how social relationships and decision-making processes work in all their contextual complexity, keeping in mind that no one set of factors will be sufficient to determine success, and no one-size-fits-all set of solutions will suit the diverse combinations of social needs and ecological conditions that confront real communities.

The evidence from our cases indicates that local resource management dynamics are far more complex and indeterminate than the models available to us convincingly explain. Agrawal (2007) comes to a similar conclusion in his review of the literature on governance of forest commons. In particular, diverse findings on the role of internal social heterogeneity and external pressures from the state and market, as well as on the relative significance of private property, communal tenure or state control in accounting for forest conditions, mean 'we still need to track down *how* context matters to commons governance in complex social situations' (2007: 126; authors' emphasis). We need more nuanced tools for analysing the complexities of the social and ecological variables at work in real communities and real environments, where change is dynamic, and multi-layered variables are tightly interdependent.

Markets, externalities, and the negotiation of reform

The questions raised by McCay and Jentoft (1998: 27) about the role of state and market in 'community failure' remain important considerations for interpreting the mixed picture presented by our case studies:

> The task is then to determine for any given case of apparent abuse of common resources, where the failures lie and what can be done about them. To do this requires exploring how property rights are understood by various parties and how those meanings are translated into behavior, custom, and law. It requires understanding the nature of conflicts over rights and responsibilities, the roles of science and other forms of expertise and of larger global processes affecting land and natural resource management throughout the world. It also requires understanding, respecting and building upon the social and political capacities of local communities, but also of [addressing] the disembedding forces of modern society.

The neoliberal celebration of the market's role in raising living standards and efficiently allocating resources through global economic integration is paralleled by equally strident criticism of its ecologically and socially destructive impacts (Heal 1998, 2000; Nadeau 2006). Within the revisionist commons literature, markets alongside the state tend to be presented as the villains of the piece, penetrating and disrupting the internal processes that might otherwise make 'communities' with common purpose out of aggregates of private interest.[26] In our case studies, markets stimulated over-exploitation of the Lake Lindu fishery by in-migrant entrepreneurs (Chapter 4); debt–credit relationships and competition drove intensification and depletion of fish stocks along Java's coast (Chapters 2 and 3); an insatiable global appetite for timber and oil palm accelerated deforestation in Kalimantan, Papua and Sumatra (Chapters 5, 6 and 7); and the leisure desires of global middle classes for exotic landscapes and cultures pressure the coastlines, padi fields and sacred sites of Bali (Chapter 8).

Typically, the marginalized poor bear the heaviest burden of the externalities of extractive intensification and resource decline. Case studies in Visser's (2004) collection and in this volume by Schiller and Fauzan, Lucas and Acciaioli (Chapters 2, 3 and 4) reveal increasing disparities between small scale artisanal fishers and highly capitalized operators with ever more 'efficient' extractive technologies at their disposal. Parallel disadvantage has been the experience of smallholder agricultural producers where primary products are grossly undervalued in the commodity market chain (Stiglitz 2002; Tudge 2004; Dunkley 2004), as we found too for community-based tourism struggling to carve a place for itself against an industry organized around mass markets (Chapter 8). Forest, marine, agricultural products, and even 'culture' itself are taken onto the open market through a process of commodification and value conversion that McCay and Jentoft (quoted above) remind us disembeds the product's relationship to its producers and their social arrangements, in ways not fundamentally different from those described by Marx and Weber more than a century ago, notwithstanding all the institutional and technological transformations that have taken place over the intervening period.[27]

The significant differential in costs and benefits of resource exploitation between the local and wider domains reflects profound asymmetries in exchange relations – with centres of power and wealth in commanding positions to determine and realize value. A Sulawesi fisher receives only US$7.50 out of the US$180 paid in a Hong Kong restaurant for the illegally caught and exported Napoleon wrasse (Lowe 2000: 239). But it is the local fisher – whether actively involved in the live reef fish trade or not[28] – who pays the most immediate cost of degraded reefs as a result of the destructive techniques used for extracting this lucrative catch on a scale that would satisfy the demands of market profitability.

This is the case for virtually all open access environmental services, discounted – and more often completely unaccounted for – within the free market regime. Because the contribution of environmental services and biodiversity are largely uncalculated in capitalist market exchanges (except by means of explicit political-regulatory interventions), the world market has been the ultimate 'free rider' on the environmental commons. On a global scale, the same pattern of growing inequities in wealth and disproportionate environmental impacts ramifies the local picture (Heal 2000; Clapp and Dauvergne 2005; Nadeau 2006). Effectively a completely disembedded 'free market' and the unregulated open access 'commons' of Hardin's tragedy are interlinked economic and ecological mirrors of one another, driven by analogous competitive, and ultimately unsustainable positive feedback dynamics.

On the other hand, alternative and regulated markets, small credit facilities, cooperatives and fair trade schemes offer to diversify the economies and reduce the environmental footprint of smallholder agriculturalists or artisanal fishers in some of these same remote areas threatened by commercial intensification. Li (2001) and Parnwell (2003) point to the importance of including marketing strategies within the community development and environmental protection agenda. And there is little question that the state still has essential regulatory and redistributive responsibilities to perform, however much these have been undercut by decades of neoliberal dismantling.

In the process of decentralizing governance, the central state in Indonesia effectively lost control over resource management, as indicated by the failure of government authorities to prevent looting of state forests and nature reserves (Chapter 2) or to limit the allocation of logging and plantation concessions by district administrations (Chapters 5 and 7). Demands by Javanese fishers (Chapter 3) to control foreign fishing vessels competing with local fishers in coastal waters have also gone unheeded. Whether the 2004 revisions of the regional autonomy legislation and efforts at reviving the moral authority of the central government enable it to assume the role of umpire in the effort to level the playing field and reverse the spiralling resource declines that have characterized the political ecology of Indonesia in recent decades remains to be seen.[29]

Better regulation of competing external interests and of destructive technologies is clearly an important quid pro quo that the state (and the international agencies increasingly taking on quasi-state governance roles in conservation policy) could offer local communities in return for negotiated rights and responsibilities within nested resource protection regimes.[30] Another is a serious commitment to supporting diversification of local livelihood options as attempted by the joint government/ADB funded Co-FISH project in Tegal (Chapter 3), several TNC conservation strategies (Chapter 4; see also Sodhi *et al.* 2008) and the NGO programme in Bali (Chapter 8).

Among emerging new environmental governance mechanisms at the international level, the potential for carbon offset and taxation schemes to support a variety of ecosystem service payment arrangements represents one means of subsidy for the hybrid economies and local experiments that offer a starting point for reconciling equity and sustainability in marginalized communities. If local custodians are to manage their components of agrarian, forest and marine ecosystems in ways that serve the wider 'common good' as well as their own needs on a sustainable basis, systems must be developed that will fairly subsidize their role in maintaining ecosystem balance. Li (2007: 365) insists: 'Conservation should be a shared burden, not one borne disproportionately by marginalized populations.'

The historical experience of 'elite capture' at every level of governance, how ever, means that policies must be devised with the view to ensure that those who bear the burden of protection will be beneficiaries of emerging new modes of environmental governance and their redistributions.[31] As Dove *et al.* (2005: 19) observe, 'the local communities responsible for the stewardship of natural resources are often politically, economically and culturally marginal and, as a result, it is unlikely that national elites will pass on to them the benefits being offered by the international conservation community.' Such concerns have led to calls for ecosystem service payments from emerging carbon trading and tax regimes to be made directly to local communities, bypassing the state (*The Age*, January 2008; see also Clémonçon 2008: 80–81).

Although the practical means of achieving environmentally sustainable along with socially equitable improvements in the local domain are highly contested, there is little doubt that pressures for intervention will only mount.[32] Governments, NGOs, transnational conservation agencies, private commercial interests and researchers have to be prepared to commit their resources to longer-term and more reciprocal engagements with local communities to overcome the serious difficulties they confront. Multi-channelled strategies can contribute to an 'institutional bricolage' approach appropriate to the dynamic institutional settings of these times. These at least provide the prospect of more creative responses to local cultural and ecological diversity that could resist the 'simplification' errors of earlier centralized planning and management (Scott 1998; Li 2002, 2007; Cleaver and Franks 2003). In the meantime, the challenges of creating accountable institutional structures that break from the vicious cycle that divorces governance practice from legal and policy principle remain.

The multi-scaled and multi-level set of interests competing for declining marine, forest, mineral and agrarian resources mean policy solutions to resolve this classic tragedy of the commons situation are by no means simple. Even the most accountable of democratic governments face the simultaneous need to conserve the resource base, to bring in budget revenue from licences and development projects, and to win votes by responding to popular demands for resource access and employment to ensure local livelihoods.

The immediate problem for Reform Era Indonesia is how to turn the vicious cycle of political corruption, inequities of power and wealth, expropriation of the environmental 'commons' and resource decline into a more far-sighted, sustainable and equitable virtuous cycle. We previously argued (Warren and McCarthy 2002: 96) in an early discussion of the issues that formed the background to this research project, that struggles over growth and distribution, individual and group rights, short and long-term sustainability were 'unlikely to be resolved in the interests of ordinary Indonesians independently of broader transformations on a global scale'. In the intervening years, the evidence on climate change has quite dramatically altered the scales affecting the place of social equity and environmental sustainability concerns vis-à-vis the narrowly framed economic growth priorities that have dominated the global policy agenda to date. It could be argued that the spate of political and economic crises, precipitated by serious asymmetries and inadequacies in global governance, are conspiring with environmental signals to produce a tipping point[33] that would force collective action in the international arena towards the serious reforms now required to move knowledge, policy and practice into alignment.

The increasing complexity of cross-scale and cross-level relationships between communities, the state, markets and other globalizing forces pose daunting challenges, but also creative opportunities, for addressing equity and sustainability issues in the local domain. Even in the remotest of our study sites, it was impossible to consider questions of resource management and decision-making practice without taking account of the effects of these wider spheres of engagement. The cases of resource contestation in contexts of environmental insecurity presented here point up the difficulties of delimiting a commons regime at any one scale or level, and the need to negotiate the sphere of 'common' identities and interests within a nested commonweal that articulates local, national and global domains. Coastal habitats may be localized, but the marine resources they support are important to national, as well as global 'communities' and vice versa. The environmental services provided by Indonesia's tropical rainforests that have critical importance for the everyday livelihoods of Kalimantan or Papuan villagers are ultimately also important to the environmental security of ordinary people in distant parts of the world. Conversely, energy use and consumption in the world's industrial capitals have far reaching consequences for even the remotest communities threatened by the impacts of climate change on their local environments. These interdependencies grant qualified legitimacy for diverse and distant claimants to some reciprocal stake in local resource management. Although the limited efforts at matching responsibilities with rights and negotiating trade-offs at different scales and levels of governance to date have proved slow, difficult, and too frequently misdirected, this seems the only legitimate way forward.

Concluding remarks

In the mixed picture presented by our studies, the community, the state and the market pose themselves as both problem and prospect for reformulating governance toward something approaching a commonweal. This leads us to draw the conclusion at this critical juncture – cognizant of salutary warnings from critical observers[34] on the limits and risks of interventionist strategies – that it is no longer a matter of whether, but of how and how soon, the articulations between these institutional spheres are reformulated which is critically important to improving prospects for the future.

In the face of escalating political, economic and ecological crises, it is imperative to develop governance frameworks that will re-embed and transform the market economy, reformulate the role of the state, and revitalize communities within emerging international regimes. This must be done in a manner that gives primacy to social equity, environmental sustainability and democratic legitimacy if we are to bring community, state and market into a virtuous cycle of accountable resource use. All of these institutional spheres claim to represent sites for constructing shared welfare and common interest. Those claims need to be negotiated into a check and balance framework, turning the rhetorics of identity and interest, transparency and accountability, into constructive real-world practices.

At the level of 'communities' we need to encourage processes that broaden forms of democratic representation, redefine relationships to land and resources to foster stewardship, and build new identities and alliances to support expanding definitions of the 'commons'. The exploratory potential of the many bottom-up NGO and community-based experiments with co-management and sustainable development programs must be seen as important if partial steps in the process of reframing the 'local' within wider spheres of interest and identity.

Beyond reforming the hierarchic state–local grid and the structural gap between private and common interest, there is need to give attention also to the nexus between formal modes of governance and the informal processes that could contribute towards (or undo) the expansion of those vital elements that make democratic governance and sustainable development serious common projects – broadening the bounds of civil society and strengthening the bridging and linking dimensions of social capital[35] – in ways that could turn contests over boundaries and authorities into productive partnerships and synergies.

A number of researchers[36] point to the need for greater collaboration between the social and natural sciences that will also accommodate the practical knowledge of their community counterparts in order to facilitate adaptive learning. Several of our case studies do suggest such a trajectory of learning and consequent efforts at negotiated governance reform at the local level. The experience of crisis led to more conservation-oriented collective action at some of the study sites where natural and cultural resources had been observably degraded by overexploitation.[37] At several sites new strategies have emerged with the engagement of conservation organisations, government agencies and academic researchers, although without as yet the systemic international framework that would be required to achieve ongoing and replicable solutions on a wide scale.

Our explorations suggest that constructing sites of common interest and shared welfare – a commonweal – across scales and competing interests will require the active forging of new alliances and new institutional arrangements in recognition of environmental realities. Consolidating precarious convergences of common interest and identity, such as those emerging in some of these Indonesian cases, will require transparent and accountable governance processes that provide internal and cross-scale/level checks and balances; that bring resource rights into line with responsibilities, and respond to the glaring social inequities that currently exist.[38] Such points of convergence and accommodation are fragile by their very nature. Where trade-offs create winners and losers, they require remediation of impacts and transformative processes for responding to conflict. Nor can the costs in time and resources of participatory processes be neglected. The high transaction costs of maintaining the symbolic identities and the participatory and negotiative practices that could sustain common interest solutions in the long term, helps to explain why the ideal of 'community' is honoured so often in its breach.

The 'commonweal' then is fragile, contingent, shifting, and its parameters constantly in need of reworking as our knowledge of social and ecological conditionalities expands along with better understandings of how externalities and trade-offs differentially impact upon local environments and the livelihoods dependent upon them. The accelerating degradation of natural resources will certainly sharpen the problems and raise the stakes identified in these Indonesian studies. This can only make more urgent the imperative of avoiding the tragedy scenario so that, instead of 'rational' actors pursuing their private ends, diverse communities work together across localities and levels of governance towards collective goals, without ignoring individual needs and rights. At present the commonweal remains little more than an immanent possibility, a rhetorical construct in need of realization. As economic and political crises converge with accelerating environmental change, there has never been more powerful impetus for committing research and governance reform to this end.

Notes

1 For an overview of shifting approaches to the sustainable development agenda launched in 1987 with the World Commission on Environment and Development Report, *Our Common Future* (1987), see Berkhout, Leach and Scoones (2003:1–31).
2 Parnwell (2005:18) traces out a model of sustainable/unsustainable livelihoods and the mix of local and external pressures and processes that swing the pendulum in one direction or the other over time. Based on community level research in northeast Thailand, Parnwell found key features in the resource degradation process analogous to features of our own case studies: spatial integration and marketization of the local economy accompanied by increased social differentiation, debt and dependency. In his process-oriented model economic and ecological crises, precipitated by state modernization and development policies, ultimately swing the balance back toward more sustainable and equitable orientations, aimed at stabilizing environmental impacts and rebuilding local institutions, social capital, community safety nets, etc. However idealized, he argues, images of a moral economy of earlier times were catalytic 'in bringing about the rehabilitation of the natural environment and a reformed social cohesiveness at least among significant segments of the study communities, as they have clawed back a certain degree of control over the development process' (2005: 6). At the same time, Parnwell also found that institutionalizing these revived moral and social commitments proved somewhat tenuous in a context still dominated by hierarchies of power and wealth and a globally dominant neoliberal policy framework (2005:17–19).
3 Indonesia, which recently surpassed Malaysia as the largest producer of crude palm oil, plans to open 300,000 hectares per year for oil palm plantations with low interest (10 per cent annual) credit incentives offered through the national banking system. Although the official policy intent is to resume undeveloped or abandoned plantation lands, it is also the case that vast areas of forest are being converted in Kalimantan and Sumatra for this purpose. The process of deforestation has been exacerbated since decentralization by districts hungry for revenue. Arguing that the establishment of oil palm plantations has been used in recent years in Indonesia 'as a pretext to clear land and take the more valuable logs', Rully Syumanda, of Indonesia's peak environmental organization, WALHI, estimates that nearly 17 million hectares of Indonesia's forests have been cleared ostensibly for oil palm plantations since the 1960s, only 6 million hectares of which have actually been cultivated (*Suara Pembaruan*, 11 May 2006, 'Indonesia Produsen Kelapa Sawit Terbesar'; Agence France Press, 11 September 2007). Furthermore, the economic benefit of oil palm to farmers in remote locations is

not assured, because of the exceptional dependency of small producers on processors. See *Kompas*, 14 May 2007, 'Petani Arso Keluhkan Rendahnya Harga Sawit', in which Papuan smallholders complain of the low prices they receive for their unprocessed palm fruit.

4 In particular, see collections that include case studies on environmental change in Indonesia edited by Zerner (2000), Persoon *et al.* (2003), Greenough and Tsing (2003), Visser (2004), Brosius *et al.* (2005), Dove *et al.* (2005), Boomgaard *et al.* (2005), Sodhi *et al.* (2008).

5 Some 53 per cent of Indonesia's animal protein supply comes from fish, compared to a global figure of some 16.5 per cent (FAO 1997, cited in Dutton 2005: 165).

6 For other case studies from the Southeast Asian region illustrating the dramatic, and in some cases apparently irreversible, decline in local fisheries, mainly as the result of over-fishing for external markets, see Lowe (2000), Visser (2004), Dutton (2005).

7 Agrawal (2007), however, points out that the complexity of factors affecting resource management confounds any simple formula for predicting successful outcomes.

8 See the collection of studies on the politics of protected areas edited by Sodhi et al. (2008).

9 See Agrawal and Gibson 2001; McCay and Jentoft 1998; Uphoff and Langholz 1998; Li (2001) and Dove *et al.* (2005) for diverging perspectives on the role and relationship of states, markets and communities in local resource management.

10 See the introductory chapter to this volume for an overview of these debates.

11 'The costs of disregarding the embeddedness factor (and in worse-case scenarios, terminating it by legislation) can be enormous even in economic terms' (Paine 1994: 193, quoted in McCay and Jentoft: 1998: 25).

12 In this sense 'outsider' status refers not only to migrants from different ethnic groups, but also to Balinese migrants from other parts of the island and even locally born Balinese where they do not continue to carry out the social and ritual obligations imposed by the village-based *adat* regime.

13 The Lindu and Kepala Telake (Chapters 4 and 5) examples of migrant inclusion suggest some of the conditions and possibilities for negotiating trade-offs and building constituencies based on convergence of interest in multi-ethnic contexts.

14 Pretty and Smith (2004: 636) regard social capital as providing important social incentives and a supportive regulatory framework for conservation, 'both to prevent free-riding and to encourage individual investments for the collective good'. Trust building, public participation, transparency and accountability practices have become key features of the environmental governance agenda.

15 In many of our cases resource entitlement claims and corresponding restrictions were qualified in specific ways by equity and sustainability considerations in consequence of involvement of government, non-government or international agencies. In the Lindu, Tegal and Jepara fisheries (Chapters 2, 3 and 4), efforts to restrict resource use were negotiated on the basis of appropriate technology or extraction limits rather than on the ethnic identity of resource users.

16 Under the previous regime, a good deal of decentralized rent seeking from natural resource extraction was already well established (see McCarthy 2006).

17 The Togean (Sulawesi) villagers in Lowe's study, who took action to enforce the law against outside fishers illegally using cyanide to capture large reef fish, were threatened by police, who exploited their position in order to get protection money from both sides of the resource struggle (Lowe 2000: 250–3).

18 Compare the Mului with the Jambi and Papua cases (Chapters 5–7) for instance.

19 In the feldspar mining dispute (Chapter 2) local considerations were counter-balanced in the district head's calculations by his interest in keeping provincial and central government as well as the company on side.

20 For discussions of the community based natural resource management model in theory and practice, see McCay 2001; Lynch and Harwell 2002; Persoon *et al.* 2003.

21 See two important collections of essays with counterposed perspectives on this subject edited by Cooke and Kothari (2004) and Hickey and Mohan (2004).
22 Despite the tight dependence of local NGOs on resourcing from their international patrons on the one hand and the acquiescence of community groups to their programmes on the other, they are not always in accord with either their patrons or clients over concrete goals or the methods for achieving them. See Farrington and Lewis (1993) and Eldridge (1995), for accounts of the role of NGOs under the New Order; see the Kalimantan, Bali and Sulawesi chapters in this volume for examples in the Reform Era period, as well as extended discussion of these cases in Bakker (2005), Warren (2005) and Acciaioli (2008); see also Hilhorst (2003) for examples from the Philippines.
23 Widely touted as the basis of an alternative 'people's economy', cooperatives have a checkered history in Indonesia. In Balinese villages, an entire network of small credit institutions (LPD) have been set up in the *adat* sphere since the late 1990s, apparently with more success than the older state established village cooperatives (KUD), although not without some of the same problems of oversight, monitoring and bad debts, that undermined the KUD system and clearly plague the cooperative 'solution' to equitable development in the majority of case studies presented here (ILO 2001; Henley 2007; Warren 2007).
24 See Lansing (2005: 89–90, 108–15) on the variations in practices associated with leadership, accountability and local governance styles among irrigation associations in Bali. His research suggests that the constant reinterpretation and balancing of finely judged and ambivalent egalitarian vs hierarchic and collective vs individual interest principles leads to a surprisingly great range of variation in democratic practice among these extraordinarily sophisticated traditional local organizations, despite their shared cultural basis and generations-long experience of cooperation. A forthcoming Australian Research Council project to be carried out by contributors to this volume intends to explore these issues in depth.
25 For a useful discussion of the contrasting approaches of Bourdieu and Putnam to the concept of 'social capital', see Portes (2000). See Arce (2003), however, who is critical of the extension of the concept of 'capital' to such social and cultural contexts: 'A more dynamic approach to the understanding of local livelihoods which stresses the interplay and mutual determination of the contestations of values and relationships and which recognizes the central place played by human action and identity rather than capital is needed' (2003: 204).
26 The final section of Chapter 8 discusses Li's (2001) critique of the tendency to treat markets as 'external' to village relations in the commons literature. It could be argued that the 'externality' of market relations is rather more a reflection of the tension of underlying principles with those of community solidarity than an assumption that we are dealing with empirically bounded entities, although we can speak of various mechanisms of exclusion or enclosure – in terms of property, citizenship or protectionist regulation for example – as aimed at constructing such boundaries with varying degrees of effectiveness: see Uphoff and Langhoz 1998; McCay 2001. See also Benda Beckmann *et al.* (2006) and Verdery and Humphrey (2004) for important discussions of the complexity of property relations and in particular the reification and fetishization of 'private property' in development policy.
27 In their essay, McCay and Jentoff (1998) take up the argument developed by substantivist economic anthropologists and historians who see a fundamental opposition between the principles of socially embedded reciprocal and redistributive exchange and the disembedded 'negative reciprocity' of the supply–demand, cost–benefit calculus of private individuals competing in the market (Polanyi 1968; Sahlins 1972).
28 Lowe (2000: 244–5) describes the hand-fishing methods that enable Sama Bajo fishers to catch live reef fish without damaging their habitat. This technique is only possible because of the deep ecological knowledge of this seagoing culture. But these traditional knowledges and technologies involve much lower yields, and do not satisfy

the maximizing, credit-providing market intermediaries or the modern consumption desires of the upwardly mobile, mainly male, youth engaged in the trade. Both those desires and the means to fulfil them are generated by the market, and are most easily satisfied by fishing technologies that destroy habitats and the future viability of the resource. In the relentless process of converting use to exchange values, globalizing markets expand and propel conditions favouring the vicious cycle of the tragedy of the commons narrative.

29 The administration of President Susilo Bambang Yudhoyono promised to crack down on corruption and illegal logging. Since 2004, the central government increased the use of the provincial police commands that remained under central control to enforce central government regulations against illegal land clearing and logging, including such practices taking place under provincial and district logging permits and plantation licences where these contravene national regulations (McCarthy 2007). This, together with new presidential orders and the revised decentralization laws of 2004, alongside the state of depletion of readily accessible timber, has reduced the epidemic of illegal logging.

30 This point is also made by Worms *et al.* in Visser (2004).

31 The other side of the equation requires that those who disproportionately benefit from resource exploitation past and present pay fair compensation. New 'progressive' taxation regimes would be needed to respond adequately to the social and environmental externalities of the current world system.

32 Ferraro (2001) and Wunder (2007) advocate direct contractual arrangements and 'performance payments' for environmental services rather than indirect conservation and development interventions, which they argue are less well targeted and send weak or ambiguous conservation signals. In other contexts regulatory initiatives to encourage sustainable development, for example through fair trade and product certification schemes that foster sustainable agriculture, forestry, fisheries or tourism, may prove more viable long term options for bringing local livelihood and conservation needs into concert. At the other end of the spectrum, Dove (2005) argues that disengagement of self-sufficient indigenous communities successfully managing their environments has to be included among the options in the schematic environmental governance repertoire. This would require up-scaling international legal protection by giving teeth to the UN Declaration on Indigenous Rights and to the International Labour Organization Conventions.

33 The IPCC Report (2007) predicts serious impacts from climate change on the resources and livelihoods of heavily populated developing countries such as Indonesia (2007: 28–30). The forest fires that destroyed some 9 million hectares in Kalimantan and Sumatra between May 1997 and March 1998 were estimated to have caused direct and indirect damage to the value of between 4 and 8 billion $US (Eaton 2005: 46–51). Livelihoods of rural villagers were affected by loss of crops and forest resources; and the smoke caused illness and injury as far away as Thailand, Malaysia and the Philippines as well as at home. High costs are predicted as a consequence of the degradation of the marine environment, like forests affected by human impacts at both global and local levels. Losses to Indonesia alone as a result of degradation of reefs from overfishing, pollution, and climate induced bleaching are estimated at 30 billion US$ over 25 years (Dutton 2005:167)

34 See for example Berkhout et al. (2003) and Dove (2005). Dove warns against relying on the same institutions that are the source of contemporary problems to resolve them. But time and scale factors make it unlikely that radically new structures and institutions will be conjured quickly enough to deal with urgent environmental issues that are already causing serious social, political and economic stress. For the coming decade, a broad raft of experiments with familiar frameworks and institutions, reformed and remodelled, will be needed to give breathing space for more substantive transformations to follow.

35 Woolcock (1998) distinguishes 'linking' (the capacity to engage with vertically related

external groups) and 'bridging' (capacity to engage with horizontal groups that have different views) types of social capital in expanding the concept beyond narrowly bounded in-group interests and identities.

36 See Dolsak and Ostram 2003; Berkes *et al.* 2003; Visser 2004; Armitage *et al.* 2007, for example.

37 Henley's (2007) review of a number of recent studies of environmental issues in Southeast Asia points to the evidence on the effect of crisis thresholds in creating the conditions for adaptive learning and reversal of destructive patterns of resource use, as does Parnwell's study referred to above (n. 2).

38 Questions of representation, identity, citizenship, and the practical capacities of individuals and collectivities to reign in fellow users, leaders and elites through formal institutions and other cultural mechanisms are as important at local level as they are within wider governance spheres. Among the Mului (Chapter 5) we saw that charismatic leadership did not translate into arbitrary decision-making power. Authority was decentred and provisional upon collective validation. But in other situations, for example among Rendani Papuans and Melayu in Jambi (Chapters 6 and 7), local cultural and organizational mechanisms were unable to prevent strategically placed leaders from acting on their own behalf at the expense of community interests.

Bibliography

Acciaioli, G. (2008) 'Strategy and subjectivity in co-management of Lore Lindu National Park (Central Sulawesi, Indonesia)', in N. Sodhi, G. Acciaioli, M. Erb, A. K-J. Tan (eds), *Biodiversity and Human Livelihoods in Protected Areas: Case Studies from the Malay Archipelago*. Cambridge: Cambridge University Press, pp. 266–88.

Agrawal, A. (2001) 'Common property institutions and sustainable governance of resources', *World Development* 29 (10): 1649–72.

Agrawal, A. (2007) 'Forests, governance, and sustainability: common property theory and its contributions', *International Journal of the Commons* 1 (1): 111–36.

Agrawal, A. and Gibson, C. (eds) (2001) *Communities and the Environment*. New Brunswick, NJ: Rutgers University Press.

Arce, A. (2003) 'Value contestations in development interventions: community development and sustainable livelihoods approaches', *Community Development Journal* 38 (3): 199–212.

Armitage, D., Berkes, F. and Doubleday, N. (eds) (2007) *Adaptive Co-management: Collaboration, Learning and Multi-level Governance*. Vancouver: UBC Press.

Aspinall, E. and Fealy, G. (eds) (2003) *Local Power and Politics in Indonesia: Decentralisation and Democratisation*. Singapore. Institute of Southeast Asian Studies.

Bakker, L. (2005) 'Resource claims between tradition and modernity: *Masyarakat Adat* Strategies in Mului (Kalimantan Timur)', *Borneo Research Bulletin* 36: 29–50.

Benda Beckmann, F., Benda Beckmann, K, and Wiber, M. (2006) *Changing Properties of Property*. New York: Berghahn.

Berkes, F., Colding, J. and Folke, C. (eds) (2004) *Navigating Social-Ecological Systems: Building Resilience for Complexity and Change*. Cambridge: Cambridge University Press.

Berkhout, F., Leach, M., and Scoones, I. (eds) (2003) *Negotiating Environmental Change; New Perspectives from Social Science*, Cheltenham: Edward Elgar.

Biller, D. (2003) *Harnessing Markets for Biodiversity: Towards Conservation and Sustainable Use*. Paris: OECD.

Boomgaard, P., Henley, D. and Osseweijer, M. (eds) (2005) *Muddied Waters: Historical and Contemporary Perspectives on Management of Forests and Fisheries in Island Southeast Asia*. Leiden: KITLV Press.

Bourdieu, P. (1990 [1980]) *The Logic of Practice*. Stanford, CA: Stanford University Press.

Brosius, J. P., Tsing, A. L. and Zerner, C. (eds) (2005) *Communities and Conservation: Histories and Politics of Community-Based Natural Resource Management.* Walnut Creek, CA: Altamira Press.

Butcher, J. (2004) *The Closing of the Frontier: A History of the Marine Fisheries of Southeast Asia c. 1850–2000.* Singapore: Institute of Southeast Asian Studies.

Clapp, J. and Dauvergne, P. (2005) *Paths to a Green World: The Political Economy of the Global Environment.* Cambridge, MA: MIT Press.

Cleaver, F. and Franks, T. (2003) 'How institutions elude design: river bank management and sustainable livelihoods', *BCID Research Paper No 12*, Alternative Water Forum, Bradford Centre for International Development.

Clémonçon, R. (2008) 'The Bali road map: A first step on the difficult journey to a post-Kyoto protocol agreement', *The Journal of Environment and Development* 17 (1): 70-94.

Cooke, B. and Kothari, U. (eds) (2004) *Participation: The New Tyranny?* London: Zed Books.

Davidson, J. S. and Henley, D. (eds) (2007) *The Revival of Tradition in Indonesian Politics.* London: Routledge.

Dolsak, N. and Ostrom, E. (eds) (2003) *The Commons in the New Millennium: Challenges and Adaptation.* Cambridge MA: MIT Press.

Dove, M. (2005) 'Use of global legal mechanisms to conserve local biogenetic resources: problems and prospects', in Dove *et al.*, *Conserving Nature in Culture: Case Studies from Southeast Asia.* New Haven, CT: Yale University, Southeast Asia Studies, pp. 279–306.

Dove, M. R., Sajise, P.E. and Doolittle, A. A. (eds) (2005) *Conserving Nature in Culture: Case Studies from Southeast Asia,* New Haven, CT: Yale University, Southeast Asia Studies.

Dunkley, G. (2004) *Free Trade: Myth Reality and Alternatives.* London: Zed Books.

Dutton, I. (2005) 'If only fish could vote', in B. P. Resosudarmo (ed.), *The Politics and Economics of Indonesia's Natural Resources.* Singapore: ISEAS, pp. 162–78.

Eaton, P. (2005) *Land Tenure, Conservation and Development in Southeast Asia.* London: Routledge Curzon.

Eldridge (1995) *Non-government Organizations and Democratic Participation in Indonesia.* Kuala Lumpur: Oxford University Press.

Ellen, R., Parkes, P. Bicker, A. (2000) *Indigenous Environmental Knowledge and its Transformations.* Amsterdam: Harwood Academic.

Ellen, R. (ed.) (2007) *Modern Crises and Traditional Strategies: Local Ecological Knowledge in Island Southeast Asia.* New York: Berghahn Books.

Erb, M., Sulistiyanto, P. and Faucher, C. (eds) (2005) *Regionalism in Post-Suharto Indonesia.* London: Routledge Curzon.

Farrington, J., and Lewis, D. (1993) *Non-governmental Organizations and the State in Asia.* London: Routledge.

Ferraro, P. J. (2001) 'Interventions and a role for conservation performance payments', *Conservation Biology* 15 (4): 990–1000.

Fox, J., Adhuri, D. and Resosudarmo, I. A. (2005) 'Unfinished edifice or Pandora's box: decentralisation and resource management in Indonesia', in B. Resosudarmo (ed.), *The Politics and Economics of Indonesia's Natural Resources.* Singapore: ISEAS, pp. 92–108.

Goldblatt, D. (1996) *Social Theory and the Environment.* Cambridge: Polity Press.

Goldman, M. (ed.) (1998) *Privatizing Nature: Political Struggles for the Global Commons.* New Brunswick, NJ: Rutgers University Press

Greenough, P. and Tsing, A. L. (eds) (2003) *Nature in the Global South: Environmental projects in South and Southeast Asia.* Durham, NC: Duke University Press.

Hadiwinata, B. S. and Pakpahan, A. K. (2004) *Fair Trade: Gerakan Perdagangan Alternatif.* Yogyakarta: Pustaka Pelajar, and Oxfam.

Heal, G. (1998) *Valuing the Future: Economic Theory and Sustainability*. New York: Columbia University Press.

Heal, G. (2000) *Nature and the Marketplace*. Washington, DC: Island Press.

Henley, D. (2007a) 'Custom and *koperasi*: the co-operative ideal in Indonesia', in J. S. Davidson and D. Henley (eds), *The Revival of Tradition in Indonesian Politics*. London: Routledge, pp. 87–112.

Henley, D. (2007b) '"The folly our descendants are least likely to forgive us": the end of nature in Southeast Asia?', *Bijdragen* 123(2/3): 440–53.

Hickey, S. and Mohan, G. (eds) (2004) *Participation: From Tyranny to Transformation?* London: Zed Books.

Hilhorst, D (2003) *The Real World of NGOs: Discourses, Diversity and Development*. London: Zed Books.

ILO (2001) *Report V (1) on the Promotion of Cooperatives*. Geneva: International Labor Organisation, http://www.ilo.org/public/english/standards/relm/ilc/ilc89/rep-v-1.htm (accessed 6 September 2005).

IPPC Report (2007) *Climate Change 2007: Synthesis Report*. Valencia, Spain: Intergovernmental Panel on Climate Change.

Jervis, P. and Richards, S. (1997) 'Public management: raising our game', *Public Money & Management* 17 (2): 9–16.

Lansing, J. S. (2006) *Perfect Order: Recognizing Complexity in Bali*. Princeton: Princeton University Press.

Li, T. M. (2000) 'Articulating indigenous identity in Indonesia: resource politics and the tribal slot', *Comparative Studies in Society and History* 42 (1): 149–79.

Li, T. M. (2001) 'Boundary work: community market and state reconsidered', in A. Agrawal and C. Gibson (eds), *Communities and the Environment*, New Brunswick, NJ: Rutgers University Press, pp. 157–79.

Li, T. M. (2002) 'Engaging simplifications: community-based natural resource management, market processes and state agendas in upland Southeast Asia', *World Development* 30 (2): 265–83.

Li, T. M. (2007) *The Will to Improve*, Durham, NC: Duke University Press.

Lowe, C. (2000) 'Global markets, local injustice in Southeast Asian seas: the live fish trade and local fishers in the Togean Islands of Sulawesi', in C. Zerner (ed.), *People Plants and Justice: The Politics of Nature Conservation*. New York: Columbia University Press, pp. 234–58.

Lynch, O. J. and Harwell, E. (2002) *Whose Natural Resources? Whose Common Good?: Towards a New Paradigm of Environmental Justice and the National Interest in Indonesia*. Jakarta: Lembaga Studi dan Advokasi Masyarakat (ELSAM).

McCarthy, J. F. (2004) 'Changing to gray: decentralization and the emergence of volatile socio-legal configurations in Central Kalimantan, Indonesia', *World Development* 32 (7): 1199–223.

McCarthy, J. F. (2006) *The Fourth Circle: A Political Ecology of Sumatra's Rainforest Frontier*. Stanford: Stanford University Press.

McCarthy, J. F. (2007) 'Local voice in shifting modes of decentralised resource control in Central Kalimantan, Indonesia', RMAP Working Paper No. 65. Canberra: Australia National University, http://rspas.anu.edu.au/papers/rmap/Wpapers/rmap_wp65.pdf>http://rspas.anu.edu.au/papers/rmap/Wpapers/rmap_wp65.pdf.

McCay, B. J. (2001) 'Community and the commons: romantic and other views', in A. Agrawal and C. Gibson (eds), *Communities and the Environment*. New Brunswick, NJ: Rutgers University Press, pp. 180–92.

McCay, B. J. and Jentoft, S. (1998) 'Market or community failure? Critical perspectives on common property research', *Human Organization* 57 (1): 21–9.

Marifa, I. (2005) 'Institutional transformation for better policy implementation and enforcement,' in B. Resosudarmo (ed.) *The Politics and Economics of Indonesia's Natural Resources*. Singapore: ISEAS.

Meynen, W. and Doornbos, M. (2004) 'Decentralising natural resource management: a recipe for sustainability and equity?', *The European Journal of Development Research* 16 (1): 235–54.

Munro, W. A. (1998). *The Moral Economy of the State: Conservation, Community Development, and State Making in Zimbabwe*. Athens, OH: Ohio University Center for International Studies Publications.

Nadeau, R. L. (2006) *The Environmental Endgame: Mainstream Economics, Ecological Disaster, and Human Survival*. New Brunswick, NJ: Rutgers University Press.

O'Riordan, T. and Stoll-Kleeman, S. (2002) *Biodiversity, Sustainability and Human Communities*. London: Earthscan.

Ostrom, E. (1990) *Governing the Commons: The Evolution of Institutions for Collective Action*. Cambridge: Cambridge University Press.

Paine, R. (1994) *Herders of the Tundra: A Portrait of Saami Reindeer Pastoralism*. Washington: Smithsonian Institution Press.

Parnwell, M. (2003) 'Shaping sustainable natural and social environments in Thailand: development, crisis and response', in Mertz O., Wadley R. and Christensen A. (eds), *Proceedings of the International Conference on Local Land Use strategies in a Globalizing World: Shaping Sustainable Social and Natural Environments*. University of Copenhagen, pp. 301–21.

Parnwell, M. (2005) 'The power to change: rebuilding sustainable livelihoods in North-East Thailand', *Journal of Transdisciplinary Environmental Studies* 4 (2): 1–21.

Persoon, G., van Est, D. and Sajise, P. (2003*) Co-Management of Natural Resources in Asia: A Comparative Perspective*. CopenhagenNIAS Press.

Pierre, J. P. and Guy, B. (2000). *Governance, Politics and the State*. Basingstoke: Macmillan.

Polanyi, K. (1968) *Primitive, Archaic and Modern Economies: Essays of Karl Polanyi*. Boston: Beacon Press.

Portes, A. (2000) 'The two meanings of social capital', *Sociological Forum* 15 (1): 1–12.

Pretty, J. and Ward, H. (2001) 'Social capital and the environment', *World Development* 29 (2): 209–27.

Pretty, J. and Smith, D. (2004) 'Social capital in biodiversity conservation and management', *Conservation Biology* 18 (3): 631–8.

Putnam, R. (1993) *Making Democracy Work,* Princeton, NJ: Princeton University Press.

Resosudarmo, B. (ed.) (2005) *The Politics and Economics of Indonesia's Natural Resources*. Singapore: ISEAS.

Ribot, J. C. and Peluso, N. L. (2002) 'A theory of access', *Rural Sociology* 62 (2): 153–81.

Sahlins, M. (1972) *Stone Age Economics*. Chicago: Aldine.

Sakai, M. (ed.) (2002) *Beyond Jakarta: Regional Autonomy and Local Society in Indonesia*. Belair South Australia: Crawford House Publishing.

Salafsky, N., Cauley, H., Balanchander, G. *et al.* (2001) 'A systematc test of an enterprise strategy for community-based biodiversity conservation', *Conservation Biology* 15 (6): 1585–95.

Schulte Nordholt, H. and van Klinken, G. (2007) *Renegotiating Boundaries: Agency, Access, and Identity in Post-Suharto Indonesia*. Leiden: KITLV Press.

Scott, J. C. (1998) *Seeing Like A State: How Certain Schemes to Improve the Human Condition Have Failed*. New Haven: Yale University Press.

Sodhi, N., Acciaioli, G., Erb, M., Tan, A. K-J. (eds) (2007), *Biodiversity and Human Livelihoods in Protected Areas: Case Studies from the Malay Archipelago*. Cambridge: Cambridge University Press.

Stiglitz, J. (2002) *Globalization and its Discontents*. London: Penguin.

Tudge, C. (2004) *So Shall We Reap: What's Gone Wrong with the World's Food and How to Fix It*. London: Penguin.

Uphoff, N. and Langholz, J. (1998) 'Incentives for avoiding the Tragedy of the Commons', *Environmental Conservation* 25 (3): 251–61.

Van Laerhoven, F. and Ostrom, E. (2007) 'Traditions and trends in the study of the commons', *International Journal of the Commons* 1 (1): 3n28.

Verdery, K. and C. Humphrey (2004): *Property in Question: Value Transformation in the Global Economy.* Oxford: Berg.

Visser, L. (ed.) (2004) *Challenging Coasts: Transdisciplinary Excursions into Integrated Coastal Zone Development.* Amsterdam: Amsterdam University Press.

Warren, C. (2005) 'Mapping common futures: customary communities, NGOs and the state in Indonesia's Reform Era', *Development and Change* 36 (1): 49–73.

Warren, C. (2007) 'Balinese adat in discourse in practice', in J. Davidson and D. Henley (eds), *The Revival of Tradition in Indonesian Politics.* London: Routledge, pp. 170–202.

Warren, C. and McCarthy, J. (2002) 'Customary regimes and collective goods in Indonesia's changing political constellation', in S. Sargeson (ed.) *Collective Goods, Collective Futures in Asia.* London: Routledge, pp. 75–101.

Woolcock, M. (1998) 'Social capital and economic development: toward a theoretical synthesis and policy framework', *Theory and Society* 27 (2): 151–208.

World Commission on Environment and Development (1987) *Our Common Future.* Oxford: Oxford University Press.

Worms, J. Ducrocq, M and Saleck, A. (2004) ' A concerted approach towards managing living resources in a marine protected area', in L. Visser (ed.) *Challenging Coasts: Transdisciplinary Excursions into Integrated Coastal Zone Development.* Amsterdam: Amsterdam University Press.

Wunder, S. (2007) 'The efficiency of payments for environmental services in tropical conservation', *Conservation Biology* 21 (1): 48–58.

Zerner, C. (ed). (2000) *People, Plants and Justice: The Politics of Nature Conservation.* New York: Columbia University Press.

Index

For Product Safety Concerns and Information please contact our EU
representative GPSR@taylorandfrancis.com
Taylor & Francis Verlag GmbH, Kaufingerstraße 24, 80331 München, Germany